Lecture Notes in Geoinformation and Cartography

For further volumes:
http://www.springer.com/series/7418

Joaquín Huerta · Sven Schade
Carlos Granell
Editors

Connecting a Digital Europe Through Location and Place

Springer

Editors
Joaquín Huerta
University Jaume I of Castellón
Castellón
Spain

Sven Schade
Carlos Granell
European Commission, Joint Research Centre
Ispra
Italy

ISSN 1863-2246 ISSN 1863-2351 (electronic)
ISBN 978-3-319-03610-6 ISBN 978-3-319-03611-3 (eBook)
DOI 10.1007/978-3-319-03611-3
Springer Cham Heidelberg New York Dordrecht London

Library of Congress Control Number: 2014940057

Printed on acid-free paper

Springer is part of Springer Science+Business Media (www.springer.com)

Preface

Since 1998, the Association of Geographic Information Laboratories for Europe (AGILE) promotes academic teaching and research on geographic information at the European level. Its annual conference reflects the variety of topics, disciplines, and actors in this research area. It provides a multidisciplinary forum for scientific knowledge production and dissemination and has gradually become the leading Geographic Information Science conference in Europe.

For the eighth consecutive year, the AGILE conference full papers are published as a book by Springer-Verlag. This year, 67 documents were submitted as full papers, of which 22 were accepted for publication in this volume after a thorough selection and review process, which resulted in an acceptance rate of approximately 33 %. As much as we congratulate the authors for the quality of their work, we thank them for their contribution to the success of the AGILE conference and book series. We also send our acknowledgments to the numerous reviewers for providing us with their thorough judgements. Their work was fundamental to select the very best papers.

Under the title *Connecting a Digital Europe Through Location and Place*, this book pays special attention to the role Geographic Information Science and Technology can play to connect European universities, research centers, industry, government, and citizens in the digital information age. The scientific papers published in this volume cover a wide range of associated topics. The first part covers innovative experiments and applications that take user generated and social network data as inputs; the second part focuses on novel algorithms and process for analyzing trajectories. The third part covers studies concerned with data mining, fusion, and integration. The fourth part includes a series of papers related to data representation, visualization, and perception. The fifth and last part comprises works on geospatial decision support services.

Organizing the program of an international conference and editing a volume of scientific papers takes time, effort, and support. The input from the AGILE Council and Committees was a tremendous asset for us and we are grateful to all members for their contributions.

We would also like to thank our sponsors (ESRI, Ubik Geospatial Solutions, Zed, ESRI España, Google, Geodan, and Prodevelop), for their kind contributions to the 17th AGILE Conference on Geographic Information Science and Springer-Verlag for their willingness to publish the accepted full papers in their academic series Springer Lecture Notes in Geoinformation and Cartography.

Castellón, March 2014
Ispra

Joaquín Huerta
Sven Schade
Carlos Granell

Organizing and Scientific Committee

Programme Committee

Programme Chair Joaquín Huerta
University Jaume I of Castellón (Spain)
Programme Co-Chair Sven Schade
European Commission, Joint Research Centre (Italy)
Programme Co-Chair Carlos Granell
European Commission, Joint Research Centre (Italy)

Local Organizing Committee

Laura Díaz, UNOPS (Spain)
Michael Gould, University Jaume I of Castellón (Spain)
Óscar Belmonte, University Jaume I of Castellón (Spain)
Marco Painho, Universidade Nova de Lisboa, ISEGI (Portugal)
Dori Apanewicz, University Jaume I of Castellón (Spain)
Ana Sanchis, University Jaume I of Castellón (Spain)
Luis Enrique Rodríguez, University Jaume I of Castellón (Spain)
Rubén Vidal, University Jaume I of Castellón (Spain)
Joaquín Torres-Sospedra, University Jaume I of Castellón (Spain)

Scientific Committee

Peter Atkinson, University of Southampton (UK)
Fernando Bação, New University of Lisbon (Portugal)
Itzhak Benenson, Tel Aviv University (Israel)
Lars Bernard, TU Dresden (Germany)
Michela Bertolotto, University College Dublin (Ireland)
Ralf Bill, Rostock University (Germany)

Thomas Blaschke, University of Salzburg (Austria)
Thomas Brinkhoff, Jade University Oldenburg (Germany)
Gilberto Camara, National Institute for Space Research (Brazil)
Sven Casteleyn, University Jaume I of Castellón (Spain)
Nicholas Chrisman, University of Laval (Canada)
Christophe Claramunt, Naval Academy Research Institute (France)
Serena Coetzee, University of Pretoria (South Africa)
David Coleman, University of New Brunswick (Canada)
Lex Coomber, University of Leicester (UK)
Oscar Corcho, Universidad Politécnica de Madrid (Spain)
Max Craglia, European Commission, Joint Research Centre (Italy)
Joep Compvoets, KU Leuven Public Governance Institute (Belgium)
Laura Díaz, UNOPS (Spain)
Anders Friis-Christensen, European Commission-Joint Research Centre (Italy)
Jérôme Gensel, University of Grenoble (France)
Yola Georgiadou, University of Twente (The Netherlands)
Michael Gould, University Jaume I of Castellón (Spain)
Carlos Granell, European Commission, Joint Research Centre (Italy)
Henning Sten Hansen, Aalborg University (Denmark)
Lars Harrie, Lund University (Sweden)
Francis Harvey, University of Minnesota (USA)
Gerard Heuvelink, Wageningen University (The Netherlands)
Stephen Hirtle, University of Pittsburgh (USA)
Hartwig Hochmair, University of Florida (USA)
Joaquín Huerta, University Jaume I of Castellón (Spain)
Bashkim Idrizi, State University of Tetova (Republic of Macedonia)
Mike Jackson, University of Nottingham (UK)
Bin Jiang, University of Gävle (Sweden)
Didier Josselin, University of Avignon (France)
Derek Karssenberg, Utrecht University (The Netherlands)
Tomi Kauppinen, Aalto University (Finland)
Marinos Kavouras, National Technical University of Athens (Greece)
Karen Kemp, University of Southern California (USA)
Alexander Kotsev, European Commission, Joint Research Centre (Italy)
Menno-Jan Kraak, University of Twente (The Netherlands)
Patrick Laube, University of Zurich (Switzerland)
Robert Laurini, INSA-Lyon (France)
Francisco J. Lopez-Pellicer, University of Zaragoza (Spain)
Joan Masó, CREAF (Spain)
Filipe Meneses, University of Minho (Portugal)
Peter Mooney, National University of Ireland Maynooth (Ireland)
Adriano Moreira, University of Minho (Portugal)
Beniamino Murgante, University of Basilicata (Italy)
Pedro Muro Medrano, University of Zaragoza (Spain)
Javier Nogueras Iso, University of Zaragoza (Spain)

Toshihiro Osaragi, Tokyo Institute of Technology (Japan)
Frank Ostermann, University of Twente (The Netherlands)
Volker Paelke, Institute of Geomatics, Castelldefels (Spain)
Marco Painho, New University of Lisbon (Portugal)
Francesco Pantisano, European Commission, Joint Research Centre (Italy)
Poulicos Prastacos, Institute of Applied and Computational Mathematics FORTH (Greece)
Ross Purves, University of Zurich (Switzerland)
Francisco Ramos, University Jaume I of Castellón (Spain)
Martin Raubal, ETH Zurich (Switzerland)
Wolfgang Reinhardt, Universität der Bunderwehr Munich (Germany)
Femke Reitsma, University of Canterbury (New Zealand)
Claus Rinner, Ryerson University (Canada)
Jorge Rocha, University of Minho (Portugal)
Anne Ruas, IGN (France)
Maribel Yasmina Santos, University of Minho (Portugal)
Tapani Sarjakoski, Finnish Geodetic Institute (Finland)
Sven Schade, European Commission, Joint Research Centre (Italy)
Christoph Schlieder, University of Bamberg (Germany)
Monika Sester, Leibniz University Hannover (Germany)
Takeshi Shirabe, Royal Institute of Technology (Sweden)
Jantien Stoter, Delft University of Technology (The Netherlands)
Maguelonne Teisseire, IRSTEA (France)
Fred Toppen, Utrecht University (The Netherlands)
Joaquín Torres-Sospedra, University Jaume I of Castellón (Spain)
Nico Van de Weghe, Ghent University (Belgium)
Jos Van Orshoven, KU Leuven (Belgium)
Danny Vandenbroucke, KU Leuven (Belgium)
Lluis Vicens, Universitat de Girona (Spain)
Luis M. Vilches Blazquez, Universidad Politécnica de Madrid (Spain)
Agnès Voisard, FU Berlin and Fraunhofer FOKUS (Germany)
Monica Wachowicz, University of New Brunswick (Canada)
Robert Weibel, University of Zurich (Switzerland)
Stephan Winter, The University of Melbourne (Australia)
Mike Worboys, University of Maine (USA)
Bisheng Yang, Wuhan University (China)
Javier Zarazaga Soria, University of Zaragoza (Spain)
Alexander Zipf, Heidelberg University (Germany)

Contents

Part I
User Generated and Social Network Data

Estimating Completeness of VGI Datasets by Analyzing Community Activity Over Time Periods

Simon Gröchenig, Richard Brunauer and Karl Rehrl

Abstract Due to the dynamic nature and heterogeneity of Volunteered Geographic Information (VGI) datasets a crucial question isu concerned with geographic data quality. Among others, one of the main quality categories addresses data completeness. Most of the previous work tackles this question by comparing VGI datasets to external reference datasets. Although such comparisons give valuable insights, questions about the quality of the external dataset and syntactic as well as semantic differences arise. This work proposes a novel approach for internal estimation of regional data completeness of VGI datasets by analyzing the changes in community activity over time periods. It builds on empirical evidence that completeness of selected feature classes in distinct geographical regions may only be achieved when community activity in the selected region runs through a well-defined sequence of activity stages beginning at the start stage, continuing with some years of growth and finally reaching saturation. For the retrospective calculation of activity stages, the annual shares of new features in combination with empirically founded heuristic rules for stage transitions are used. As a proof-of-concept the approach is applied to the OpenStreetMap History dataset by analyzing activity stages for 12 representative metropolitan areas. Results give empirical evidence that reaching the saturation stage is an adequate indication for a certain degree of data completeness in the selected regions. Results also show similarities and differences of community activity in the different cities, revealing that community activity stages follow similar rules but with significant temporal variances.

S. Gröchenig · R. Brunauer (✉) · K. Rehrl
Salzburg Research Forschungsgesellschaft mbH, Jakob-Haringer-Straße 5, 5020 Salzburg, Austria
e-mail: richard.brunauer@salzburgresearch.at

S. Gröchenig
e-mail: simon.groechenig@salzburgresearch.at

K. Rehrl
e-mail: karl.rehrl@salzburgresearch.at

J. Huerta et al. (eds.), *Connecting a Digital Europe Through Location and Place*, Lecture Notes in Geoinformation and Cartography, DOI: 10.1007/978-3-319-03611-3_1,

1 Introduction

Volunteered Geographic Information (VGI) denotes one of the most promising and interesting developments in the field of geographic information science. Since the coining of the term by Goodchild (2007) researchers all over the world have started to scientifically investigate the phenomenon. One of the most crucial questions in VGI research is concerned with the assessment of geographic data quality of VGI datasets (ISO 2011; Goodchild and Li 2012). One of the outstanding categories of geographic data quality addresses completeness. Although completeness estimations of geographic datasets are not new, the VGI movement raises some new aspects such as inherent heterogeneity, high regional differences or frequent changes. In previous work researchers have addressed the assessment of data completeness in VGI datasets with well-known approaches like comparisons with external reference datasets (Haklay et al. 2010; Mondzech and Sester 2011; Zielstra and Hochmair 2011). Due to the success and open license of OpenStreetMap (OSM) (Haklay and Weber 2008) the project has been focus of most previous studies. The number of features, total lengths of linear features or the overlapping area of buffered features are compared. Although significant progress has been achieved, comparisons with external reference datasets have certain disadvantages such as the incertitude concerning completeness of the reference datasets, the missing of global availability or legal restrictions as well as high fees (Hecht et al. 2013). To overcome these disadvantages this work introduces a novel approach aiming at internal evaluation of data completeness. The presented approach analyzes the community activity over time periods in order to determine whether a certain level of completeness has been reached in a selected region. For estimating the completeness level in a region the approach derives the three activity stages *Start*, *Growth* and *Saturation* from the annual increase of geographic features being mapped by volunteering community members. The measure for completeness estimation is based on the hypotheses that completeness in a region can only be achieved when community activity passes a well-defined sequence of activity stages.

The remainder of this chapter is organized as follows: The next section discusses related work. It is followed by a section on the theoretical aspects of assessing completeness of VGI datasets. Section 4 outlines the novel approach for internal estimation of data completeness. Section 5 introduces the dataset used for proof-of-concept evaluation. Section 6 presents and discusses results and finally, Sect. 7 concludes the work.

2 Related Work

Goodchild and Li (2012) propose three approaches for quality assurance of VGI datasets: the crowdsourcing approach (i) relies on the community to check each other's contribution, the social approach (ii) gives people the responsibility of moderating the mapping process and the geographic approach (iii) deals with correctness of spatial data. For measuring data quality of already mapped features related work considers ISO 19157 (2011) where standardized quality measures for geographic information are defined. Related work mainly addresses the quality categories *completeness* and *positional accuracy* for the most prominent open VGI dataset OSM.

Haklay (2010) compared the OSM street network (motorways, A- and B-roads) of London with the federal dataset provided by Ordnance Survey. He concluded that on average 80 % of the streets are already mapped. Neis et al. (2012) compared the OSM street network of Germany with the commercial data provided by TomTom. They showed that OSM has a longer street network for pedestrians while TomTom is more detailed at rural street networks for cars. Moreover, authors revealed that urban street networks developed earlier than rural ones. Similarly, Zielstra and Hochmair (2011) compared the street network in selected cities in Germany and in the US with three reference datasets, namely Tiger, NAVTEQ and TomTom. Girres and Touya (2010) conducted a similar study for French roads, rivers and lakes. They determined a relative completeness of 45 % for roads, 83 % for lakes and 8 % for rivers compared to the French IGN dataset. Hecht et al. (2013) compared the OSM buildings with the ALKIS/ATKIS datasets for selected regions in Germany and concluded that less than 30 % of all buildings have been mapped.

While previous approaches pursue external data quality measures, the following studies focus on internal measures without relying on reference datasets. One of the first internal quality assessments was done by Mooney et al. (2010) who examined the geometry of polygons. Neis et al. (2013) analyzed the development of OSM data in 12 metropolitan areas distributed all over the world. According to their analysis of active users, European cities show a more active OSM community. Furthermore, authors analyzed the creation date and latest update of all features. They found that more than 20 % of all features have been created in 2012 and used this as indicator, that the dataset is not complete, yet. Corcoran et al. (2013) proofed that the growth of OSM street networks follows the development pattern of street networks in the real world defined by Strano et al. (2012). This pattern describes that the exploration phase (when new areas are mapped) is followed by a densification phase (when more details are added). Barron et al. (2014) developed a tool to analyze 25 indicators for assessing OSM data quality. Arsanjani et al. (2013) simulate the OSM mapping development for upcoming years based on development in previous years.

From examining previous work it can be concluded that only few studies address internal completeness measures of VGI datasets. To the knowledge of the authors there is no approach analyzing the development of the community activity over time periods for estimating regional data completeness.

3 Estimating Completeness of VGI Datasets

The International Organization for Standardization defines in ISO 19157 "Geographic information—Data quality" (ISO 2011) five data quality categories for geographic information, namely *completeness*, *logical consistency*, *positional accuracy*, *thematic accuracy* and *temporal quality* where completeness, which is addressed in this chapter, is defined as:

> [...] the presence and absence of features, their attributes and relationships. It consists of two data quality elements: commission—excess data present in a dataset; and omission—data absent from a dataset.

The completeness of a dataset depends on the presence of features in the dataset and on the correspondence between these features and the objects or properties in the real world. The measure does not depend on the positional accuracy or on the level of detail of the features. Completeness is a property of a geographical dataset and restricted to a geographical area and a purpose. The purpose defines the set of feature classes which are investigated. Hence we define completeness as:

> The *completeness measure* of the geographical dataset D, where D is defined by geographical region R and for purpose P, depends on the degree of correspondence between the existence of objects and properties in the real world and the presence of their representing features in dataset D.

However, the degree of correspondence (i) cannot be measured directly and the value (ii) cannot be calculated from the geographical dataset alone. Thus, completeness is commonly estimated by comparing two geographical datasets where the reference dataset is used instead of the real world. The comparisons of datasets with reference datasets or with the real world are so-called "external approaches" of quality evaluations, while internal approaches estimate data quality by calculating quality parameters from the dataset itself (ISO 2011). Internal approaches have to use well-defined rules to derive completeness indicators. The adequacy of such rules has to be proofed empirically. For VGI, the following three rules for estimating data completeness can be applied:

1. **Community activity and contributions** One possibility to estimate internal data completeness is to conduct an analysis of community contributions to VGI datasets (Neis et al. 2012; Steinmann et al. 2013a, b). Characteristic of VGI datasets are frequently appearing, disappearing or changing features. Since these changes are assigned to their contributors the current development of mapping activity of the community may always be treated as indicator for data completeness (e.g. Neis et al. (2013)).
2. **Hierarchical relationships between feature classes** In VGI feature classes are typically mapped according to their importance and appearance. For example, motorways are usually mapped before lower-level streets (Neis et al. 2012). An approach for estimating completeness may consider such hierarchical structures (e.g. Corcoran et al. (2013)). Thus, the temporal appearance of feature classes or feature class combinations may be used as completeness indicator.
3. **Relations between neighboring, sub- and super-regions** Completeness assessments have not to be treated as regionally isolated tasks. For example, it seems obvious that complete regions are more likely to be in a cluster of complete neighbors or that they contain at least complete sub-regions (Arsanjani et al. 2013). Together with the results of other rules relationships between spatially close and equally developed regions could be used as characteristic completeness indicator.

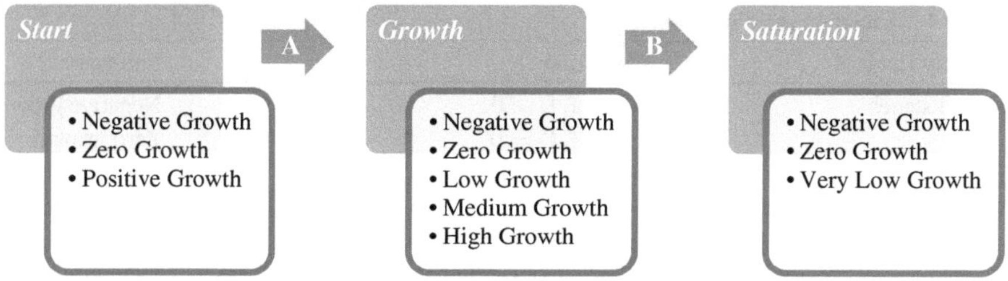

Fig. 1 Activity stages for analyzing the completeness of VGI datasets

In this work we outline an approach for completeness estimation which is a combination of rule types (1)–(3). As a first step, the growth rates of features in a dataset are analyzed to derive annual stages of community activity (rule type (1)). The activity stages represent an empirically determined mapping progress where the last stage is supposed to be a proper candidate indicator for completeness. Additionally, a detailed analysis of community activities is conducted by regarding rules from type (2) and (3). The results are used to gather additional evidence that completeness in a certain region has been reached or is near to be reached.

4 Deriving Community Activity Stages

Activity stages describe the contributors' activity by analyzing the annual changes to features in a dataset. The stages describe an ideally unidirectional development of the activities: at the start of a community activity only a few contributors are contributing to the dataset, afterwards more contributors are joining the activity and start contributing data before the mapping activity ceases since a certain level of data completeness has been reached. The development of these activities is described with the three stages *Start*, *Growth* and *Saturation* (Fig. 1). Within a certain stage community activity may change to more detailed sub-stages. The transitions between main stages follow distinct rules and are typically unidirectional. For the presented analysis the activity stage *Saturation* is the most relevant one. It occurs in the final years of a development in case that no more new (or just few) features are created.

The definition of stage transitions and sub-stage classifications is based on a growth value. For time interval i, region r and feature class f the growth value g is defined as the difference between the number of created features c and the number of deleted features d

$$g(i, r, f) := c(i, r, f) - d(i, r, f) \tag{1}$$

The progress value p is defined for time interval i, region r and feature class f as the fraction of the growth value from the overall growth value over the whole analyzed time interval $I\,(i \subseteq I)$

Table 1 Transition rules for sub-stages

Activity stage	Condition	Sub-stage
For all stages	$p(i, r, f, I) < 0$	Negative growth
	$p(i, r, f, I) = 0$	Zero growth
Start	$0 < p(i, r, f, I)$	Positive growth
Growth	$0 < p(i, r, f, I) \leq 0.25$	Low growth
	$0.25 < p(i, r, f, I) \leq 0.75$	Medium growth
	$0.75 < p(i, r, f, I)$	High growth
Saturation	$0 < p(i, r, f, I) \leq 0.03$	Very low growth

$$p(i, r, f, I) := g(i, r, f) / g(I, r, f) \quad (2)$$

A progress value of 0.36 indicates that 36 % of all features have been created in the respective time period. Transition rules between sub-stages are shown in Table 1.

In Fig. 1 the rules for the unidirectional transitions *A* and *B* between the activity stages are empirically defined as heuristic rules. For transition *A* from *Start* to *Growth*, two or more active contributors within a distinct region are required. Transition *B* from *Growth* to *Saturation* requires the progress within a time period to be very low (less than 3 %), whereas the cumulated progress value is greater than 0.97 and the number of years with active contributions is greater than two. Due to the retrospective calculation of growth values the resulting activity stages and sub-stages are subject to change. Since community effort is continuously changing, significant annual growth may occur although there was only minor growth during the previous years. It should be noted that any re-evaluation of the dataset with additional data may result in other activity stages for the previous years. It should also be noted that transition rules, although being derived from empirical evidence, should be treated as 'subject to change' since additional analyses could reveal the necessity of adjustments.

5 OSM History as Evaluation Dataset

To evaluate the proposed measure for data completeness, it is applied to the historic changes of the OSM data (OSM History). The OSM History has been selected since it includes all versions of all features starting from 2006 until the current date (for this analysis the file from 5th Feb. 2013 has been used). Since the calculation of activity stages is based on the definition of the growth value 4.1 and progress value 4.2 the historic data has to be prepared. Data preparation is based on the algorithm proposed in Rehrl et al. (2012) and results in a list of annually aggregated growth shares based on the total number of created and deleted features for the selected year. In addition to growth shares, the total number of active contributors is calculated for each year. Annual growths are considered as well-suited temporal units by avoiding seasonal variability. The geographic scope of the analysis has been set to the same

12 metropolitan areas (equal delimitation) as proposed in Neis et al. (2013). Besides fostering comparability of results the selected areas are considered well-suited due to worldwide distribution, cultural diversity and homogeneous settlement structure with a large number of geographic features and feature classes. Moreover, it has been previously found that urban communities are commonly more developed and more active (Neis et al. 2012). For proving the results of the 12 metropolitan areas, the three Austrian cities Vienna, Linz and Salzburg have been added. At least the mapping of the street network has been estimated "complete" by the local OSM community (OSM Wiki 2013b) and the authors' local knowledge confirms this estimation. Thus, the results for the Austrian cities are used as ground-truth for the proposed completeness measure.

Beside geographic scopes, completeness measures have to be focused on different feature classes. While OSM does not follow strict rules for classifying features, the proposed keys and values in the OSM Wiki may be used for selecting feature classes (OSM Wiki 2013a). As previously found, the feature classes denoted by the keys *highway* and *building* are significantly more developed in comparison to all other classes (Steinmann et al. 2013a). Due the high development it may be assumed that both classes have passed several years of mapping activity in all of the selected regions.

In OSM the key *highway* comprises all kinds of features related to the street network. This includes motorways, roads, residential streets, tracks, paths and footways. The highway key is also used for point features like traffic lights, turning points or pedestrian crossings. Due to the heterogeneous nature of the feature class it is suggested to analyze sub-classes separately.

The key *building* is used for mapping each kind of buildings. The value specifies the type of building (e.g. residential buildings, hotels or churches). In contrast to highways, the building class is homogenously structured and thus may be analyzed as a whole. In addition to the footprint and the building type, additional information such as addresses may be attributed to buildings. Since address information is typically mapped after building footprints, a separate analysis is suggested.

For the evaluation of the completeness measure, four different feature classes are selected: (i) the class *street* subsumes the OSM *highway* sub-classes *primary*, *secondary*, *tertiary*, *living_street*, *residential* and *unclassified*, (ii) the class *path* subsumes the sub-classes *path*, *footway*, *cycleway* and *steps*, (iii) the class *building* regards all features having the key *building* and finally (iv) the class *house number* regards all features having the key *addr:housenumber*. While the classes *street* and *building* are mainly used for calculating community activity, the classes *path* and *house number* are used as additional indicators to estimate the level of completeness.

6 Results and Discussion

This section presents selected evaluation results and discusses the results in the context of the following criteria: (i) impacts of different spatial resolutions, (ii) time series of activity stages to highlight the transitions from *Start* to *Growth* to *Saturation*,

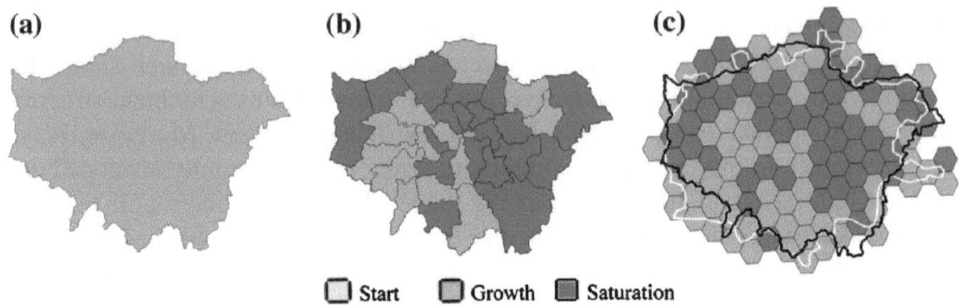

Fig. 2 Activity stages for London using three spatial resolutions (feature class: street; year: 2012): **a** shows the activity stage for Greater London. **b** shows activity stages for the 32 boroughs plus the City of London. **c** shows activity stages as hexagon grid consisting of cells with a diameter of 5 km; administrative boundary (*black line*); metropolitan area (*white line*) defined by Neis et al. (2013)

(iii) comparisons of activity stages for the selected metropolitan areas for one year, (iv) comparisons of activity stages between selected feature classes and (v) comparisons of spatial activity stage patterns of the last year.

Figure 2 shows the impact of different spatial resolutions on the calculation of activity stages for the London metropolitan area based on the same delimitation used by Neis et al. (2013). Firstly, the algorithm is applied to Greater London resulting in one conflated activity stage, in the middle the 32 London Boroughs plus the City of London are analyzed separately resulting in different activity stages and on the right the metropolitan area is subdivided by a hexagon grid with a cell diameter of five kilometers. According to Hagenauer and Helbich (2012) the shapes of hexagon grids follow urban patterns best. While the former two resolutions are bound to administrative boundaries the third one ignores boundaries. The benefit of using a grid resolution can be found in the worldwide applicability as well as in the comparability of different world regions. Analyses based on administrative boundaries cannot be compared due to variances in size and shape. For example, Great Britain is subdivided by different administrative structures with totally different sizes. Moreover it has been previously found that coarse-grained spatial resolutions with larger regions conflate individual results, which could get apparent with more fine-grained resolutions (Haklay et al. 2010). The London example from Fig. 2 confirms this finding for activity stages as the spatial resolution of Greater London conflates the different activity stages of the boroughs. However, activity stages for unpopulated areas should be specifically addressed due to lower mapping activity. Indeed it should be noted that larger evaluation units (administrative boundaries) may be useful for more general analyses. Table 2 summarizes advantages and disadvantages of the three proposed spatial resolutions with emphasis on analyzing community activity. The remainder of this work builds on the hexagon approach.

Figure 3 shows the sequential changes of community activity stages (see rule (1) in Sect. 3) for the feature class *street* in the metropolitan area of London during the years 2006–2012 using hexagon cells with 5 km diameter. The annual results shown in the hexagon maps are summarized in the bottom right diagram. While in the year

Table 2 Advantages and disadvantages of different spatial resolutions

	Greater London	Boroughs + City	Hexagons
Advantage	Fewer test regions; overview analysis	Residential areas are considered, no areas without population or infrastructure	All polygons have the same size and emphasize; world-wide comparability; detailed analysis of homogenous topographies (e.g. big cities)
Disadvantage	Places with different activity stages are conflated; no detailed conclusions are possible	Places with different activity stages are conflated; detailed conclusions only with additional contextual knowledge; large polygons are more emphasized in visualization	Areas with low contribution level; especially unpopulated areas; hexagons do not fit administrative boundaries

2006 71 cells are still in *Start*, in 2007 119 out of 120 cells have proceeded to *Growth* which is an indication for rising community activity in all parts of London. Since London has been the incubator city of the OSM project, activity stages are temporally ahead in comparison to other cities. Community activity most likely starts in the city center and moves towards the suburbs subsequently (see Cairo in Fig. 5 and Buenos Aires in Fig. 6). In 2010, the first two cells reached *Saturation*. In 2012, a majority of cells has reached *Saturation* which can be interpreted as indication that a certain level of completeness has been achieved.

To address the question whether saturated cells are also complete cells the next analysis regards the hierarchical structure of the feature classes (see rule (2) in Sect. 3). Tables 3 and 4 compare the activity stages of 12 metropolitan areas and three Austrian reference cities for the year 2012. While Table 3 shows the results for feature classes *street* and *path*, Table 4 has its focus on comparing the classes *building* and *house number*. The tables show (i) the number of hexagon cells per metropolitan area or city, (ii) the absolute numbers of cells which are in *Start* and *Saturation*, respectively, and charts showing the relative share of the three activity stages and (iii) ratios between the numbers of created features between the related feature classes. Results in Table 3 emphasize that streets are mapped before paths, while Table 4 indicates that buildings are mapped before house numbers. In case of streets in Berlin, 34 of 85 hexagons have already reached *Saturation* by the end of 2012 while for the paths only 14 hexagons have achieved the final stage. Results indicate that cities with similar activity patterns exist. In Table 3, the cities Berlin,

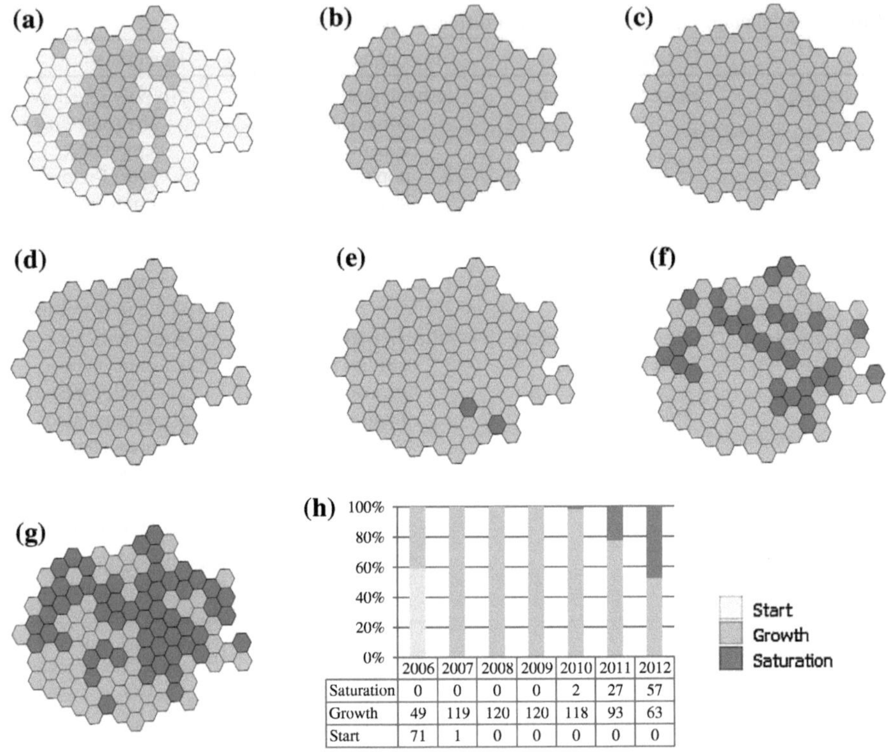

	2006	2007	2008	2009	2010	2011	2012
Saturation	0	0	0	0	2	27	57
Growth	49	119	120	120	118	93	63
Start	71	1	0	0	0	0	0

Fig. 3 Activity stages (2006–2012) for feature class *street* for the metropolitan area of London showing the mapping progress. **a** 2006, **b** 2007, **c** 2008, **d** 2009, **e** 2010, **f** 2011, **g** 2012, **h** Shares of activity stages of the 120 hexagon cells of London as time series from 2006–2012

London, Los Angeles, Moscow, Paris and Sydney have a faster mapping progress than the other cities. These cities also have the most advanced ratio towards mapping paths which is an indication for a high level of street completeness. Contrary, Johannesburg has a very low ratio and a high proportion of saturated cells, which is an indicator for a temporary inactive community than for completeness. For buildings and house numbers, the cities Buenos Aires, Cairo and Istanbul have the slowest activity progress. Due to Japan's different address scheme, only a very low number of house numbers has been mapped in Osaka which results in an adverse ratio (similar for Seoul). This finding gives indication that slow progress can also be the result of cultural variations. In case of Paris it has to be stressed that the city benefited from a major building import in 2010. In both tables the results for the three Austrian reference cities Linz, Salzburg and Vienna reveal the supposed advanced mapping progress with Linz as the most saturated city for all feature classes which reflects own and community observations. The overall results confirm previous results from Neis et al. (2013) that European cities have a more active OSM community compared to other cities.

Table 3 Comparisons between 12 metropolitan areas + 3 reference cities based on activity stages of the year 2012 for the feature classes street and path

Highway 2012	Hex. #	Street			Path			Ratio s:p
		Start	Sat.	%	Start	Sat.	%	
Berlin	85	0	34		0	14		1 : 1.5
Buenos Aires	177	8	52		134	7		13.0 : 1
Cairo	105	21	34		93	3		19.5 : 1
Istanbul	110	19	21		91	1		18.7 : 1
Johannesburg	167	12	80		140	4		24.7 : 1
London	120	0	57		0	7		1.2 : 1
Los Angeles	369	6	157		186	23		4.9 : 1
Moscow	300	1	113		30	14		1 : 1.9
Osaka	224	18	11		80	3		14.6 : 1
Paris	188	0	44		10	12		1.5 : 1
Seoul	147	39	15		112	3		12.4 : 1
Sydney	126	1	4		12	21		8.0 : 1
Linz	14	0	9		0	5		1 : 7.7
Salzburg	13	0	2		0	0		1 : 1.5
Vienna	41	0	9		0	3		1 : 1.9

Start Growth Saturation

The ratio s:p is the proportion between the number of created *street* (s) and *path* features (p)

A more detailed comparison demonstrates the shift in mapping progress between different hierarchically structured features. Figure 4 shows two time series for London and Paris. For London, the time series compares the progress of feature classes *street* and *path*, while for Paris, the progress of class *building* is compared to the progress of class *house number*. Both cities have been selected due to their advanced progress for the respective feature classes (see Tables 3 and 4). Figure 4 illustrates the shares of activity stages based on the hexagons for the years between 2006 and 2012. The diagrams outline that a shift between the related feature classes is observable. Streets are mapped before paths while buildings are mapped before house numbers. A possible reason for that phenomenon is that the focus of mapping interests follows a hierarchical order being determined by hierarchical relationships between feature classes. For example, most building footprints are mapped before house numbers are added. Based on this observations it may be valid to assess saturated hexagons more likely as complete if the observed area followed the typical hierarchical mapping schema, too. This trend can also be observed in the hexagon maps visualized in Figs. 5 and 6 and confirms the findings of Corcoran et al. (2013) about the exploration and densification phase.

Table 4 Comparisons between 12 metropolitan areas + 3 reference cities based on activity stages of the year 2012 for the feature classes building and house number

Building 2012	Hex. #	Building			House number			Ratio b:hn
		Start	Sat.	%	Start	Sat.	%	
Berlin	85	1	18		4	3		1.6 : 1
Buenos Aires	177	126	0		160	1		3.3 : 1
Cairo	105	79	5		100	1		21.7 : 1
Istanbul	110	78	1		98	0		27.3 : 1
Johannesburg	167	93	1		166	0		2.4 : 1
London	120	8	13		56	3		3.0 : 1
Los Angeles	369	178	25		249	8		5.8 : 1
Moscow	300	26	39		119	7		3.9 : 1
Osaka	224	137	0		220	0		1657 : 1
Paris	188	11	70		83	8		7.6 : 1
Seoul	147	128	0		140	0		11.6 : 1
Sydney	126	55	9		101	0		3.1 : 1
Linz	14	0	6		0	4		4.3 : 1
Salzburg	13	0	0		1	1		2.1 : 1
Vienna	41	0	0		1	3		1.6 : 1

Start Growth Saturation

The ratio b:hn is the proportion between the number of created *building* (b) and *house number* features (hn)

Concerning relationships between neighboring cells (see rule (3) in Sect. 3), the last comparison outlines the activity stage for the year 2012 for cities with a fast and a slow mapping progress. Figure 5 compares the results of London and Cairo for the feature classes *street* and *path* using hexagon maps. While London depicts advanced progress with respect to the mapping of streets, path mapping is still predominately stuck in *Growth*. Similarly, street mapping activity in Cairo is more advanced than path mapping. In contrast to London, Cairo still has 20 % of all hexagon cells for the street class and 89 % of the cells for the path class in *Start*. The high percentage of cells in *Start* for both feature classes indicates a low level of completeness for Cairo.

Figure 6 shows differences in building and house number mapping for the metropolitan areas of Paris and Buenos Aires. As Table 4 indicates, Paris has a high proportion of cells in *Saturation* while Buenos Aires still has many cells in *Start*. Again, more progress has been identified for buildings compared to house numbers. Single and distributed *Saturation* hexagons, as for paths in London (Fig. 5) and house numbers in Paris (Fig. 6), should be treated carefully. Those hexagons can also indicate temporal inactivity at the beginning of the mapping progress.

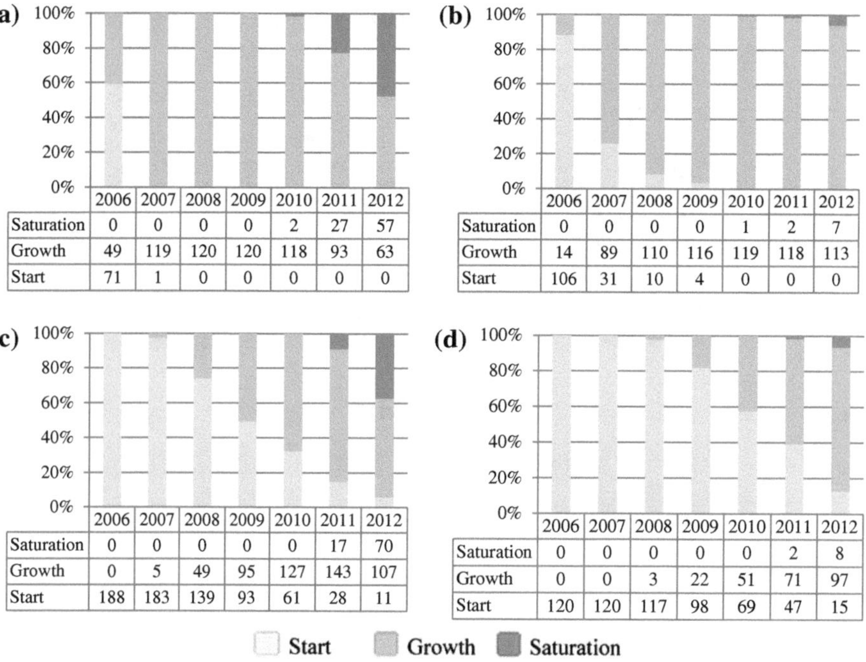

(a)

	2006	2007	2008	2009	2010	2011	2012
Saturation	0	0	0	0	2	27	57
Growth	49	119	120	120	118	93	63
Start	71	1	0	0	0	0	0

(b)

	2006	2007	2008	2009	2010	2011	2012
Saturation	0	0	0	0	1	2	7
Growth	14	89	110	116	119	118	113
Start	106	31	10	4	0	0	0

(c)

	2006	2007	2008	2009	2010	2011	2012
Saturation	0	0	0	0	0	17	70
Growth	0	5	49	95	127	143	107
Start	188	183	139	93	61	28	11

(d)

	2006	2007	2008	2009	2010	2011	2012
Saturation	0	0	0	0	0	2	8
Growth	0	0	3	22	51	71	97
Start	120	120	117	98	69	47	15

Fig. 4 Mapping progress based on activity stages (2006–2012) for London and Paris. **a** London—*street*, **b** London—*path*, **c** Paris—*building*, **d** Paris—*house number*

To summarize, it can be concluded that combining different rules for interpreting saturated cells leads to more accurate estimations. The presented examples have especially shown that (i) an appropriate selection of different spatial resolutions, (ii) the consideration of hierarchical structures between feature classes and (iii) the consideration of spatial distributions provide a proper analysis method for completeness.

However, the proposed approach is based on the simple hypothesis that when contributors cease to create features in a region, a sufficient level of completeness has been reached. A drawback of the method comes from the fact that low mapping activity can also be the result of non-ideal community developments (Suh et al. 2009). Thus, a critical evaluation of resulting activity stages is necessary for adequate estimations.

7 Conclusions and Outlook

In this chapter we proposed a new method for analyzing changes in VGI datasets to determine community activity stages in order to estimate regional completeness. The presented results show that local community activities provide sufficient information

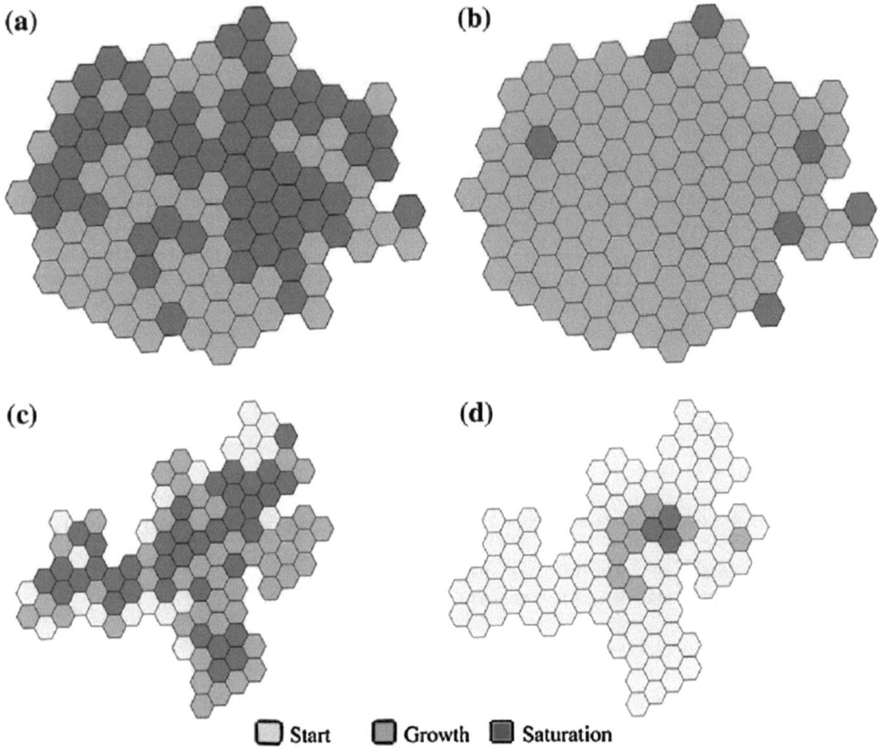

Fig. 5 Comparison of two different progress patterns in London and Cairo based on calculated activity stages for the year 2012 using the feature classes *street* and *path*. **a** London—*street*, **b** London—*path*, **c** Cairo—*street*, **d** Cairo—*path*

for assessing regions as complete. Several indicators show that regions with a low local community activity are estimated to be "regional complete", too.

The examples outlined above lead to plausible indications that the level of regional completeness can be derived from the temporal progress of community activity. Together with a detailed analysis of spatial distributions of activity stages a more accurate estimation between inactive and complete can be achieved. Furthermore, a selective analysis regarding the mapping progress of hierarchically structured feature classes, e.g. for streets and paths, facilitates the understanding and estimation of completeness. The proposed method can be easily adapted to different time periods, temporal resolutions, spatial resolutions and feature classes in order to provide deeper insights into the mapping progress of VGI communities.

To achieve reasonable results, the interpretation of activity stages for estimating completeness requires consideration of multiple aspects such as different hexagon sizes or related feature types. This applies especially for regions with a small or young VGI community or for regions with different cultural or topographic characteristics.

An open issue for further analyses would be to investigate more diverse geographic regions on applicability and generalization of the approach. For example

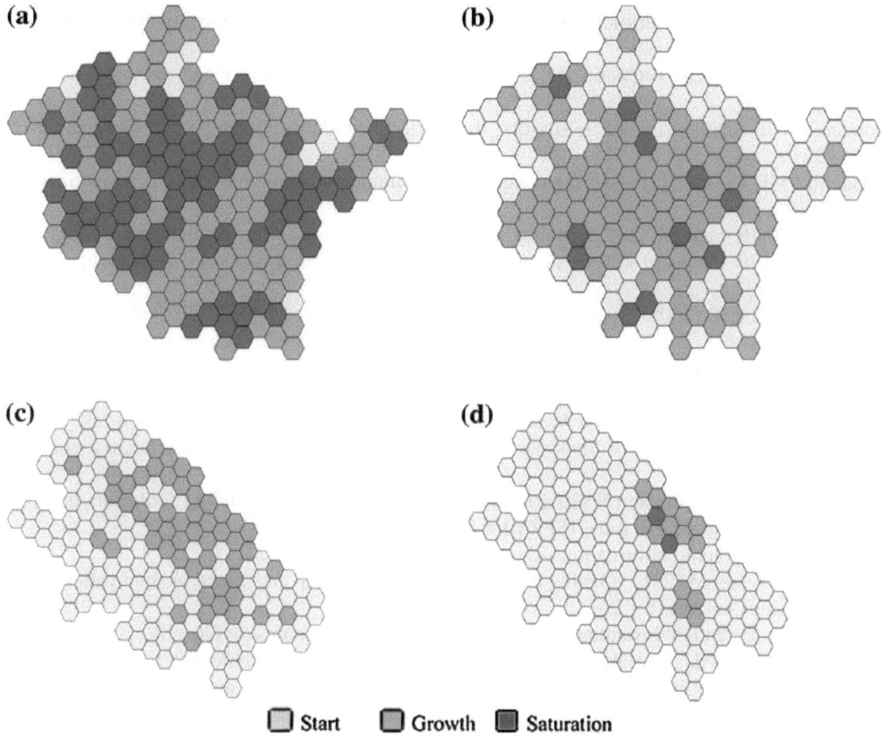

Fig. 6 Comparison between two different progress patterns in Paris and Buenos Aires based on calculated activity stages for the year 2012 using the feature classes *building* and *house number*. **a** Paris—*building*, **b** Paris—*house number*, **c** Buenos Aires—*building*, **d** Buenos Aires—*house number*

rural regions, sparsely populated regions or other feature classes than streets and buildings would be proper candidates. Finally, externally estimated complete regions may be considered as candidates for deriving and validating additional inference rules for completeness estimations. These rules can be used for automated completeness assessments in the future.

Acknowledgments This work was partly funded by the Austrian Federal Ministry for Transport, Innovation and Technology. We thank Pascal Neis for providing the delimitations of the twelve world regions defined in Neis et al. (2013).

References

Arsanjani JJ, Helbich M, Bakillah M, Loos L (2013) The emergence and evolution of OpenStreetMap: a cellular automata approach. Int J Digit Earth 1–30

Barron C, Neis P, Zipf A (2014) A comprehensive framework for intrinsic OpenStreetMap quality analysis. Trans GIS. doi:10.1111/tgis.12073

Corcoran P, Mooney P, Bertolotto M (2013) Analysing the growth of OpenStreetMap networks. Spat Stat 3:21–32

Girres J, Touya G (2010) Quality assessment of the french OpenStreetMap dataset. Trans GIS 14(4):435–459

Goodchild MF (2007) Citizens as sensors: the world of volunteered geography. GeoJournal 69: 211–221

Goodchild MF, Li L (2012) Assuring the quality of volunteered geographic information. Spat Stat 1:110–120

Hagenauer J, Helbich M (2012) Mining urban land-use patterns from volunteered geographic information by means of genetic algorithms and artificial neural networks. Int J Geogr Inf Sci 26(6):963–982

Haklay M, Weber P (2008) OpenStreetMap: user-generated street maps. IEEE Pervasive Comput 7(4):12–18

Haklay M (2010) How good is volunteered geographical information? a comparative study of OpenStreetMap and ordnance survey datasets. Environ Plan B, Plan Des 37(4):682–703

Haklay M, Basiouka S, Antoniou V, Ather A (2010) How many volunteers does it take to map an area well? the validity of linus' law to volunteered geographic information. Cartographic J 47(4):315–322

Hecht R, Kunze C, Hahmann S (2013) Measuring completeness of building footprints in Open-StreetMap over space and time. ISPRS Int J Geo-Inf 2(4):1066–1091

ISO (2011) Geographic information—data quality (ISO/DIS 19157:2011)

Mondzech J, Sester M (2011) Quality analysis of OpenStreetMap data based on application needs. Cartographica 46(2):115–126

Mooney P, Corcoran P, Winstanley AC (2010) Towards quality metrics for OpenStreetMap. In: 18th ACM SIGSPATIAL international conference on advances in geographic information systems

Neis P, Zielstra D, Zipf A (2012) The street network evolution of crowdsourced maps: Open-StreetMap in Germany 2007–2011. Future Internet 4(1):1–21

Neis P, Zielstra D, Zipf A (2013) Comparison of volunteered geographic information data contributions and community development for selected world regions. Future Internet 5(2):282–300

OSM Wiki (2013a) Map features. http://wiki.openstreetmap.org/wiki/Map_Features. Accessed 14 Nov 2013

OSM Wiki (2013b) Vienna OSM coverage. http://wiki.openstreetmap.org/wiki/Vienna_OSM_Coverage. Accessed 03 Dec 2013

Rehrl K, Gröchenig S, Hochmair H, Leitinger S, Steinmann R, Wagner A (2012) A conceptual model for analyzing contribution patterns in the context of VGI. In: LBS 2012–9th symposium on location based services. Springer, Berlin

Steinmann R, Brunauer R, Gröchenig S, Rehrl K (2013a) Wie aktiv sind freiwillige Mapper? In: Angewandte Geoinformatik 2013. Beiträge zum 25. AGIT-Symbosium Salzburg, pp 173–182

Steinmann R, Gröchenig S, Rehrl K, Brunauer R (2013b) Contribution profiles of voluntary mappers in OpenStreetMap. In: Online proceedings of the international workshop on action and interaction in volunteered geographic information, 16th AGILE conference

Strano E, Nicosia V, Porta S, Barthélemy M (2012) Elementary processes governing the evolution of road networks. Sci Rep 2:296

Suh B, Convertino G, Chi EH, Pirolli P (2009) The singularity is not near: slowing growth of Wikipedia. In: WikiSym '09 proceedings of the 5th international symposium on Wikis and open collaboration

Zielstra D, Hochmair HH (2011) A comparative study of pedestrian accessibility to transit stations using free and proprietary network data. J Transp Res Board 2117:145–152

Estimation of Building Types on OpenStreetMap Based on Urban Morphology Analysis

Hongchao Fan, Alexander Zipf and Qing Fu

Abstract Buildings are man-made structures and serve several needs of society. Hence, they have a significant socio-economic relevance. From this point of view, building types should be strongly correlated to the shape and sized of their footprints on the one hand. On the other hand, building types are very impacted by the contextual configuration among building footprints. Based on this hypothesis, a novel approach is introduced to estimate building types of building footprints data on OpenStreetMap. The proposed approach has been tested for the building footprints data on OSM in Heidelberg, Germany. An overall accuracy of 85.77 % can be achieved. Residential buildings can be labeled with accuracy of more than 90 %. Besides, the proposed approach can distinguish industrial buildings and accessory buildings for storage with high accuracies. However, public buildings and commercial buildings are difficult to be estimated, since their footprints reveal a large diversity in shape and size.

1 Introduction

Nowadays, OpenstreetMap (OSM) is considered as one of the most successful and popular VGI (Volunteered Geographic Information) projects, with a global cast of volunteers. Currently, there are more than 1.5 million registered members

H. Fan (✉) · A. Zipf
Chair of GIScience, Heidelberg University, Berlinerstr.48, 69120 Heidelberg, Germany
e-mail: hongchao.fan@uni-heidelberg.de

A. Zipf
e-mail: zipf@uni-heidelberg.de

Q. Fu
College of Surveying and Geomatics, Tongji University, Sipinglu 1239,
Shanghai 200092, China
e-mail: fuqing96031@gmail.com

J. Huerta et al. (eds.), *Connecting a Digital Europe Through Location and Place*,
Lecture Notes in Geoinformation and Cartography, DOI: 10.1007/978-3-319-03611-3_2,

(OSM 2014a) and OSM is growing rapidly. Sparked by the availability of high-resolution imagery from Bing since 2010, there has been an increase in building information in OSM, proving that volunteers do not only contribute roads or POIs to the database. According to the statistics (the values are derived from our internal OSM database which is updated daily), on November 20, 2013, the number of buildings in OSM was over 77 million. In Germany, there are almost 9 million objects with "building=yes" to the same time point.

Currently, building footprints data in OSM is mainly used for reconstructing 3D buildings. At present there are several projects which generate and visualize 3D buildings from OSM: OSM-3D, OSM Buildings, Glosm, OSM2World, etc. And applications based on these projects i.e. 3D navigation on mobile devices, web-based visualization, and simulation etc. are getting increased. The most of 3D buildings in these projects are rendered as polyhedral, extruded footprints with flat roofs, whereby the height information of a number of buildings are directly taken from the attribute of building footprints or converted from the number of stories, while the majority of 3D buildings own random heights. In OSM-3D, many buildings are modeled in LoD2 (Level of Detail according to CityGML) in case there are indications for their roof types (Goetz and Zipf 2012). In further, Goetz (2013) proposed a conception to generate buildings in LoD3 and LoD4 in CityGML. Besides, buildings in different LoDs from other sources can be uploaded via OpenBuildingModels and visualized in OSM-3D. But the buildings for uploading have to be adapted with the corresponding building footprints in OSM (Uden and Zipf 2012).

The applications based on the abovementioned projects i.e. 3D navigation on mobile devices (Li et al. 2012), web-based visualization, and simulation etc. are getting increased. 3D buildings could play important roles not only for the 3D visualization but also for other interoperability like spatial analyses and/or queries in the 3D environment. For this purpose, it is crucial to know the types of buildings. According to the latest research on quality assessment of OSM building footprints data, OSM building footprints data has a high completeness in terms of area covered, while there is limited attributive information such as building types (Fan et al. 2013).

Unfortunately, there are very few researches on building footprints data enrichment. To the best of the authors' knowledge, only one detailed study has been conducted by Huang et al. (2013). In their work, an automatic method is developed by using the geometric and topological features in footprint data, in order to enhance the maps with the building usage information. The work utilized the knowledge that building types are strongly characterized by the geometric features of footprints. In addition, simple neighborhood relation was taken into account to improve the classification. However, size/area of building footprints was not considered. Another related work is proposed by Henn et al. (2012) to derive the architectural types of buildings based on coarse LoD1 block models with vertical walls and flat roofs by employing Support Vector Machines (SVMs), whereby, geometric features such as length, width, area, and degree of perpendicularity of building footprints, types of buildings, as well as height information of buildings are required for the classification process.

In this work, a novel approach is introduced to estimate building types of building footprints data in OpenStreetMap. The approach is proposed based on urban morphology analysis by using ATKIS building footprints data in five city districts in Heidelberg. First of all, the correlations among geometries (mainly area and rectangularity) of building footprints and their types are derived. Then it investigates the similarity in building types for the buildings whose footprints are similar in shape and size. Based on the results of this urban morphology analysis, a set of rules are established to estimate buildings types in OpenStreetMap. Prior to the rule-based estimation of building types, the existing records of building types in OSM are classified into six types: (i) residential buildings, (ii) industrial buildings, (iii) commercial buildings, (iv) public buildings, (v) accessory buildings for storage and (vi) accessory buildings for supply. The building footprints on OSM are then labeled to one of these six types using the proposed method.

The reminder of this chapter is structured as follows. Section 2 describes the algorithm for detecting similar building footprints in urban area. Section 3 presents the urban morphology analysis. Section 4 describes the algorithm for estimation of building types on OSM. Section 5 presents the experimental results for the OSM data set in Heidelberg and evaluates the results by using ATKIS data. Finally, Sect. 6 concludes the whole work and gives some works in the future.

2 Finding Similar Building Footprints in Urban Area

As indicated in the introduction, we would like to investigate if the building types are similar when these buildings are similar in shape and size. For this purpose, building footprints have to be clustered into a number of groups at first, whereas building footprints in each cluster have similar shape and size.

Most of the existing research works about finding similar building footprints focus on urban building clustering, because building footprints with similar shapes and size form patterns according to Gestalt theory. The most common approaches to find similar building footprints are to define polygon of building footprints using a set of parameters, i.e. minimum distance, area of visible scope, area ratio, edge number ratio, smallest minimum bounding rectangle (SMBR), directional Voronoi diagram (DVD) and so on (Yan et al. 2008). Similar footprints then have similar parameters, for instance, Qi and Li (2008) calculated the similarity of two footprints by comparing their intersected and united areas. These approaches, however, can only find approximately similar building footprints. Therefore, they are more suitable for building footprints with less detailed geometries. In further, most of these approaches are very sensitive to orientations.

In the field of computer vision, polygon curve representation is used to measure similarity of polygons with high accuracy. Arkin et al. (1991) introduced turning function to represent polygons. As shown on the left of Fig. 1, let C be the polygon.

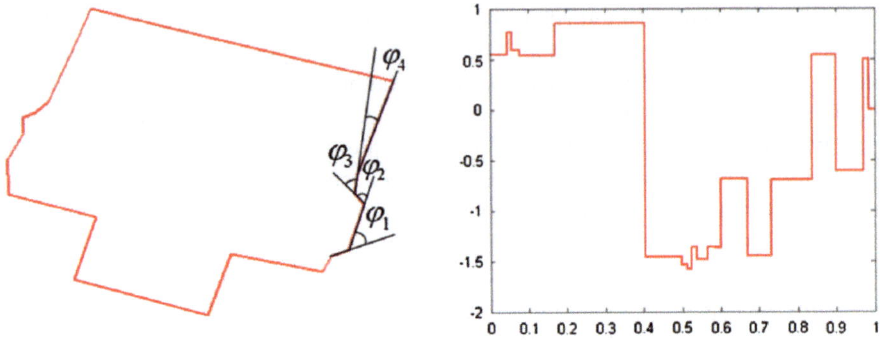

Fig. 1 Turning function representation of polygon

The tangent angle at the starting vertex is $\theta_1 = \varphi_1$. Then θ_i can be calculated as $\theta_i = \theta_{i-1} + \varphi_i$. The right of Fig. 1 shows the change of tangent angles (y-axis) along the normalized accumulated length of the polygon sides (x-axis). From this point of view, the tangent angle can be treated as a function of the normalized accumulated length $\boldsymbol{T}_c(\mathbf{l})$. It can be called tangent function or turning function.

The turning function $T_c(\mathbf{l})$ measures the angle of the counter-clockwise tangent as a function of the normalized accumulated length l. The cumulative angle increases with left hand turns and decreases with right hand turns. This kind of representation is invariant to rotation, because it contains no orientation information. Furthermore, it is invariant to scaling, since the normalized length makes it independent to the polygon size.

Then similarity of two polygons can be derived based on the L_2-norm of their turning functions.

$$\mathrm{S(A, B) = d(A, B) = \|T_A - T_B\|_2 = \left(\int_0^1 (T_A(l) - T_B(l))\right)^{\frac{1}{2}}} \tag{1}$$

Note that S(A, B) denotes actually the dissimilarity between A and B. The smaller S(A, B) is, the more similar are the two polygons. In the case A is identical to B, there is S(A, B) = 0.

However, the similarity measurement above is strongly affected by the starting point shift of the curve, because there is translation of the turning function when shifting the starting points. In the presented work, this problem is solved by resampling turning functions into power spectrum. Then similarity can be measured by comparing Fourier Descriptors of the power spectrum, as suggested by Lee et al. (2003).

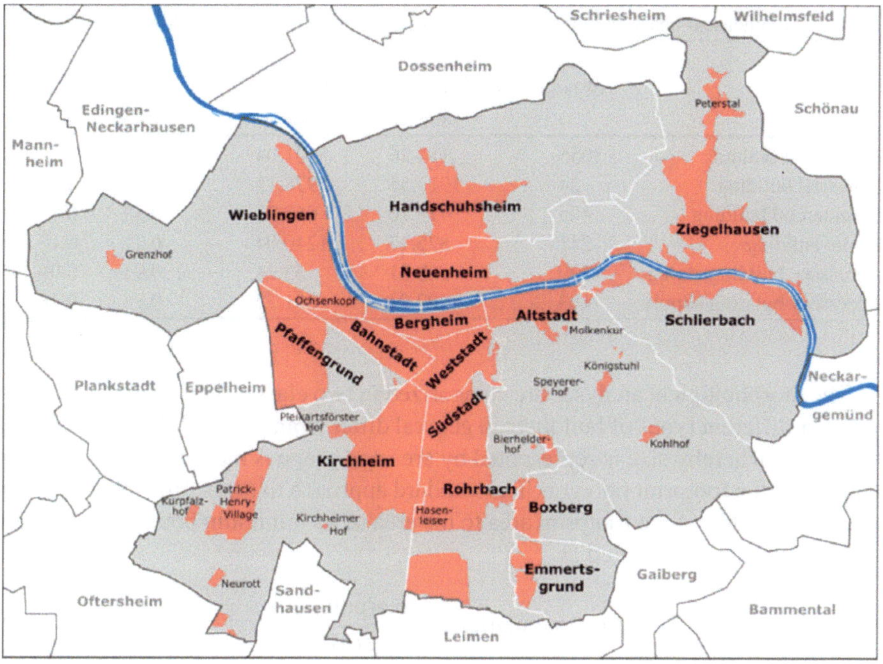

Fig. 2 Administrative districts in Heidelberg

3 Impact of Urban Morphology on Types of Buildings

Buildings are man-made structures and serve several needs of society. Hence, buildings have a significant socio-economic relevance. Their dimensions and architectures are strongly characterized by the purposes for that buildings are used, namely, the types of buildings. On the 2D map, these impacts are reflected by the shapes and sizes of building footprints. For instance, footprints of residential buildings normally have small area and relative simple shape—most of them are in form of rectangle, while public buildings normally have large and complicated footprints. In this work, the morphology of building footprints are analyzed using the ATKIS building footprints data in five city districts of the German city Heidelberg, namely, Altstadt, Bergheim, Boxberg, Emmertsgrund and Südstadt (Fig. 2). ATKIS stands for Amtliches Topographisch-Kartographisches Informationsystem—Authorative Topographic-Cartographic Information System. It is a common project of the Working Committees of the Survey Administrations of the States of the Federal Republic of Germany (AdV) (Grünreich 2000). It contains information on settlements, roads, railways, vegetation, waterways, and more. Building footprints data in ATKIS is represented by polylines with building height, types, address and other attribute information.

Table 1 Geometrical analysis of building footprints for different types of buildings

	Number of buildings	Area		Rectangularity	
		Mean value (m^2)	Standard deviation	Mean value	Standard deviation
Residential building	3055	154.26	101.04	0.88	0.11
Industrial building	24	1912.55	1978.13	0.92	0.12
Commercial building	405	479.10	660.71	0.83	0.15
Public building	211	875.17	1240.04	0.72	0.17
Accessory building storage	1390	38.21	33.24	0.95	0.09
Accessory building supply	103	09.04	851.72	0.83	0.18

The morphological analyses are categorized in two classes. Firstly, it investigated whether different types of buildings in general differ from size and shape of building footprints, whereby size is represented by area and shape is measured by the rectangularity of the footprint polygon. The standard approach to measuring rectangularity is to use the ratio of the footprint area to the area of its minimum bounding rectangle (MBR), which is calculated as

$$\text{rectangularity} = \frac{\text{area(polygon)}}{\text{area(MBR)}} \tag{2}$$

The larger the *rectangularity* is, the simpler the building is in architectural style. Oppositely, the smaller the rectangularity is, the more complicated the building is constructed in architecture.

Table 1 summarizes the results of the statistical analysis on building footprints of six types of buildings respectively. It shows that building types are characterized by the area and rectangularity of building footprints, although the built area of a certain type of buildings varies as much as the average area of this type of buildings. In terms of built area, industrial buildings and accessory buildings for storage can easily be differentiated from other types of buildings, because industrial buildings are normally very large and accessory buildings for storages are normally very small. Residential buildings are majority in a city. The most of residential buildings have built area of 100–200 m^2. And their footprints are relatively simple with respect to architectural style. In the contrast, public buildings are normally large and the rectangularity of their footprints is low, because many public buildings are treated as landmarks due to their extraordinary dimension and architectural styles in their local environment.

The second measurement of the morphological analysis is proposed according to the first law of geography (Tobler 1970). It says: "Everything is related to everything else, but near things are more related than distant things". In this work, the analysis is conducted based on the following hypothesis:

- Buildings share attributes i.e. types, heights, structure of roof and facades, if their footprints are similar in shape and size.

Table 2 Context analysis on types of building footprints in city district

	Number of buildings	Number of clusters	Average similarity of building type in clusters (%)
Altstadt	1978	534	79.17
Bergheim	877	282	79.36
Boxberg	545	111	94.18
Emmertsgrund	333	41	97.50
Südstadt	1189	133	89.88

Table 3 Context analysis on types of building footprints in urban block

	Number of urban blocks	Average similarity of attribute in clusters (%)
Altstadt	97	95.42
Bergheim	39	95.96
Boxberg	32	95.86
Emmertsgrund	12	100
Südstadt	47	94.53

- For buildings with similar footprints, the closer they are located to each other, the more likely they share attributes such as type.

In the second morphological analysis, building footprints in a city district are classified into a number of clusters according to their shapes and sizes, as described in Sect. 2. As shown in Table 2, a cluster in city center (Altstadt) contains 4 buildings in average, while in the suburban region (Emmertsgrund and Südstadt) a cluster contains approximately 10 buildings in average. This means that building footprints in city center reveal a larger diversity than those in suburban regions. In each cluster, the similarity of building types is calculated based on Eq. (3).

$$\text{Similarity}_{\text{type}} = \frac{\text{number of the most frequent building type}}{\text{numberof building footprints in the cluster}} \tag{3}$$

Comparing the average similarities in the five city districts, the types of buildings are about 80 % similar in city center while they are more than 90 % similar in suburban regions, if the building footprints are similar in shape and size.

The similarity of building type in a cluster is getting higher, when decreasing the search area from city district to urban block which is defined as the smallest area that is surrounded by streets. Table 3 indicates that the possibility that buildings have same types is more than 95 %, if they are located in the same urban block and their footprints are similar in shape and size.

4 Estimation of Building Types by Rule-Based Approach

Based on the morphological analysis in Sect. 3, for a building footprint $foot_A$ on OpenStreetMap, its type can be estimated as follows:

Step 1: Start with a building footprint $foot_A$ on OpenStreetMap. Keep the building type of $foot_A$, if there is an attribute for the building type.

Step 2: If there is no information about building type for $foot_A$

(a) In the same urban block, find building footprints $foot_{1,2,\ldots N}$ with similar shapes and sizes to $foot_A$. If there is information of building type in these building, take the majority type as the type of $foot_A$. Otherwise,
(b) In the whole area, find building footprints $foot_{1,2,\ldots N}$ with similar shapes and sizes to $foot_A$. If there is information of building type in these building, take the majority type as the type of $foot_A$. Otherwise,
(c) Estimate the building type using the results of the statistical analysis of building footprints in Table 1.

- If $|Area_{foot_A} - 150| \leq 50\,m^2$, $foot_A$ is a residential building.
- If $Area_{foot_A} \geq 2000\,m^2$ and $AI_{foot_A} \geq 0.9$, $foot_A$ is an industrial building.
- If $|Area_{foot_A} - 450| \leq 50\,m^2$, $foot_A$ is a commercial building.
- If $Area_{foot_A} \geq 750\,m^2$ and $AI_{foot_A} < 0.78$, $foot_A$ is a public building.
- If $Area_{foot_A} \leq 50\,m^2$, $foot_A$ is an accessory building for storage.
- If $|Area_{foot_A} - 800| \leq 50\,m^2$and $AI_{foot_A} \geq 0.83$, $foot_A$ is a an accessory building for supply.

(d) In an urban block, if there is an industrial building with $Area_{foot_A} \geq 2000\,m^2$ and $AI_{foot_A} \geq 0.9$, then change all the buildings in the urban block as industrial buildings.

5 Experiments and Evaluation

The proposed approach is implemented and tested for the OSM data set of Heidelberg in Germany. The test area covers $46.38\,km^2$. The whole area is composed of 14 administrative districts which are called as city districts in this work, as shown in Fig. 2. There are 32836 buildings in ATKIS while 14335 buildings in OSM, because many buildings in OSM are recorded as blocks of buildings in the real world on the one hand. On the other hand, a number of buildings are difficult to be recognized and mapped on OSM due to occlusion by vegetation and other buildings surrounding them. The ATKIS footprints in five city districts are used for morphology analysis, as described in Sect. 3, while the rest 9 city districts are used for the experiments and evaluation.

In this section, a preprocessing step is introduced to filter the building footprints which are recorded as block of buildings in OSM. Then the existing building types are

Table 4 Possible relations between building footprints in two data sets

Relation	1:1	1:0	1:n
Illustration	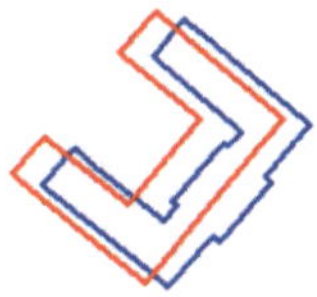	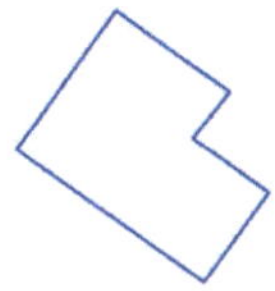	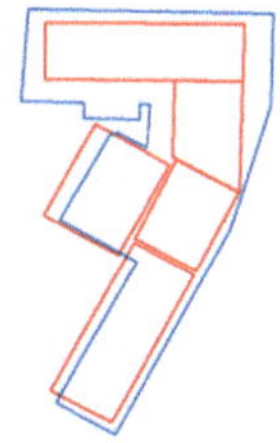
Relation	n:1	0:1	n:m
Illustration	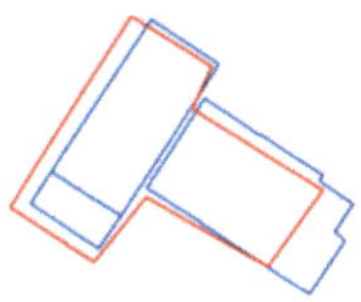	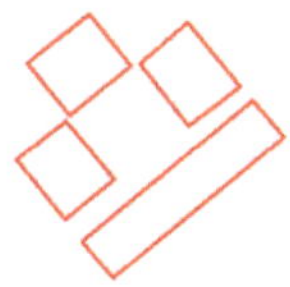	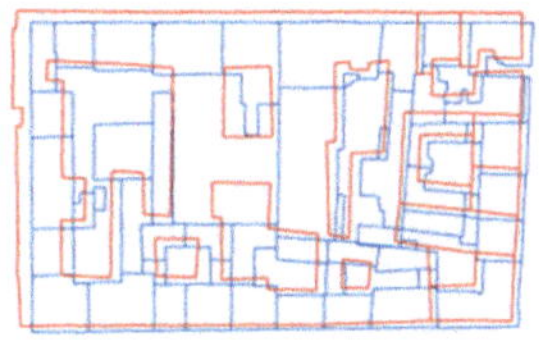

classified into six types and used as input data for the process of estimation building types based on urban morphology analysis. The results are presented and evaluated by using authority data of ATKIS building footprints.

5.1 Pre-processing

- Finding semantically correctly recorded buildings in OSM.

There might be 1:1, 1:n, 1:0, 0:1, n:1, and n:m relations between OSM building footprints and those in reference data, as shown in Table 4, whereby footprints in two data sets are distinguished in red and blue colors. While footprints in OSM are visualized in red color, footprints in ATKIS data are in blue.

According to the OGC standard of CityGML building models (Gröger et al. 2008), semantic hierarchy and geometrical level of details (LoD) relate them inherently. Hence, only 1:1 buildings are recorded semantically correctly in OSM.

In this work, building footprints are selected for the test of type estimation, when they have 1:1 relation to ATKIS building footprints. Since the most of building footprints in OSM have been digitalized according to the Bing Map images (http://www.bing.com/maps) (Goetz and Zipf 2012; OSM 2014b,c), there is normally offset between footprints in OSM and the reference data due to the distortion caused by oblique view of the used sensors. Considering this factor, large buildings in OSM have larger percentage of area overlap with their correspondence in the reference data, while small and high buildings might have smaller percentage of area overlap with their correspondence. The threshold of the judgment depends actually strongly upon the parameters of the Bing map images used for digitalization in OSM. In their work, Rutzinger et al. (2009) found out that the correspondence might be caused by

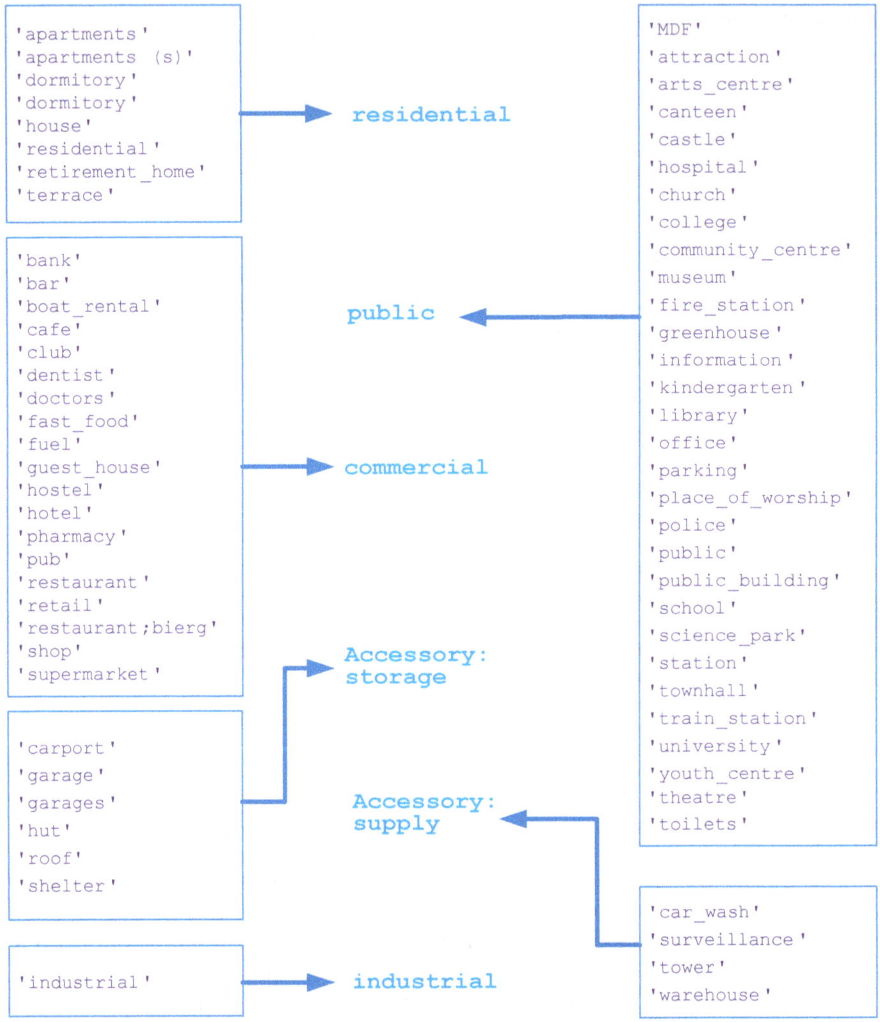

Fig. 3 classification of building types in OSM

their neighboring building if the overlapped area is less than 30 %. Therefore, the threshold of the overlapping is set as 30 %. If

$$\frac{\text{Area}_{\text{overlap}}}{\min\left(\text{Area}(\text{foot}_{\text{osm\_i}}), \text{Area}(\text{foot}_{\text{ATKIS\_j}})\right)} > 30\,\% \tag{4}$$

then the footprints Area($\text{foot}_{\text{osm\_i}}$ and $\text{foot}_{\text{ATKIS\_j}}$ are matched. A 1:1 relation is identified when a footprint in OSM can only be matched to one footprint in ATKIS.

- Classification of OSM building types into the pre-defined types.

Because users are allowed to define building types by themselves (OSM 2014c), there arise a large number of building types with high duplication and ambiguity. In this work, these building types are classified into six types (see Fig. 3): (i) residential buildings, (ii) industrial buildings, (iii) commercial buildings, (iv) public buildings, (v) accessory buildings for storage and (vi) accessory buildings for supply.

5.2 Experimental Results and Evaluation

In our test field, 12382 of 14335 building footprints in OSM were selected using the method in Sect. 5.1, because they have 1:1 relation to ATKIS footprints and are regarded as semantically correctly mapped on OSM. Among these selected building footprints, 2027 buildings are recorded with types of buildings on OSM. The building types of these 2027 building footprints are compared with those of ATKIS data. The results are listed in Table 5. It shows that: (i) the overall accuracy of the building type recording in Heidelberg on OSM is 88.86 %; (ii) approximately 90 % residential buildings are mapped with correct types, as well as industrial, public and storage buildings; (iii) commercial buildings and accessory building for supply are badly recorded with types, because they are normally difficult to be differentiated from residential and public buildings.

Although the types of 2027 buildings in Heidelberg are recorded with errors (with an accuracy of 88.86 %), they are treated as correct and used as seeds in the process of type estimation, as described in Sect. 5. Table 6 summarizes the results of type estimation for the Heidelberg data on OSM. Residential buildings are estimated with high quality (with accuracy of 92.21 %). There are several reasons. First of all, residential buildings are majority in the city. Even all the buildings are labeled as residential buildings; accuracy over 60 % can be achieved. Secondly, in the input OSM data (i) the majority of the buildings recorded with types on OSM are residential buildings; (ii) the residential buildings are recorded with high accuracy (93.13 %); and (iii) the buildings recorded as residential building are distributed almost everywhere in Heidelberg. Therefore, they contribute much with respect of context effect during the process of type estimation.

Note in the evaluation of type estimation in Heidelberg, the accuracy of type recording in the input OSM data is not considered.

Despite of residential buildings, the accessory buildings for storage are estimated with high accuracy, too, because they can be easily distinguished from other buildings due to their characteristics in shape and size, namely, almost rectangular and in small size of area. Similarly, around 70 % of industrial buildings are correctly estimated because of the simple architectural shape and large size in footprints. However, commercial buildings, public buildings and accessory building for supply are not so good estimated, because they reveal huge diversity in shape and size and therefore can be easily confused with each other. Many of these buildings are estimated as residential buildings, because buildings similar to residential buildings can also be used for commercial, public or (energy) supply buildings.

Table 5 Confusion matrix of the type record in Heidelberg on OSM

	Building types in ATKIS data: true data						
	Residential buildings	Industrial buildings	Commercial buildings	Public buildings	Accessory building storage	Accessory building supply	
Estimated building types	Residential buildings	1303	1	20	5	13	17
	Industrial buildings	4	21	0	0	0	3
	Commercial buildings	25	0	35	6	5	6
	Public buildings	47	0	16	207	8	19
	Accessory building storage	20	1	0	0	229	4
	Accessory building supply	0	0	3	0	2	7
Total in column	1399	24	74	218	257	56	
Producer accuracy (%)	93.13	87.50	47.30	94.95	89.11	12.50	
Overall accuracy (%)	88.86						

Table 6 Confusion matrix of the type estimation for OSM data in Heidelberg

	Building types in ATKIS data: true data						
	Residential buildings	Industrial buildings	Commercial buildings	Public buildings	Accessory building storage	Accessory building supply	
Estimated building types	Residential buildings	9060	17	198	116	6	53
	Industrial buildings	0	111	7	7	0	8
	Commercial buildings	371	9	316	109	16	25
	Public buildings	254	12	126	304	5	3
	Accessory building storage	4	0	6	6	1242	54
	Accessory building supply	136	11	5	52	239	152
Total in column	9825	160	658	594	1508	295	
Producer accuracy (%)	92.21	69.38	48.02	51.18	82.36	51.53	
Overall accuracy (%)	85.77						

Table 7 The accuracy of type estimation in the nine city districts in Heidelberg

	Producer accuracy	Overall accuracy					
	Residential (%)	Industrial (%)	Commercial (%)	Public (%)	Storage (%)	Supply (%)	
Handschuhsheim	94.12	100	52.63	57.14	91.48	71.42	89.27
Kirchheim	97.22	79.31	45.22	78.26	85.12	92.98	88.61
Neuenheim	96.62	60	66.67	48.89	81.55	37.04	89.08
Paffaffengrund	92.56	100	43.55	41.67	88.10	52.22	83.84
Rohrbach	98.30	–	41.66	27.65	78.66	27.53	90.85
Schlierbach	95.30	0	0	38.46	78.62	32.14	86.41
Weststadt	95.47	0	77.12	58.82	87.12	63.93	91.71
Wieblingen	95.99	95.12	8.60	25.93	91.86	48.70	81.38
Ziegelhausen	93.98	0	82.61	66.67	95.23	72.46	90.66

For the nine city districts in Table 7, the process of type estimation is conducted again, and respectively. In most of city districts, better results are obtained than using the whole Heidelberg data as input data, since not only the overall accuracies are higher than that of using the whole Heidelberg data as input data, but also the producer's accuracies are higher. The reason is that the search area is reduced from the whole city to city district in the step 2b of the process of type estimation (Sect. 4). This verifies the hypothesis in Sect. 3: buildings with similar footprints might share attributive information, and the closer the similar footprints are located, the more likely they have same attributes.

6 Conclusion and Future Works

In this work, a rule-based approach for building type estimation is proposed with the attempt of data enrichment for OpenStreetMap footprints data. The rules are derived based on two hypotheses. First of all, building types are very impacted by the shapes and sizes of building footprints. Secondly, for buildings with similar footprints in shape and size, the more closely they are located to each other, the more likely they have the same building type. These two hypotheses are proofed by using authority footprints data in four city districts in Heidelberg by means of urban morphology analysis. The proposed approach is tested for OSM building footprints data in Heidelberg at first. The overall accuracy for the type estimation is 85.77 %. With respect to the individual type of buildings, residential buildings are estimated with high accuracy, as well as industrial and accessory buildings for storage. Only about 50 % of commercial buildings, public buildings and accessory building for supply are correctly estimated, because they reveal huge diversity in shape and size and therefore can be easily confused with each other. Many of these buildings are

Fig. 4 Residential buildings on OSM before type estimation

estimated as residential buildings, because buildings similar to residential buildings can also be used for commercial, public or (energy) supply buildings. When reducing the input area to city district level, better results can be yielded. Building types are estimated with both higher overall accuracy and producer's accuracies. This proofs our second hypothesis as well.

In comparison with the original accuracy of building type recording on the input data of OSM, the proposed approach achieve higher accuracy, in terms of both the overall accuracy and producer's accuracies. Besides, in the evaluation of type estimation in Heidelberg, the accuracy of type recording in the input OSM data is not considered. If this kind of impact could be calculated, even higher accuracy can be achieved by the proposed approach.

There are still many error estimations, especially for the commercial buildings, public buildings and accessory buildings for storage, because only 50 % of these kinds of buildings can be correctly estimated with type. As mentioned above, one reason is that these kinds of buildings reveal huge diversity in shape and size. Other reasons could be that: (i) the types in the input OSM data are labeled with relative low accuracies, (ii) there are ambiguous classifications when classifying a lot of building types on OSM into six building types, and (iii) similarly, there are confusion classifications when classifying a lot of building types on authority data into six buildings types.

In the future, the abovementioned building type classification will be investigated for better correspondences in the two databases. Secondly, the proposed approach will be tested for other cities in Germany, in order to investigate the dependency of parameters upon different cities in different regions. Thirdly, the algorithm of Support Vector Machine (SVM) will be used for the type estimation, whereby the authority data and buildings with known information of types will be used as training data respectively. In further, the results by using SVM will be compared with the proposed approach presented in this work.

Acknowledgments This work was supported in part by the Klaus Tschira Foundation and the National Natural Science Foundation of China under Project: 41101443.

References

Arkin EM, Chew LP, Huttenlocher DP, Kedem K, Mitchell JSB (1991) An efficiently computable metric for comparing polygonal shapes. IEEE Trans Pattern Anal Mach Intell 13:3

Fan H, Zipf A, Fu Q, Neis P (2013) Quality assessment for building footprints data on OpenStreetMap. Int J Geogr Inf Sci. doi:10.1080/13658816.2013.867495

Goetz M (2013) Towards generating highly detailed 3D CityGML models from OpenStreetMap. Int J Geogr Inf Sci. doi:10.1080/13658816.2012.721552

Goetz M, Zipf A (2012) OpenStreetMap in 3D—detailed insights on the current situation in Germany. In: AGILE 2012. Avignon, France

Gröger G, Kolbe TH, Czerwinski A, Nagel C (2008) OpenGIS city geography markup language (CityGML) encoding standard—version 1.0.0. OGC Doc. No. 08–007r1

Henn A, Roemer C, Groeger G (2012) Automtic classification of building types in 3d city models using svms for semantic enrichment of low resolution building data. Geoinformatica 16:281–306

Huang H, Kieler B, Sester M (2013) Urban building usage labeling by geometric and context analyses of the footprint data. In: Proceeding of 26th international cartographic conference (ICC), Dresden, Germany

Lee DJ, Antani S, Long LR (2003) Similarity measurement using polygon curve representation and fourier descriptors for shape-based vertebral image retrieval. In: Proceeding of International Society for Optics and Photonics, Medical Imaging, Bellingham, 1283–1291

Li Q, Li Z (2008) An approach to building grouping based on hierarchical constraints. In: The international archieves of the photogrammetry. Remote sensing and spatial information science, vol XXXVII. Part B2, Beijing

OSM (2014a) Stats—OpenStreetMap Wiki. http://wiki.openstreetmap.org/wiki/Statistics. Accessed 19 Feb 2014

OSM (2014b) Bing—OpenStreetMap Wiki. http://wiki.openstreetmap.org/wiki/Bing. Accessed 19 Feb 2014

OSM (2014c) Buildings–OpenStreetMap Wiki. http://wiki.openstreetmap.org/wiki/Buildings. Accessed 19 Feb 2014

Rutzinger M, Rottensteiner F, Pfeifer N (2009) A comparison of evaluation techniques for building extraction from airborne laser scanning. IEEE J Sel Top Appl Earth Obs Remote Sens 2(1):11–20

Tobler W (1970) A computer movie simulating urban growth in the detroit region. Econ Geogr 46(2):234–240

Uden M, Zipf A (2012) open building models–towards a platform for crowdsourcing virtual 3D cities. In: 7th 3D GeoInfo conference. Quebec City, QC, Canada

Yan H, Weibel R, Yang B (2008) A multi-parameter approach to automated building grouping and generalization. Geoinformatica 12(1):73–89. doi:10.1007/s10707-007-0020-5 http://dx.doi.org/10.1007/s10707-007-0020-5

Qualitative Representations of Extended Spatial Objects in Sketch Maps

Sahib Jan, Angela Schwering, Malumbo Chipofya and Talakisew Binor

Abstract With the advent of Volunteered Geographic Information (VGI) the amount and accessibility of the spatial information such as sketched information produced by layperson increased drastically. In many geo-spatial applications, sketch maps are considered an intuitive user interaction modality. In sketch maps, the spatial objects and their relationships enable users to communicate and reason about their actions in the physical environment. The information people draw in sketch maps are distorted, schematized, and incomplete. Thus, processing spatial information from sketch maps and making it available in information systems requires suitable representation and alignment approaches. As typically only qualitative relations are preserved in sketch maps, performing alignment and matching with geo-referenced maps on qualitative level has been suggested. In this study, we analyzed different qualitative representations and proposed a set of plausible representations to formalize the topology and orientation information of extended objects in sketch maps. Using the proposed representations, the qualitative relations among depicted objects are extracted in the form of Qualitative Constraint Networks (QCNs). Next, the obtained QCNs from the sketch maps are compared with QCN derived from the metric maps to determine the degree to which the information is identical. If the representations are suitable, the QCNs of both maps should be identical to a high degree. The consistency of obtaining QCNs allows the alignment and integration of spatial information from sketch maps into Geographic Information Systems (GISs).

S. Jan (✉) · A. Schwering · M. Chipofy · T. Binor
Institute for Geoinformatics, University of Muenster, Münster, Germany
e-mail: Sahib.jan@uni-muenster.de

A. Schwering
e-mail: schwering@uni-muenster.de

M. Chipofy
e-mail: mchipofya@uni-muenster.de

T. Binor
e-mail: talakisew@uni-muenster.de

J. Huerta et al. (eds.), *Connecting a Digital Europe Through Location and Place*,
Lecture Notes in Geoinformation and Cartography, DOI: 10.1007/978-3-319-03611-3_3,

Keywords Qualitative representations · Extended objects · Sketch maps · Qualitative constraints networks · Qualitative alignment

1 Introduction

Hand-drawn sketch maps have extensively been used to investigate how humans memorize spatial knowledge. Cognitive maps and cognitive collages (Tolman 1948; Tversky 1993) have been suggested as metaphors to describe mental organization of geographic information. They relate the concrete and detailed spatial information from the physical environment to abstract and conceptual information stored in our brain (Casakin et al. 2000). Sketch maps are used to externalize the individual's mental image of the environment. They contain objects which represent real world geographic features, relations between these objects, and oftentimes symbolic and textual annotations (Blaser 1998). These spatial objects and their relationships enable us to use sketch maps to communicate about our environments and to reason about our actions in those environments. In this way, sketch maps provide an intuitive user interaction modality for many geo-spatial applications (Egenhofer 1996; Nedas and Egenhofer 2008; Wallgrün et al. 2010). Especially with the advent of Volunteered Geographic information (VGI) (Goodchild 2007), sketch maps may be the key to contribute spatial information in Geographical Information Systems (GIS) without taking into account the technical barriers imposed by traditional GIS as noted by Goodchild (2007).

The information represented in sketch maps reflects the user's spatial knowledge that is based on observations rather than on measurements. However, Humans' cognitive maps are typically distorted, schematized, incomplete, and generalized, thus the information in sketch maps is equally distorted, schematized, incomplete, and generalized (Tversky 1992, 2003; Huynh and Doherty 2007). Cognitive errors documented in Wang and Schwering (2009), Schwering and Wang (2011) are neither random nor solely due to human ignorance. In sketch maps, people present a few significant objects and their configuration in terms of qualitative relations (Wang and Schwering 2009; Schwering and Wang 2011). In many GIS applications (Egenhofer 1996; Nedas and Egenhofer 2008; Wallgrün et al. 2010), these relations are used to represent and reason about spatial configurations between depicted objects. However, processing spatial information from sketch maps and making it available in information systems requires computational approaches to represent, align, and integrate the sketched spatial information.

During the last two decades, a series of qualitative spatial calculi have been proposed in the area of Qualitative Spatial Reasoning (QSR) (Freksa 1993), focusing on different aspect of space such as representations for the topological relations (Randell et al. 1992; Cohn et al. 1997), orderings (Allen 1983; Schlieder 1995; Osmani 1999), directions (Frank 1996; Renz and Mitra 2004), relative position of points (Moratz et al. 2000, 2005; Renz and Mitra 2004) and others. These representations provide general and sound reasoning mechanisms based on spatial configurations in terms of qualitative relations. Wang et al. (2010, 2011) identify a set of qualitative aspects in sketch maps throughout a series of experiments. These qualitative aspects represent spatial configurations between depicted objects in terms of relations. In our previous study (Jan et al. 2013), we propose a set of coarsened representations to formalize the ordering aspect of spatial objects in sketch maps.

This study extends our previous work on qualitative representations of spatial objects in sketch maps. In this study, we propose qualitative representations to formalize spatial configurations between extended objects such as containment (topology) of landmarks in city-block, their orientations, and the topology of city-blocks themselves. We identify these representations being robust against schematizations, distortions, and other cognitive effects (Tversky 1992, 2003; Huynh and Doherty 2007) found in sketch maps. Using the proposed representations, we extracted qualitative information of extended objects in the form of Qualitative Constraint Networks (QCNs) (Wallgrün et al. 2010; Chipofya et al. 2013). Next, the obtained QCNs from the sketch and metric maps are tabularized to evaluate the proposed representations. The evaluation is done by testing the accuracy of qualitative relations between landmarks and city-blocks from sketch maps with the qualitative relations of corresponding spatial objects in metric maps—generated from OpenStreet Map. The tested sketch maps (28 in total) are from two different locations (area about 1.04 and 2.10 km^2) in Münster, Germany. All the sketch maps are generated by different participants and most of them were holding an academic degree at University of Muenster, Germany. Though none of the participants were residents of the predefined locations, all of them were familiar with the locations by frequent visits by foot or vehicle. During the experiment, participants were asked to produce sketch maps of predefined locations as detailed as possible but only from memory.

The results of the evaluation show that the proposed representations are suitable to formalize the qualitative information of extended objects. They provide high accuracy of identical relations between objects from sketch and metric maps. The highly identical qualitative relations will allow users to align and integrate spatial information from sketch maps into geographic information systems (GISs) as VGI (Goodchild 2007).

The remainder of this chapter is structured as follows: In the following section, we briefly introduce related work. In Sect. 3, we introduce extended objects found in sketch maps. Representations are proposed to formalize the qualitative information of extended objects in Sect. 4, which are evaluated with respect to accuracy of qualitative information in Sect. 5. Section 6 concludes the chapter with an outlook on future work.

2 Related Work

In qualitative representations, everyday descriptions of distinguishing the relative direction (left, right), distance (near, far), and topology (disjoint, overlap) are used to identify associations or correspondences between scenes. During the last two decades, several approaches attempt to capture spatial configurations between objects qualitatively. Egenhofer (1996, 1997) propose Spatial-Query-by-Sketch to query spatial databases using a sketch-based interface. It focuses on specifying spatial relations by drawing them. The approach uses five types of spatial relations such as coarse, detail topological relations, metric refinements, and coarse and detailed cardinal directions relations to capture spatial configurations between depicted objects.

Forbus et al. (2003) develop a sketch understanding system, nuSketch, which is a battle-space search system that focuses on qualitative topological reasoning. The system uses both qualitative topological relations and quantitative information to construct spatial configurations between depicted entities. Nedas and Egenhofer (2008) propose a similarity measure to compare two spatial scenes by identifying similarities between (i) objects in the two scenes, (ii) similarity between the binary relations among spatial objects such as buildings and lakes, and (iii) the ratio of the total number of objects to the number of objects that has been matched—or equivalently, not matched.

Similarly, there are several approaches on how graph-like structure can be represented qualitatively. For example, Wallgrün et al. (2010) propose an approach for qualitative matching of geo-/non-referenced datasets using qualitative relations between spatial objects. In Chipofya et al. (2013), we propose a simple model for matching qualitatively described spatial scenes extracted from sketch maps. The qualitative direction relations over points in the plane depend on the angles formed by the points, where angles that yield the same direction relation belong to a common direction sector bounded by different angles.

There are two types of qualitative representations that allows for defining sectors with different angles: the STAR calculi (Renz and Mitra 2004) for absolute directions and $\mathcal{OPRA}$ calculi (Moratz et al. 2005) for relative directions. In STAR calculus, the direction sectors are same for every point p in the plane, while the sectors in $\mathcal{OPRA}$ depend on the orientation of p. Lücke et al. (2011) propose a qualitative approach for navigating in the street-network. They use Oriented Point Relation Algebra ($\mathcal{OPRA}$) (Moratz et al. 2005) together with Klippel's turn directions (Klippel and Montello 2007) for navigating in the street-network. Renz and Wölfl (2010) use STAR calculi (Renz and Mitra 2004) for representing direction sectors in order to have a consistent sector arrangement for every intersection node in the route-network.

All the above cited approaches use the method of representing spatial configurations with some abstract qualitative relations. They share motivation with our work and use similar methods of representing spatial configurations in sketch maps. However, they did not consider the influences of human spatial cognition and the effects of cognitive distortions (Tversky 1992, 2003; Huynh and Doherty 2007) in the qualitative representation and alignment of spatial objects. Since spatial objects' outlines in freehand sketches are imprecise, the qualitative representation of spatial objects with imprecise boundaries leads to different qualitative relations when compare with relations in geo-referenced maps.

In this study, a set of plausible qualitative representations is proposed to formalize topology and orientation information of extended objects in sketch maps.

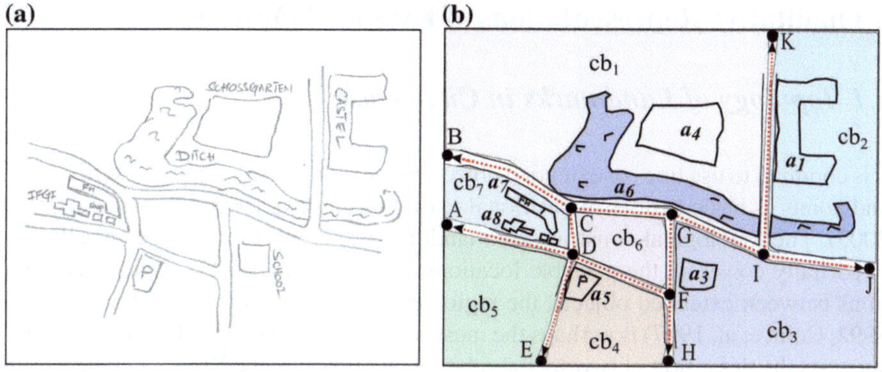

Fig. 1 **a** The unprocessed sketch map with depicted street segments and landmarks. **b** Street segments, landmarks, and city-blocks in the processed sketch map

3 Extended Objects in Sketch Maps

3.1 Landmarks

According to Blaser (2000), landmarks and road entities are the most frequently depicted spatial objects in the sketch map. In freehand sketches, landmarks are vectorized and approximated by polygons. They are represented as multiple intersecting or non-intersecting strokes. Sketcher considers main street segments that lead to frequently visited or important landmarks and few side street segments which contain landmarks along them (Huynh and Doherty 2007). In sketch maps, landmarks represent spatial entities such as water bodies, buildings, and parks (see Fig. 1a).

3.2 City-Blocks

City-blocks are important areal features for sketch map alignment. We define city-blocks as the smallest area, completely surrounded by street segments. In our representation, a city-block plays the role of a container for other spatial objects such as buildings, water bodies, and parks (see Fig. 1b). People do not always sketch complete city-blocks. They may sketch a network of the streets without any loop because they have omitted other street segments, in particular at the edge of sketch medium. Therefore, sketch maps do not contain many closed city-blocks. In order to maximize the number of city-blocks, we consider them as areas bounded not only by the street segments, but also by the medium-boundary. Therefore, all incomplete street segments with endpoints towards the medium-boundary are extended. This is done until either the medium-boundary or other street segments extension is encountered (see Fig. 1b).

4 Qualitative Representations of Extended Objects

4.1 Topology of Landmarks in City-Blocks

It is common to use lines or extended objects as basic entities in topological reasoning and points as basic entities in positional reasoning (Freksa 1993; Moratz et al. 2000, 2005). The topological constraints on landmarks and city-blocks together allow us to partially constrain the possible locations of the landmarks. For topological relations between extended objects, the region connected calculus RCC (Randell et al. 1992; Cohn et al. 1997) is perhaps the most well-known topological formalism. RCC supports the definition of two spatial relation algebras, i.e. the RCC5 and the RCC8. These two algebras make a small number of five and eight topological distinctions between regions.

For the topology of landmarks in the city-blocks, we analyze different qualitative representations that support extended objects as entity types such as region connected calculus RCC (Randell et al. 1992; Cohn et al. 1997) and string based topological representation (Li and Liu 2010). In sketch maps, city-blocks are non-overlapping regions, while landmarks may overlap several city-blocks (see Fig. 2a). Since spatial objects' outlines in freehand sketches are imprecise, the distinction between overlapping and disconnected boundaries becomes less important when landmarks are involved. Similarly, for the topology of landmarks with respect to city-blocks, the distinction between completely inside and sharing boundaries become less important. Therefore, we propose RCC5 to capture the topological relations between landmarks and city-blocks. RCC5 base relations consists of DR ("discrete"), PO ("partially-overlap"), PP ("proper-part"), PPi ("proper-part inverse"), and EQ ("equal"). The RCC5 provides topological relations at an abstract level, which overcomes the effects of schematization and distortion of landmark's boundaries in qualitative representation and alignment.

Figure 2a shows, the landmarks and detected city-blocks (delineated by street segments and page-boundary) in the sketch map. Using RCC5, the topology about landmarks with respect to city-blocks are represented as follows: landmarks a_1 is *proper-part* of the city-block (cb_2), landmark a_2 *partially overlaps* on city-blocks (cb_1 and cb_2), and landmarks a_4 is *proper-part* of the city-block (cb_1), while landmarks them self are disconnected from each other. Figure 2b shows the constraint network for topological relations between landmarks with respect to city-blocks in the sketch map.

4.2 Topology of City-Blocks

Triangulation of City-blocks. In freehand sketches, the city-blocks are mixed (concave and convex) regions surrounded by street segments and medium-boundary. The qualitative representation of these mix regions increases topological relations

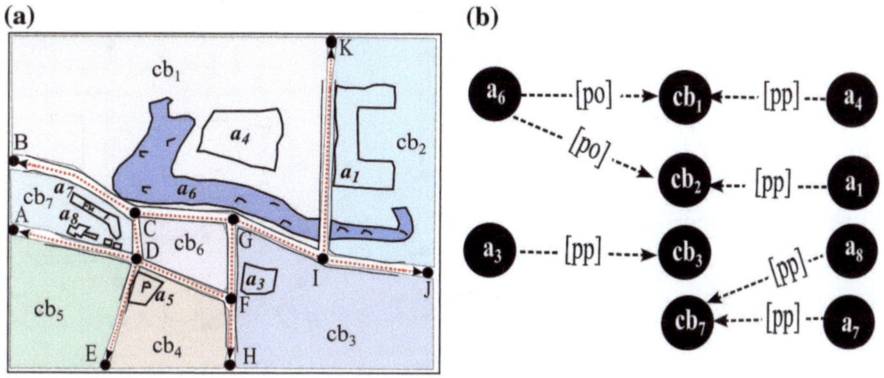

Fig. 2 **a** Landmarks and city-blocks in the sketch map. **b** QCN for the topological relations of landmarks with respect to city blocks using the RCC5 representation

significantly because there are many uncountable topologically different regions in the plane (containing infinite holes and connected components).

However, we need to restrict our representation of city-blocks to specific regions, such as simple regions (homomorphic to a closed disk), convex regions or rectangles. In many practical applications, arbitrarily shaped spatial objects are approximated by their convex hulls. Randell et al. (1992) represent the problem of representing qualitative relations of concave objects with the help of their convex hulls. Bennet et al. (1998) consider regions as the union of convex polygons. However, the qualitative representation using convex hulls of approximated city-blocks lead to different topological relations when compared with the relations of city-blocks in corresponding metric maps.

For the qualitative representation of city-blocks, we restrict ourselves to convex regions. Convexity plays a central role in computational geometry, geographical information science, and several other disciplines. For qualitative representation, concave city-blocks from sketch and metric maps are detected using an algorithm and decomposed them into a set of triangles (convex regions), known as a triangulation (Eberly 2002). In triangulation, concave polygon of n vertices is decomposed into $n-2$ triangles with the help of the ear-clipping algorithm (Eberly 2002). For example, we have concave city-blocks cb_1 and cb_3 in sketch map; the vertices of bounded street segments are used to decompose them into sets of triangles (see Fig. 3b).

Formally, the decomposition of concave city-blocks can be described as follows: A city-block A is a region surrounded by street segments and medium-boundary. It is divided into a set of triangles, called parts, such that the union of all parts constitutes A itself (Eq. 1), and all parts are mutually exclusive (Eq. 2).

$$\mathrm{A} = \bigcup_{\mathrm{i}=0}^{\mathrm{n}} \mathrm{A_i} \qquad \text{With } \mathrm{n} \geq 1 \tag{1}$$

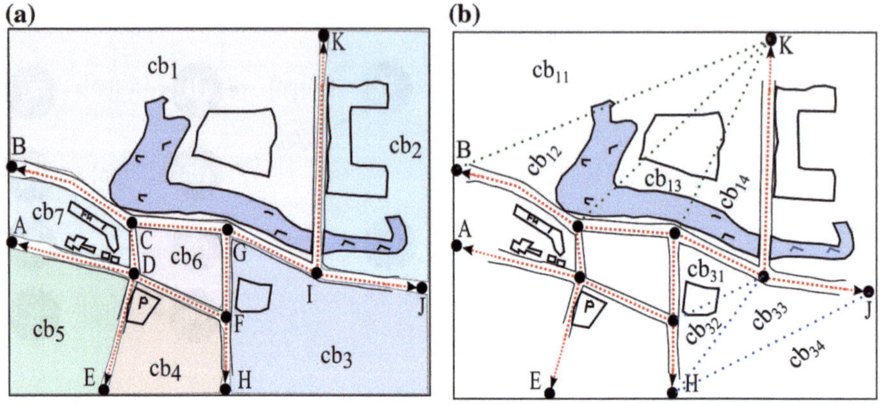

Fig. 3 **a** Concave city-blocks cb_1 and cb_3 bounded by street segments and page-boundary. **b** Decomposition of the city-blocks into sets of *triangles*

$$\forall A_i, j | i \neq j\colon A_i \cap A_j = \emptyset \tag{2}$$

Qualitative Representation. In order to formalize topological relations among city-blocks, we analyze region based qualitative representations (Randell et al. 1992; Li and Liu 2010). Since street information is incomplete in sketch maps, we find aggregated city-blocks covering larger areas quite often. Using RCC8 (Randell et al. 1992), we find the topological relations such as disconnected (DC), externally connected (EC) between city-blocks very often. In sketch maps, city-blocks are externally connected by street segments or connected diagonally at junctions. Using RCC8 (Randell et al. 1992), the relation EC represents the connectivity of the city-blocks without differentiating their connectivity by street segments or junctions. For example, the city-block cb_6 is EC with city-blocks cb_1, cb_7, cb_4, and cb_3 by street segments (see Fig. 3a). It provides the same EC relation for the connectivity between city-block cb_6 and cb_5 which are connected diagonally at junction D. The aggregation of street segments in the sketch map leads to different topological relations among city-blocks when compare with topological relations between corresponding city-blocks in metric map. Therefore, it is important to make the distinction between these two types of externally connected scenarios when city-blocks are involved.

We propose the qualitative representation known as a model for topological relations between convex regions (Li and Liu 2010). Using representation, the atomic topological relation between two convex regions can be uniquely represented as a circular string. It provides a complete classification for topological relations between regions. Using circular string, two configurations are topologically equivalent if they have the same string representation. If we have two non-equal convex region (a, b), the topological relations $\alpha_{a,b}$ between interiors (°) and boundaries (∂) of a and b is represented by a circular string over {u, v, x, y}. If a region $a \neq b$ and a is not contained in the interior of b, each maximally connected component (mcc)

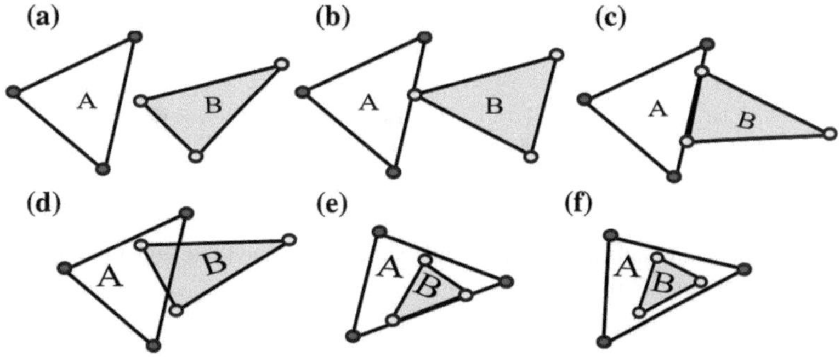

Fig. 4 Topological relations between convex regions using the *circular* strings {ε, u, v, x, and y}. **a** A (ε)B, **b** A(x)B, **c** A(y)B, **d** B(uxvx)A, **e** B(vy)A, **f** B(u)A

of $a^\circ \cap \partial b$ or $\partial a \cap b^\circ$ is homomorphic to the open interval (0,1) and each mcc of $\partial a \cap \partial b$ is single point or line (Li and Liu 2010). The circular string represents following possible intersections.

u represents mcc of $\left(a^\circ \cap \partial b\right)$
v represents mcc of $\left(\partial a \cap b^\circ\right)$
x represents 0—dimensional mcc of $(\partial a \cap \partial b)$
y represents 1—dimensional mcc of $(\partial a \cap \partial b)$

The circular strings {(ε), (**u**), (**v**), (**x**), (**y**)} represent the atomic topological relations DC, NTPP, NTPP^{-i}, and two refine sub-relations for the EC of RCC (Randell et al. 1992). The refine sub-relations for EC distinguish the topological relations between regions which are externally connected by lines or points. Similarly, the combinations of strings represent PO, TPP, and TPP^{-i} of RCC (Randell et al. 1992). Figure 4 shows, the possible topological relations between two regions using circular strings.

As shown in Fig. 3b, we have two concave city-blocks cb_1 and cb_3 in the sketch map. The decomposition of these concave city-blocks provide sets of triangles $cb_1 = \{cb_{11}, cb_{12}, \ldots, cb_{14}\}$ and $cb_3 = \{cb_{31}, cb_{32}, \ldots, cb_{34}\}$, which are basically convex sub-regions of the given concave city-blocks. If A_i and B_i represents a set of triangles in the city-block cb_1 and cb_3, then the topological relations between city-blocks can be inferred using the possible topological relations between triangles.

If the boundary of at least one triangle in A_i intersects with the boundary of B_i and the intersection of the interiors of A_i and B_i is empty, then the boundary–boundary intersection between two city-blocks is non-empty (Eq. 3). Similarly, if the boundaries and interiors of all triangles in A_i and B_i have empty intersections, then the intersection between city-blocks is also empty (Eq. 4).

$$\partial \bigcup_{i=0}^{n} A_i \cap \partial \bigcup_{i=0}^{n} B_i = \neg\emptyset \wedge \bigcup_{i=0}^{n} \overset{\circ}{A_i} \cap \bigcup_{i=0}^{n} \overset{\circ}{B_i} = \emptyset \rightarrow \text{A} \cap \text{B} = \neg\emptyset \qquad (3)$$

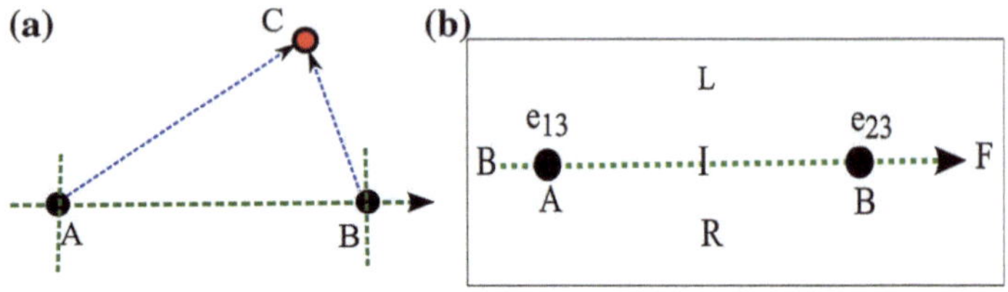

Fig. 5 **a** The orientation information of C with respect to oriented line going through A and B. **b** The basic relations of $\mathcal{LR}$ where $A \neq B$

$$\partial \bigcup_{i=0}^{n} A_i \cap \partial \bigcup_{i=0}^{n} B_i = \emptyset \wedge \bigcup_{i=0}^{n} \mathring{A_i} \cap \bigcup_{i=0}^{n} \mathring{B_i} = \emptyset \rightarrow \mathrm{A} \cap \mathrm{B} = \emptyset \quad (4)$$

Using constraints among interiors and boundaries of triangles, we formalize the set of possible topological relations between aggregated city-blocks in the sketch maps and relations between city-blocks in the metric maps. These relations can be used to infer new knowledge and can be combined for more than two parts such that $\mathbf{R_i} \otimes \mathbf{R_j} \otimes \mathbf{R_k}$ derive the relation between two pairs of adjacent regions (Tryfona and Egenhofer 1997).

4.3 Orientation of Landmarks

A person's ability to establish his or her location in an environment is termed spatial orientation (Correa de Jesus 1994). From the cognitive point of view, the spatial orientation is considered as the capability to form a cognitive map (Golledge et al. 1996). Human beings always distinguish between spatial objects ahead of them or behind their back, spatial objects on their right and on their left-side, when they proceed along the path (Scivos and Nebel 2004).

To formalize the orientation information about adjacent landmarks with respect to street segments, different qualitative representations for relative orientations are investigated such as $\mathcal{LR}$ calculus (Scivos and Nebel 2004), Single Cross Calculus (SCC) (Freksa 1992), and Double Cross Calculus (DCC) (Freksa 1992). We exclude the DCC as it is not closed under composition and permutation and there exists no finite refinement of the base relations with such a closure property (Scivos and Nebel 2001). We propose $\mathcal{LR}$ calculus (Scivos and Nebel 2004), an enhanced and refined version of the FlipFlop Calculus (Ligozat 1993). The $\mathcal{LR}$ calculus deals with point type entities in the plane R^2. It describes the position of a point C with respect to two other points A (the origin) and B (the relatum) as illustrated in Fig. 5a.

For configurations with $A \neq B$, the following base relations are distinguished using $\mathcal{LR}$ calculus: C can be to the *left* or to the *right* of the oriented line going through A and B, or C can be placed on the line resulting in one of the five relations *inside, front, back, start* or *end* (see Fig. 5b). The $\mathcal{LR}$ calculus (Scivos and Nebel

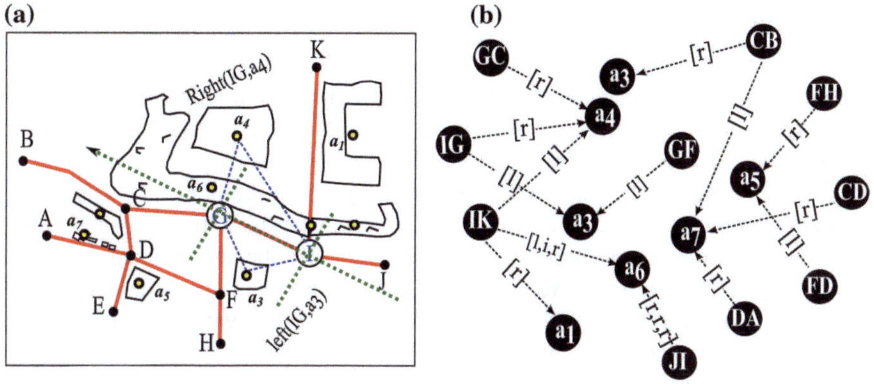

Fig. 6 **a** Orientation of landmark a_4 and a_3 in the sketch map using the $\mathcal{LR}$ calculus, **b** QCN for relative orientation of the adjacent landmarks with respect to street segments

2004) introduces relations *dou* (A = B ≠ C) and *tri* (A = B = C) as additional relations. Overall, the $\mathcal{LR}$ calculus provides nine relations. The orientation relation of an object using $\mathcal{LR}$ is represented as A, B (reILR) C.

For positional reasoning, it is common to use points as basic entities (Freksa 1993; Moratz et al. 2000). To fulfill the requirements of proposed representation, landmarks are approximated by the centroids of their minimum bounding boxes. The landmarks which are stretched over multiple city-blocks, the centroids of their sub-regions in each city-block are considered. For example, landmark a_6 is a water body that stretched over city-block cb_1 and cb_2 (see Fig. 6a). We have three approximated points, one point for each sub-region and a point on the street segment that intersects the landmark. The start-and end-junctions of street segments are used as origin and relatum points and the orientation information about adjacent landmarks is extracted. The $\mathcal{LR}$ calculus provides the orientation information of landmarks at an abstract level, which overcomes the effects of schematizations and distortions of reference street segments in the relative orientation of adjacent landmarks.

For example in Fig. 6a, we have street segments with their start- and end-junctions such as JI, IK, and IG. The orientation information about adjacent landmarks with respect to oriented street segments is represented as $r(I, G : a_4)$, where relation r represents the orientation of the landmark a_4 with respect to a street segment IG. Similarly, the orientation information about landmarks a_6 (water body stretched over multiple city-blocks) is represented as $l, i, r(I, K : a_6)$. The qualitative relation left (l), inside (i), and right (r) represents the relative orientation of a_6 with respect to reference street segment IK. Figure 6b shows the orientation of landmarks with respect to street segments in the form of QCN.

5 Evaluation

In this section, the proposed representations are evaluated by aligning aforementioned 28 sketch maps with corresponding metric maps. The sketch maps are generated by different participants at University of Münster, Germany. Spatial aggregation is ubiquitous in the sketch maps world. Particularly, street segments and city-blocks are highly aggregated spatial features (see Fig. 7a). After our participants have completed drawing the sketch maps, we asked them to describe and indicate the spatial objects on the metric maps. We also asked participants to indicate the corresponding street segment for every sketched street segment, thus we got information on how streets were aggregated. As city-blocks are delineated by street segments, these street segments are used as reference objects to identify corresponding city-blocks in metric maps on an aggregated level.

In order to evaluate the proposed representations, the topology and orientation information of extended objects is extracted. The qualitative information is derived from the geometric representations of sketch and metric maps manually. Since, the proposed representations for the topological relations are binary and the representation for the orientation of landmarks is ternary, we extracted both binary and ternary qualitative relations in the form of QCNs.[1]

Next, the obtained QCNs from the sketch maps are compared with QCNs derived from the metric maps to determine the degree to which information is identical. If the representations are suitable, the QCNs of sketch maps and metric maps should be identical to a high degree. In order to align depicted landmarks and city-blocks, we identify the possible pairing of nodes from one QCN with those in the other QCN. The hypothesis of matching city-blocks and landmarks are generated based on a visual analysis, where we consider all depicted landmarks and city-blocks from sketch maps and identify their corresponding spatial objects in metric maps.

5.1 Topology of Landmarks in City-Blocks

As proposed above, the qualitative representation RCC5 is used to formalize the topological relations between landmarks and city-blocks. The QCN derived from sketch maps are tabularized and compare with the corresponding QCN derived from metric maps. From a visual analysis of the tabularized QCNs, we find that the qualitative constraints using the RCC5 have an average accuracy rate of 99.87 % (see Table 1). While using the RCC8, we have an accuracy rate of 99.02 % for both locations. In sketch maps, most of the landmarks are depicted correctly with respect to city-blocks. However, we fine some mismatched topological relations using the RCC8, because

---

[1] Complete QCN comparison files and images are available at http://ifgibox.de/s_jan001/QualitativeRepresentation_ExtendedObjects/.

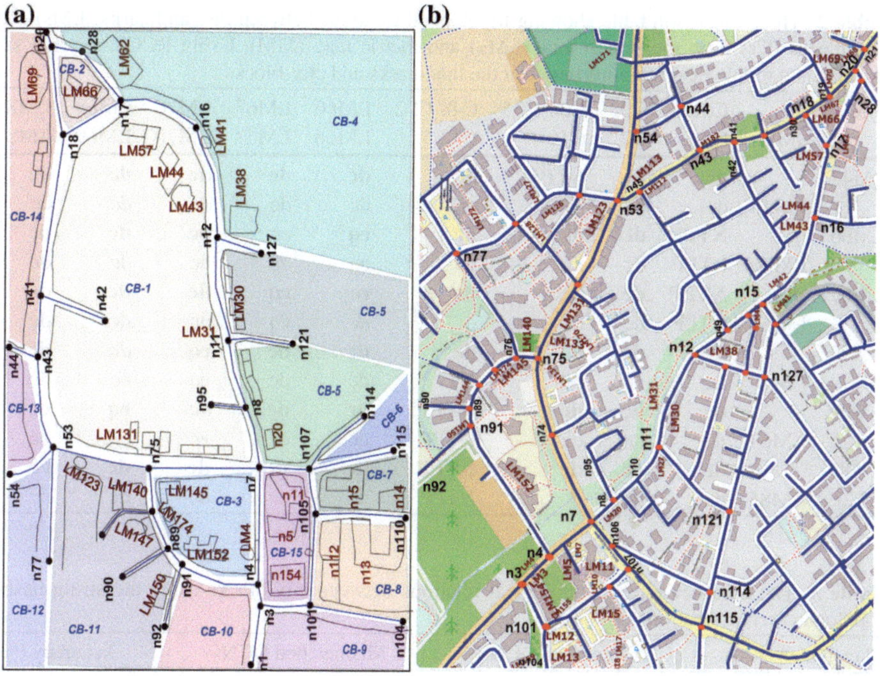

Fig. 7 **a** Landmarks and city-blocks in the sketch map. **b** Corresponding spatial objects in the metric map of the location-II

Table 1 Accuracy of topological relations between landmarks and the relations of landmarks with respect to city-blocks in the sketch and metric maps using the RCC5 representation

Sketch maps (28)	Total # of QCNs	Mismatched QCNs	Accuracy (%)
Location-I	3543	2	99.96
Location-II	22275	24	99.78
Average			99.87

topological relations such as EC, DC, TPP, and TPP^{-I} in RCC8 require precise boundary information of landmarks, which is not possible in freehand sketches.

Table 2 shows the inconsistent topological relations between landmarks and city-blocks from sketch and metric maps using the RCC8. For example, the landmark LM66 is disconnected with LM67 in the sketch map (SM_7), while the same landmarks are externally connected with each other in the metric map. Similarly, the landmark LM42 is *tangential proper-part* with respect to city-block (C10) in the sketch map. The same landmark is *non-tangential proper-part* with respect to corresponding city-block in the metric map.

The high accuracy of topological relations indicates that the proposed representation overcomes the effects of schematizations and distortions of landmark's boundaries in qualitative representation and alignment.

Table 2 The comparison table showing topological relations between the landmarks (LMs) and city-blocks (Cs) from the sketch map (SM_7) and metric map (MM). Using RCC8, we find six mismatched topological relations between the landmarks and city-blocks

MM		C61	C23	C24-C35	C48-C52	LM66	LM67	LM70	LM69	LM84
	SM_7	C9	C2	C10	C12	LM66	LM67	LM70	LM69	LM84
LM42		**dc**	**dc**	***NTPP***	**dc**	**dc**	**dc**	**dc**	**dc**	**dc**
	LM42	dc	dc	*TPP*	dc	dc	dc	dc	dc	dc
LM66		**NTPP**	**dc**	**dc**	**dc**	**eq**	***ec***	**dc**	**dc**	**dc**
	LM66	NTPP	dc	dc	dc	eq	*dc*	dc	dc	dc
LM67		**NTPP**	**dc**	**dc**	**dc**	***ec***	**eq**	**dc**	**dc**	**dc**
	LM67	NTPP	dc	dc	dc	*dc*	eq	dc	dc	dc
LM70		**dc**	**dc**	**dc**	**dc**	**dc**	**dc**	**eq**	***dc***	**dc**
	LM70	dc	dc	dc	NTPP	dc	dc	eq	*ec*	dc
LM69		**dc**	**dc**	**dc**	**NTPP**	**dc**	**dc**	***dc***	**eq**	**dc**
	LM69	dc	dc	dc	NTPP	dc	dc	*ec*	eq	dc
LM85		**dc**	**dc**	**dc**	**NTPP**	**dc**	**dc**	**dc**	**dc**	***ec***
	LM85	dc	dc	dc	NTPP	dc	dc	dc	dc	*dc*

Table 3 The accuracy rate of topological relations between city-blocks using the string based topological representation

Sketch maps (28)	Total # of QCNs	Mismatched QCNs	Accuracy (%)
Location-I	1210	40	96.94
Location-II	3631	42	98.65
Average			97.79

5.2 Topology of City-Blocks

To formalize the topology of the city-blocks, we use string based representational model (Li and Liu 2010). For both locations, the string based topological relations between city-blocks are extracted in the form of QCNs and compare with QCNs derived from the metric maps. We find the string based topological relations between city-blocks have an average accuracy rate of 97.79 % for both locations (see Table 3). While using the RCC8, we have an accuracy rate of 98.75 %. Since the EC relation of the RCC8 loses the important distinction of external connectivity, the proposed string based representation seems a promising way to formalize topological relations between city-blocks.

5.3 Orientation of Landmarks

We use $\mathcal{LR}$ calculus (Scivos and Nebel 2004) to formalize the orientation information about the adjacent landmarks. The street segments are used as reference objects to localize nearby landmarks. Nearness is defined via the distance in a Voronoi

Table 4 The accuracy of orientation information about landmarks with respect to street segments in the sketch maps using the $\mathcal{LR}$ calculus

Sketch maps	Total # of QCNs	Mismatched QCNs	Accuracy (%)
Location-I	272	2	99.43
Location-II	719	13	98.27
Average			98.85

diagram (Aurenhammer 1991). A landmark is considered adjacent, if its footprint is in the Voronoi diagram of the reference object. For both locations, the orientation information about landmarks is extracted from the sketch and metric maps. The orientation information about landmarks using $\mathcal{LR}$ calculus has an average accuracy rate of 98.85 % (see Table 4).

We compare these results to QCNs obtained from the Single Cross Calculus (SCC) (Freksa 1993). Using the SCC, the average accuracy rate drops to 93.67 %. Therefore, the $\mathcal{LR}$ representation is suitable to formalize the orientation information of landmarks in sketch maps.

6 Conclusion and Future Work

Sketch maps represent the physical environment in a highly schematized and therefore distorted way. To formalize the topology and orientation information of extended objects, we analyze the existing qualitative representations in the area of Qualitative Spatial Reasoning (QSR). We propose a set of qualitative representations that are robust against cognitive distortions. The proposed representations are evaluated by comparing the QCN matrices derived from the sketch maps and corresponding metric maps. Overall evaluation shows that the representations are suitable to extract high degree of identical information, and thus are reliable for the alignment of spatial objects from sketch maps with geo-referenced maps. This way, the additional sketched information such as information on the usage of the buildings, bakeries, and completely un-mapped information about landmarks and areas can be transferred to the geographic information system as volunteered information.

In the present study, we handle the aggregation and detection of city-blocks manually. So, the future work comprises, in part, the automatic detection of the city-blocks and the methods for handling aggregations. In evaluation, we use the sketch maps of an urban spatial structure. In the future, we will investigate the relevance of the proposed representations for the alignment of sketch maps from the rural areas. We evaluated proposed representations by comparing QCNs manually. The problem of QCN matching is ongoing research work. Evaluation of the representations using the matching model is a part of our future work.

Acknowledgments Research for this chapter is based upon work supported by the Deutsche Forschungsgemeinschaft (DFG) under grant for the SketchMapia project (584820).

References

Allen JF (1983) Maintaining knowledge about temporal intervals. Commun ACM 26:832–843

Aurenhammer F (1991) Voronoi diagrams: a survey of a fundamental geometric data structure ACM computing surveys. ACM Comput Surv 23:345–405

Bennett B, Isli A, Cohn AG (1998) A system handling RCC-8 queries on 2D regions representable in the closure algebra of half-planes. In: Proceedings of the 11th international conference on industrial and engineering applications of artificial intelligence and expert systems IEA-98-AIE Benicàssim, Castellón, Spain, 1–4 June 1998

Blaser A (1998) Geo-spatial sketches. Technical report. Department of Spatial Information Science and Engineering and National Center for Geographic Information and Analysis, University of Maine, Maine, USA

Blaser A (2000) A study of people's sketching habits in GIS. Spat Cogn Comput 2:393–419. doi:10.1023/A:1015555919781

Casakin H, Barkowsky T, Alexander K, Christian F (2000) Schematic maps as wayfinding aids. In: Freksa C et al (eds) Spatial cognition II. Lecture notes in computer science, vol 1849. Springer, Berlin, pp 54–71

Chipofya M, Schwering A, Binor T (2013) Matching qualitative spatial scene descriptions 'a la Tabu. In: Proceedings of the 12th Mexican international conference on artificial intelligence, MICAI 2013, Mexico City, Mexico, 24–30 Nov 2013

Cohn A, Bennett B, Gooday J, Gotts N (1997) Qualitative spatial representation and reasoning with the region connection calculus. GeoInformatica 316:275–316

Correa de Jesus S (1994) Environmental communication: design planning for wayfinding. Des Issues 10:32–51

Eberly D (2002) Triangulation by ear clipping. Magic Software, Inc 2002

Egenhofer MJ (1996) Spatial-query-by-sketch. In: Burnett M, Citrin W (eds) IEEE symposium on visual languages, vol 96. IEEE, Boulder, Colorado, pp 60–67

Egenhofer MJ (1997) Query processing in spatial-query-by-sketch. J Vis Lang Comput 8:403–424

Forbus K, Usher J, Chapman V (2003) Qualitative spatial reasoning about sketch maps. In: Proceedings of the fifteenth annual conference on innovative applications of artificial intelligence, Acapulco, Mexico, 12–14 Aug 2003

Frank AU (1996) Qualitative spatial reasoning: cardinal directions as an example. Int J Geogr Inf Syst 10:269–290. doi:10.1080/02693799608902079

Freksa C (1993) Dimensions of qualitative spatial reasoning. In: Carreté NP, Singh MG (eds) Proceeding III MACS-international workshop on qualitative reasoning and decision technologies—QUARDET'93, Barcelona, 1993

Freksa C (1992) Using orientation information for qualitative spatial reasoning. In: Frank AU, Campari I, Formentini U (eds) Theories and methods of spatio-temporal reasoning in geographic space. Lecture notes in computer science, vol. 639. Springer, Berlin, pp 162–178

Golledge R, Klatzky R, Loomis M (1996) Cognitive mapping and wayfinding by adults without vision. In Portugali J (ed) The construction of cognitive maps. GeoJournal Library, vol. 32. Springer, Netherlands, pp 215–246

Goodchild MF (2007) Citizens as sensors: the world of volunteered geography. GeoJournal 49:211–221

Huynh NT, Doherty ST (2007) Digital sketch-map drawing as an instrument to collect data about spatial cognition. Cartographica Int J Geogr Inf Geovisualization 42:285–296. doi:10.3138/carto.42.4.285

Jan S, Schwering A, Wang J, Chipofya M (2013) Ordering: a reliable qualitative information for the alignment of sketch and metric maps. In: Proceedings of the IEEE 12th international conference on cognitive informatics and cognitive computing (ICCI*CC13), New York, USA, 2013
Klippel A, Montello DR (2007) Linguistic and nonlinguistic turn direction concepts. In: Winter S, Kuipers B, Duckham M, Kulik L (eds) Spatial information theory. Lecture notes in computer science, vol. 4736. Springer, Berlin, pp 354–372
Li S, Liu W (2010) Topological relations between convex regions. In: Proceedings of the 24th AAAI conference on artificial intelligence (AAAI-10), Atlanta, Georgia, USA, 11–15 July 2010
Ligozat G (1993) Qualitative triangulation for spatial reasoning. In: Frank AU, Campari I (eds) Spatial information theory a theoretical basis for GIS, European conference, COSIT'93. Lecture notes in computer science, vol 716, Marciana Marina, Elba Island, Italy. Springer, Berlin
Lücke D, Mossakowski T, Moratz R (2011) Streets to the OPRA—finding your destination with imprecise knowledge. In: Renz J, Cohn AG, Wölfi S (eds) IJCAI workshop on benchmarks and applications of spatial reasoning, vol 27. Barcelona, Spain, pp 25–32
Moratz R, Dylla F, Frommberger L (2005) A relative orientation algebra with adjustable granularity. In: Proceedings of the workshop on agents in real-time and dynamic environments (IJCAI05), Edinburgh, Scohtland, 2005
Moratz R, Renz J, Wolter D (2000) Qualitative spatial reasoning about line segments. In: Horn W (ed) Proceedings of the 14th European conference on artificial intelligence (ECAI'00), Berlin
Nedas KA, Egenhofer MJ (2008) Spatial-scene similarity queries. Trans GIS 12:661–681
Osmani A (1999) Introduction to reasoning about cyclic intervals. In: Imam I, Kodratoff Y, El-Dessouki A, Ali M (eds) Multiple approaches to intelligent systems. Lecture notes in computer science, vol. 1611. Springer, Berlin, pp 698–706
Randell DA, Cui Z, Cohn AG (1992) A spatial logic based on regions and connection. In: Proceedings of the 3rd international conference on knowledge representation and reasoning. Morgan Kaufmann, San Mateo, 1992
Renz J, Mitra D (2004) Qualitative direction calculi with arbitrary granularity. In: Zhang C, Guesgen HW, Yeap WK (eds) PRICAI-04. Lecture notes in computer science, vol 3157. Springer, Berlin, pp 65–74
Renz J, Wölfl S (2010) A qualitative representation of route networks. Frontiers in artificial intelligence and applications (ECAI), vol 215. IOS Press, Amsterdam, pp 1091–1092. doi:10.3233/978-1-60750-606-5-1091
Schlieder C (1995) Reasoning about ordering. In: Frank AU, Kuhn W (eds) Spatial information theory: a theoretical basis for GIS, international conference COSIT '95, semmering, Austria, 21–23 Sept 1995
Schwering A, Wang J (2011) SketchMapia: a framework for qualitative mapping of sketch maps and metric maps. In: Las Navas 20th anniversary meeting on cognitive and linguistic aspects of geographic spaces. Las Navas del Marques, Avila, Spain, 4–8 July 2010
Scivos A, Nebel B (2004) The finest of its class: the natural, point-based ternary calculus LR for qualitative spatial reasoning. In: Freksa C et al (2005) Spatial cognition IV. Reasoning, action, interaction: international conference spatial cognition. Lecture notes in computer science, vol 3343. Springer, Berlin, pp 283–303
Scivos A, Nebel B (2001) Double-crossing: decidability and computational complexity of a qualitative calculus for navigation. In: Montello DR (ed) Spatial information theory, COSIT 2001. Lecture notes in computer science, vol 2205. Springer, Berlin, pp 431–446
Tolman EC (1948) Cognitive maps in rats and men. Psychol Rev 55:189–208
Tryfona N, Egenhofer M (1997) Consistency among parts and aggregates: a computational model. Trans GIS 1:189–206
Tversky B (1993) Cognitive maps, cognitive collages, and spatial mental models. In: Frank AU, Campari I (eds) Spatial information theory: a theoretical basis for GIS, proceedings COSIT '93. Lecture notes in computer science, vol 716. Springer, Berlin, pp 14–24
Tversky B (1992) Distortions in cognitive maps. Geoforum 23:131–138. doi:10.1016/0016-7185(92)90011-R

Tversky B (2003) Structures of mental spaces: how people think about space. Environ Behav 35: 66–80

Wallgrün OJ, Wolter D, Richter K-F (2010) Qualitative matching of spatial information. In: 18th SIGSPATIAL international conference on advances in geographic information systems, ACM, USA, 2–5 Nov 2010

Wang J, Muelligann C, Schwering A (2011) An empirical study on relevant aspects for sketch map alignment. In: Proceedings of the 14th AGILE international conference on geographic information science (AGILE 2011), Utrecht, Netherlands, 2011

Wang J, Mülligann C, Schwering A (2010) A study on empirically relevant aspects for qualitative alignment of sketch maps. In: Proceedings of the sixth international conference on geographic information science (GIScience). Zurich, Switzerland, 2010

Wang J, Schwering A (2009) The accuracy of sketched spatial relations: how cognitive errors affect sketch representation. Presenting spatial information: granularity, relevance, and integration. workshop at COSIT 2009, AberWrac'h, France, 21–25 Sept 2009

Exploring the Geographical Relations Between Social Media and Flood Phenomena to Improve Situational Awareness

A Study About the River Elbe Flood in June 2013

Benjamin Herfort, João Porto de Albuquerque, Svend-Jonas Schelhorn and Alexander Zipf

Abstract Recent research has shown that social media platforms like twitter can provide relevant information to improve situation awareness during emergencies. Previous work is mostly concentrated on the classification and analysis of tweets utilizing crowdsourcing or machine learning techniques. However, managing the high volume and velocity of social media messages still remains challenging. In order to enhance information extraction from social media, this chapter presents a new approach that relies upon the geographical relations between twitter data and flood phenomena. Our approach uses specific geographical features like hydrological data and digital elevation models to prioritize crisis-relevant twitter messages. We apply this approach to examine the River Elbe Flood in Germany in June 2013. The results show that our approach based on geographical relations can enhance information extraction from volunteered geographic information, thus being valuable for both crisis response and preventive flood monitoring.

Keywords Social media · Volunteered geographic information · Disaster management · Flood · Situation awareness · Emergency management

1 Introduction

Managing an emergency puts high demands on authorities and crisis management organizations. Collecting as much information as possible about the unfolding crisis and making sense of that information in a timely manner is critical to subsidise relief efforts. One of the main challenges in emergency management is thus achieving

B. Herfort · J. P. de Albuquerque · S.-J. Schelhorn · A. Zipf
Chair of GIScience, Heidelberg University, Heidelberg, Germany

J. P. de Albuquerque (✉)
Department of Computer Systems, ICMC, University of Sao Paulo, Sao Carlos, Brazil
e-mail: jporto@icmc.usp.br

J. Huerta et al. (eds.), *Connecting a Digital Europe Through Location and Place*,
Lecture Notes in Geoinformation and Cartography, DOI: 10.1007/978-3-319-03611-3_4,

situation awareness, which can be defined as "the perception of elements in the environment within a volume of time and space, the comprehension of their meaning, and the projection of their status in the near future" (Endsley 1995).

Social media platforms like Twitter, Flicker or Instagram are growingly used by crisis-affected individuals. Hence, they are used to share local knowledge that can be a vital source of crisis-relevant information. Although this topic has been a subject of research in the recent years (Vieweg et al. 2010; Kongthon et al. 2012; Sakaki et al. 2010; Terpstra et al. 2012; Imran et al. 2013), the process of collecting and analysing social media information has to be further improved and evaluated to offer better insights and information that really contributes to situation awareness.

Scientific research on crisis management and social media has concentrated on filtering and classifying microblog posts, e.g. tweets, applying crowdsourcing (i.e. manual message classification by volunteers) (Gao et al. 2011; Rogstadius et al. 2011; Lofi et al. 2012) or natural language processing and machine learning (Sakaki et al. 2010; Terpstra et al. 2012; Imran et al. 2013). Nevertheless, a crucial problem remains unsolved. During a crisis, the volume and the velocity of posted tweets are extremely high. Distinguishing messages that contain critical information from off-topic messages in an efficient and reliable way is the basic requirement for any feasible approach for dealing with the information overload. However, existing approaches are only partially successful in this regard.

The use of existing and well-studied geographical models about natural hazards hold a non-exploited potential to tackle the open problem of handling social media information during a crisis. In the end, this could lead to crisis-relevant and actionable information, thus contributing to situation awareness and better decision-making. There is initial work in this field (e.g. Triglav-Čekada and Radovan 2013), but this is not as comprehensively studied as other approaches. This chapter thus seeks to contribute with a new approach for leveraging existing geographical knowledge of the flood phenomena in order to improve the usefulness of VGI in crisis management.

In the pursuit of this goal, we apply a geographical approach to prioritize crisis-relevant information from social media. Our methodology is based on specific geographical relations of flood phenomena, for example hydrological features and models of terrain and affected areas. Furthermore, we conduct a case study for the River Elbe Flood in Germany in June 2013 to validate our approach. Combining information from tweets, water level measurements and digital elevation models, we thus seek to answer the three research questions as follows.

- RQ1. Does the spatiotemporal distribution of flood-related tweets match the spatiotemporal distribution of the flood phenomenon?
- RQ2. Does the spatial distribution of flood-related tweets differ dependent on their content?
- RQ3. Is distance to flood phenomena a useful parameter to prioritize social media messages in order to improve situation awareness?

The remainder of this chapter is organized as follows. In Sect. 2, related work on social media and volunteered geographic information in the context of crisis management is presented. In Sect. 3 we present our research approach. Information about the case study and the different datasets we employed is given in Sect. 4. In Sects. 5 and 6 we present the methodology and the results of our study. Finally, Sect. 7 concludes the chapter by discussing our findings and possibilities for future work.

2 State of the Art

Social media is significantly influencing social interactions. Social media is a "disruptive technology" (Hiltz and Plotnick 2013), especially because it provides an alternative to traditional authoritative information from governmental institutions like civil protection or mapping agencies (Goodchild and Glennon 2010). The ability to communicate and share geographical data through simple, freely-available tools that can be quickly learnt without demanding professional or scientific background is key to this development. This is described by the terms Neography, Volunteered Geographic Information (VGI) and Crowdsourcing (Goodchild and Glennon 2010; Gao et al. 2011; Hudson-Smith et al. 2009).

Social media has also become a potentially useful tool during crises. Citizens adapted social media applications like social sites, document management, multimedia sharing, microblogging and geo-location systems to suit their crisis management needs. Twitter, for instance, enables victims to quickly connect with the rest of the world, and so can help to minimize the effects of catastrophes and supports disaster relief (Kaewkitipong et al. 2012). On the one hand, social media offers a new communication channel for government agencies to reach the media and informing affected citizens (Chatfield and Brajawidagda 2013). Therefore, the pervasive use of social media causes significant implications for emergency management practice and policy (Palen 2008).

Many studies have examined the way of extracting crisis-relevant information from social media messages (e.g. Yin et al. 2012; MacEachren et al. 2011; Imran et al. 2013). In particular, scientific research focused on twitter messages, so called tweets, using either crowdsourcing or machine learning techniques. Sakaki et al. (2010) investigated the real-time nature of twitter and were able to detect crisis-related twitter messages using a support vector machine (SVM). Kongthon et al. (2012) analysed twitter messages about the flood that affected Thailand in 2011, concluding that, due to its up-to-the-minute character, analysis and classification of twitter messages can be useful in coordinating resources and efforts and in preparing and planning for disaster relief. Imran et al. (2013) tested an automatic method for filtering crisis-relevant social media messages vis-à-vis a crowdsourcing approach, i.e. based on manual classification by volunteers. Their results show that machine learning can be utilized to extract structured information from unstructured text-based twitter messages.

Vieweg et al. (2010) analysed twitter messages referring to the Red River Floods in spring 2009. Graham et al. (2012) analysed the use of Twitter during the UK floods in November 2012, by mapping geo-referenced tweets mentioning the words "flood" and visually checking whether the distribution of tweets corresponds to rainfall data and official flood alerts. The authors conclude that the digital trails of twitter messages are mostly matched to official data on floods and metereological precipitation. Triglav-Čekada and Radovan (2013) gathered information about the November 2012 floods in Slovenia from VGI. Their research shows that volunteered image gathering is a comparable alternative to satellite imagery.

Furthermore, there are several studies that examine the use of social media as a tool for improving situation awareness during crises (Yin et al. 2012). The so-called "crisis maps", as exemplified by the Ushahidi platform (Okolloh 2009), are the most recent entrants to the social media field (Goolsby 2010). Meier (2012) compares the value that live crisis maps can provide for situation awareness with a bird's-eye view of an unfolding event. MacEachren et al. (2011) develop and implement tools for visually-enabled information foraging and sense-making.

Most of the extant research is focused on analysing data from social media as a stand-alone information source, although situation awareness should arise from the combination of different data sources. Gao et al. (2011) state that scientific data could augment VGI to provide more detailed insights on information requirements and needs during a disaster. The integration and fusion with official and scientific data sources could lead to progress in validating and verifying information gathered from social media and thus improve the fitness-for-use of VGI as a source for crisis relevant information. This is the direction pursued in the present study, which is applied to the case of the floods in the river Elbe basin in Germany in 2013.

3 Research Approach

This chapter addresses the problem of enhancing the extraction of useful information from VGI and social media for improving situation awareness during emergencies. In contrast to the approaches reviewed in the previous section, which resort to either crowdsourcing or machine learning, our approach is based on the geographical relations between flood phenomena and social media messages.

Inspired by Tobler's first law of geography (Tobler 1970), we assume that near things are more related than distant things. Regarding crisis events, this implies that the spatiotemporal characteristics of the catastrophe affect the spatiotemporal characteristics of VGI and social media messages. As such, our approach seeks to leverage existing knowledge about the spatiotemporal characteristics of flood phenomena to improve information extraction from social media. In doing this, the hypothesis posed here is that social media messages which are closer to the flooded areas are 'more related' to the unfolding event, thus being more useful for improving situation awareness.

Our approach explores the relations between spatial information from twitter messages and the knowledge about flood phenomena both from hydrology and official

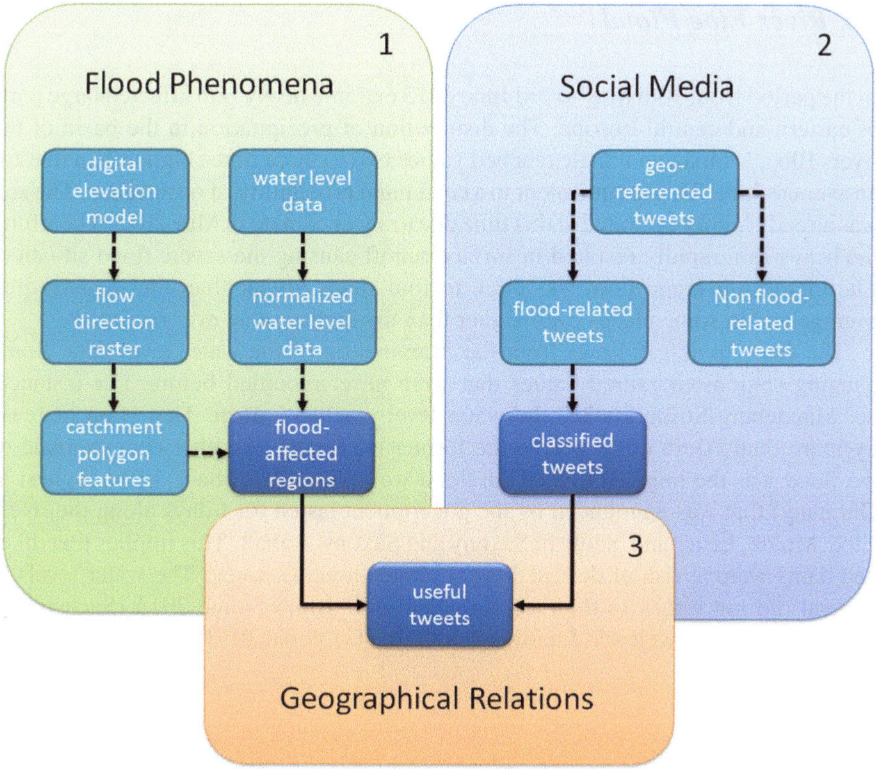

Fig. 1 Research approach

sensor data. The goal is to test our hypothesis that the distance to flood phenomena is a useful resource to prioritize messages for improving situation awareness.

Figure 1 schematically depicts our approach. It is divided into three main components: (1) gathering information on flood phenomena, i.e. flood-affected regions; (2) gathering information from social media, i.e. georeferenced twitter messages; (3) analysing the geographical relations between the information on flood phenomena (1) and social media messages (2) to assess the usefulness of tweets.

In this chapter, this approach is applied to analyse the use of Twitter during the River Elbe flood in 2013.

4 Description of the Case Study and Datasets

This section provides a description of our case study followed by an explanation of the datasets we employed.

4.1 River Elbe Flood

In the period from 30th May to 3rd June 2013 extreme heavy rain affected large parts of eastern and central Europe. The distribution of precipitation in the basin of the rivers Elbe, Moldau and Saale reached values two to three times higher than that for an average June. This is equivalent to a centennial probability of occurrence. The soil was already highly saturated at this time due to a wet climate in May 2013. Therefore, the heavy rain rapidly resulted in surface runoff causing the severe flood situation. The monthly average flow was three to four times higher than the longstanding average and in some places even higher than the higher value ever recorded.

The same finding follows from the examination of the water level data. Some gauging stations measured values that were never recorded before. For instance, at "Magdeburg-Strombrücke" the water level reached 7.46 m. That is an increase by more than 70 cm compared to the former maximum. Another characteristic of the flood was the huge stretch of the flood wave. The alert phase 4 (the highest in Germany) that was announced by the government lasted for 6 days along the rivers Elbe, Mulde, Elster and Neiße in Saxony and Saxony-Anhalt. This implies that dikes and dams were at risk of destruction for almost a whole week. The water levels in general did not return to their normal state until 16th of June 2013 (Sächsisches Landesamt für Umwelt and Landwirtschaft und Geologie 2013).

4.2 Datasets

The Twitter dataset contains of 60.524 geo-referenced short text messages ("tweets") within the territory of Germany. Each message consists of 140 Unicode characters at a maximum. Besides the actual text message string every tweet contains several additional fields representing metadata, such as a UTC time when the tweet was created, entities like *hashtags* (i.e. keywords preceded by #) and URLs, as well as an integer representation of the unique ID or information about the user who posted the tweet. The geographic location of a tweet is described in the metadata field "coordinates". The inner coordinates array is formatted as geoJSON.[1]

Users can geo-reference messages in Twitter either manually (e.g. by entering the name of a city in the field "location") or automatically via a client application that access the coordinates of a GPS receiver. Unfortunately, only a small fraction (3 % is the estimated average) of tweets are currently georeferenced by users, and this consists of a limitation for analysis approaches based on the location like the current study.

Twitter offers a number of Application Programming Interfaces (APIs), which can be used for automatically recovering data. For this study, we queried the Twitter streaming API using the 1 % garden hose access, during the period from 08th June 2013 1.30 pm to 10th June 2013 midnight, and collected every geo-referenced tweet

[1] https://dev.twitter.com/docs/platform-objects/tweets

within a bounding box covering Germany. Afterwards we further filtered tweets by their location and excluded those outside the territory of Germany.

In addition to the twitter dataset, we also gathered official water level data from 54 monitoring stations along the rivers Elbe and Saale provided by the German Federal Waterways and Shipping Administration and the German Federal Institute for Hydrology through the German online gauge system "Pegel Online" via *web feature service*.[2] In this manner, our second dataset includes information about the location of each measurement station, the current water level, the average flood water level over a time period from 1st November 2000 to 31st October 2010, and the highest water level ever recorded. The water level measurements were provided in a 15-minute resolution for the whole period analysed.

As a third dataset, we used HydroSHEDS drainage direction information derived from elevation data of the Shuttle Radar Topography Mission (SRTM) at 3 arc-second resolution in order to compute hydrographical features of the river Elbe basin. This includes information about flow accumulation, stream network and catchment boundaries (Lehner et al. 2008).

5 Methodology

This section describes the detailed methodology used in this chapter, by further elaborating the procedures used to apply the approach described in Sect. 3 and schematically depicted in Fig. 1. The next section explains the steps conducted in preparing the datasets employed (Sect. 4.2), followed in Sect. 5.2 by the description of the analytical procedures used.

5.1 Data Preparation

The first step of our data preparation consisted of defining the flood-affected regions based on the digital elevation model (for catchment areas) and on official data (river water levels). Starting with the HydroSHEDS flow direction raster, based on SRTM elevation data, we computed catchment polygon features for each location where two streams flow together using the ArcHydro Toolset for ArcGIS. The detailed workflow is shown in Fig. 2. This way of proceeding guarantees that any cell within a catchment drains into the same stream. Catchments therefore contain no more than one stream by definition.

In the next step, we analysed the water level data collected from 54 water level measurement stations along the rivers Elbe and Saale. To assess the severity of the flood at the gauge station, we computed the difference between the daily maximum

[2] http://www.pegelonline.wsv.de/webservice/wfsAktuell

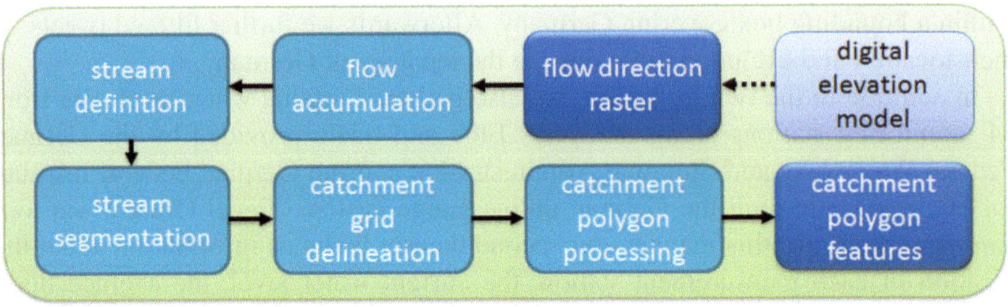

Fig. 2 Catchment processing workflow

water level and the average flood water level for the time period from 1st November 2000 to 31st October 2010.

In the third step, we combined both information on catchments and water levels based on the location of the monitoring stations. The normalized water level values were then matched to the corresponding catchment regions. If more than one water level measurement station was found to be within one given catchment region, we calculated the arithmetic mean of the values measured by those corresponding stations. If the computed flood level for a catchment exceeded the average flood water level by more than 100 cm it was considered "flood-affected".

The fourth step involved performing a content analysis of the twitter messages to identify messages that contain useful information. For doing so, we first filtered the twitter messages, sorting them out into the categories "flood-related" and "non-related". This was accomplished using keyword filtering as common practice in the analysis of twitter messages (e.g. Graham et al. 2012; Kongthon et al. 2012; Vieweg et al. 2010). Tweets containing the keywords in German "Hochwasser", "Flut", "Überschwemmung" ("Hochwasser", "Flut" and "Überschwemmung" are the German words meaning "flood") and the English word "flood", regardless of case-sensitivity, were considered "flood-related". The selection of these keywords was based on the definition of the German dictionary "Duden" for the word "Hochwasser". Furthermore, we included the additional words "Deich" (dike) and "Sandsack" (sandbag), which were found to be common in reports in the media.

Finally, we classified the flood-related tweets into thematic categories, based on a manual content analysis. The content-based classification of messages requires a well-defined set of categories, which heavily depends on the crisis context analysed, i.e. it varies for each crisis phenomenon and event. We analysed the categories proposed by Imran et al. (2013) ("caution and advice", "information source", "donation", "causalities and damages", "unknown") and Vieweg et al. (2010) (warning, preparatory activity, fire line/hazard location, flood level, weather, wind, visibility, road conditions, advice, evacuation information, volunteer information, animal management, and damage/injury reports). However, neither of the previous sets of categories was well suited for our case study, the River Elbe flood. We thus used these previous works as a guideline and adapted them to derive an own set of categories that we considered necessary.

Table 1 Thematic categories based on content analysis

Category	Description
"Volunteer actions" (VA)	Tweets related to flood combating Example: "Keine Ahnung, bin auf der Sandsackfüllstation in der Listemannstr. #Magdeburg #Hochwasser" ("I have no clue. I'm at the sandbag filling point at Listemann-street. #Magdeburg #flood")
"Media" (M)	Tweets related to media coverage, politicians and political events Example: "schaue mir das Hochwasser am Fernseher an. schrecklich. und dann gibt es auch noch Plünderer. unglaublich. #SpiegelTV" ("Watching the flood on TV, horrible, there are even looters, unbelievable. 'SpiegelTV")
"Traffic conditions" (TC)	Tweets related to road or rail traffic, traffic jams or other restraints Example: "Der #ice644 von Berlin nach Köln/Bonn soll übrigens fahren—aktuell aber zehn Minuten Verspätung. #hochwasser" ("By the way, the ice644 will go from Berlin to Cologne/Bonn, current delay 10 min #flood")
"Flood level" (FL)	Tweets related to hydrological or physical measurements, not only quantitative ("719 cm") but also qualitative information ("water level sinks") Example: "aktuelles Foto aus #Lostau: 08.30 Uhr Pegel MD Strombrücke: 719 cm #hochwasser http://t.co/uv3NkMMcIw" ("Latest photos from #Lostau: 08.30 am water level MD Strombrücke 719 cm #flood http://t.co/uv3NkMMcIw")
"Other" (O)	Tweets not related to any of the previous categories Example: "Ich wünsche den #Hochwasser betroffenen weiterhin alles Gute, und trotz alledem allen einen schönen #Sonntag" (To all #flood-affected people: Let's hope for the best. Despite all that, have a nice #Sunday")

We grouped flood-related twitter messages into five categories: "volunteer actions" (VA), "media" (M), "traffic conditions" (TC), "flood level" (FL) and "other" (O). Table 1 presents a detailed description of the categories and their characteristics.

5.2 Data Analysis Procedure

The analysis of our data was guided by our three research questions (see Sect. 1). For answering the first research question, we sought to determine whether the spatiotemporal distribution of flood-related tweets matches the spatiotemporal distribution of the flood phenomenon. For doing this, we first generated a density map by executing a kernel density function using ArcGIS software, in order to allow a visual analysis of the spatial distribution of tweets for the time period analysed (8th–10th June 2013). In the following step, we calculated the distance between each tweet and the nearest flood-affected catchment. In order to test if flood-related tweets are closer to the flooded areas than non-related tweets, we computed and compared the average distance of the two groups (flood-related tweets *versus* non-related tweets) using an independent sample t-test.

For answering our second research question, i.e. whether the spatial distribution of flood-related tweets differ depending on their content, we again firstly performed a visual analysis by producing density maps using the kernel density function. Next, we calculated the average distance for each of the categories of Table 1 and performed a post-hoc analysis (LSD) to test the mean distances depending on the categorization for statistical significance.

The final step of our study consisted of answering our third and overall research question by assessing to what extend the distance of messages to flood phenomena is a useful parameter for prioritizing social media messages in order to improve situation awareness. For doing this, we verified if the categories whose messages are closer to flood-affected areas are more useful to improve situation awareness than the categories with more distant messages. In this analysis, we considered social media messages that contain information which is not available through other sources as being more useful to improve situation awareness. As such, the criteria for defining the usefulness of social media messages that we adopt consists of the capacity to enrich and complement other information sources.

6 Results

The results of our study are presented in the following sections. The next section provides an exploratory description of the data collated, serving as a basis for the detailed analysis based on our research questions (Sect. 5.2).

6.1 Data Description

Figure 3 shows flood-affected catchments and the severity of the flooding calculated from digital elevation data and water level data for the time period from 8th to 10th June 2013. Comparing the three maps, one can visualize the shift of the flood peak from the upper reaches (southeast) in the map of 8th June to the lower reaches (north) in the map of 10th June. As such, on 8th June 2013 the catchments along the river Elbe in the federal state of Saxony were most affected, whilst the lower reaches of the river Elbe were not affected until 10th June 2013.

The results of the first classification of twitter messages based on keywords are listed in Table 2. Overall we examined 60,524 tweets within the territory of Germany. The majority (99.34 %) of them do not contain the keywords. These tweets were marked as "non-related". For the period from 8th to 10th June 2013 we selected 398 tweets containing the keywords and marked these tweets as "flood-related".

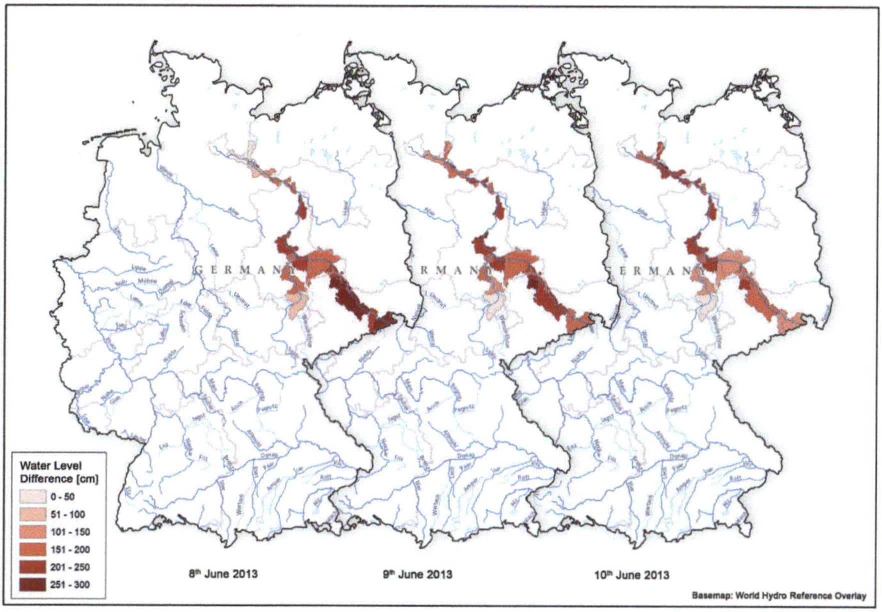

Fig. 3 Spatiotemporal distribution of flood-affected catchments based on official water level information

Table 2 Classification of twitter messages using keyword-based filtering

Period	8th–10th June 2013	8th June 2013	9th June 2013	10th June 2013
# all tweets	60,524 (100 %)	14,286 (100 %)	23,093 (100 %)	23,145 (100 %)
# flood-related tweets	398 (0.66 %)	75 (0.52 %)	197 (0.85 %)	126 (0.54 %)
# non-related tweets	60,126 (99.34 %)	14,211 (99.55 %)	22,296 (99.15%)	23,019 (99.46 %)

Table 3 Classification of twitter messages based on content analysis

Period	8th–10th June 2013	8th June 2013	9th June 2013	10th June 2013
# all tweets	398 (100 %)	75 (100 %)	197 (100 %)	126 (100 %)
# VA	113 (28.39 %)	24 (32.00 %)	59 (29.95 %)	30 (23.81 %)
# M	57 (14.32 %)	8 (10.67 %)	30 (15.23 %)	19 (15.08 %)
# TC	30 (7.53 %)	4 (5.33 %)	8 (4.06 %)	18 (14.29 %)
# FL	32 (8.04 %)	5 (6.67 %)	15 (7.61 %)	12 (9.52 %)
# O	126 (31.66 %)	34 (45.33 %)	85 (43.15 %)	47 (37.30 %)

Table 3 shows the distribution of flood-related tweets based on content classification. More than a quarter (28.39 %) of all flood-related tweets contain information referring to volunteer actions, whereas flood-related tweets referring to media, traffic conditions or flood level reach a much less share. About 30 % of the flood-related tweets were classified as "other" and therefore do not contain any viable information.

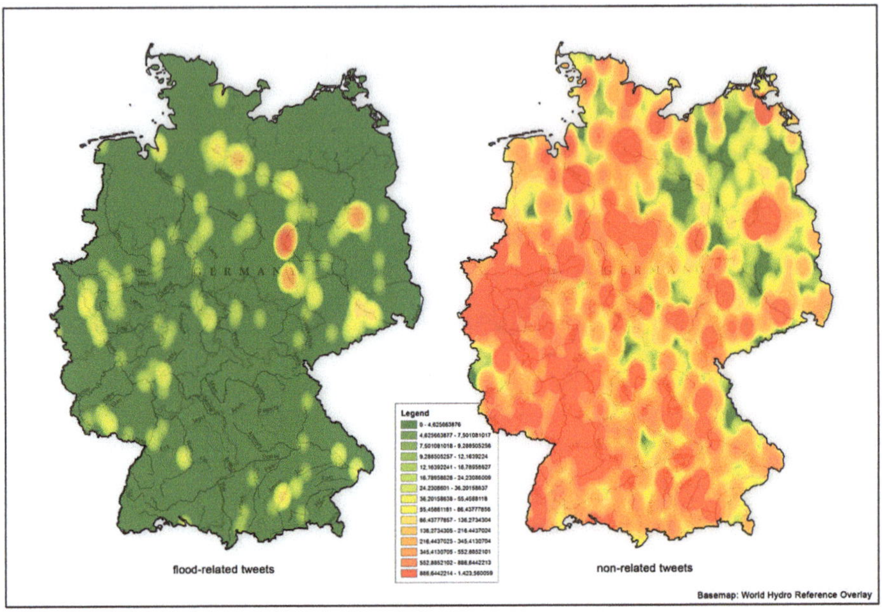

Fig. 4 Spatial distribution of flood-related and non-related tweets

6.2 Analysis

The analysis of the results achieved is presented as follows by addressing each of our three research questions in turn.

6.2.1 RQ1: Does the Spatiotemporal Distribution of Flood-Related Tweets Match the Spatiotemporal Distribution of the Flood Phenomenon?

Firstly, we examined the spatial distribution of flood-related and non-related twitter messages to review whether they follow the spatiotemporal distribution of the flood phenomenon. Figure 4 shows the density of tweets depending on keyword classification. Flood related tweets (on the left side) show peaks in the regions of Magdeburg, Berlin and Halle. Overall flood-related tweets appear only in a few parts of Germany. Non-related tweets (on the right side) concentrate in dense populated regions, e.g. urban areas like Berlin, Hamburg, Munich and the Ruhr area. The tweets cover almost all of Germany, except for some regions in the federal states of Brandenburg and Mecklenburg-Hither Pomerania.

Comparing the spatial distribution of flood-related tweets to the spatial distribution of flood-affected catchments (see Figs. 3 and 4) one can notice similarities the first look. Not the location of all flood-related tweets, but at least of a considerable amount of them does correspond to the location of flood-affected catchments. To further

Table 4 Average distances to flood-affected catchments

	# tweets	Average distance [km]	Standard deviation
Non-related	60,126	221	125
Flood-related	398	78	121

examine this relationship we statistically analysed the distance of all tweets to flood-affected catchments (Table 4).

Using the t-test, we found out that the distance to flood-affected catchments for flood-related twitter messages was statistically significantly lower (78 ± 121 km) compared to non-related twitter messages (221 ± 125 km), $t(60522) = 22.674$, $p = 0.000$.

This implies that the locations of flood-related twitter messages and flood-affected catchments match to a certain extent. In particular this means that mostly people in regions affected by the flooding or people close to these regions posted twitter messages referring to the flood.

That is remarkable as there are for instance far more tweets posted in greater distance to flood-affected regions compared to the number of tweets posted in the proximity to flood-affected regions and as such as that media coverage about the River Elbe Flood was enormous since it was one of the most severe floods ever recorded in Germany. Regarding these circumstances one would have expected a great amount of tweets referring to the flood posted in the urban areas like Munich, Hamburg or the Ruhr area. However, that was not the case. The majority of tweet referring to the flooding were posted by locals.

6.2.2 RQ2: Does the Spatial Distribution of Flood-Related Tweets Differ Depending on Their Content?

Considering that most flood-related tweets were posted by locals it seems probable that these messages contain local knowledge only available to people on site. To review this assumption we analysed the spatial distribution of flood-related tweets depending on their content.

Figure 5 visualizes the spatial distribution of tweets referring to the flooding according to the content based classification. At a first look, one can notice that tweets classified as "other" do not follow the spatial distribution of flood-affected catchments in a particular way. Tweets containing information about "volunteer actions" and "flood level" show a spatial distribution that is similar to the spatial distribution of flood-affected catchments. On the contrary tweets containing information about "media" or "traffic conditions" do not show such a match. To review our observations we statistically analysed the distance of all flood-related tweets to flood-affected catchments (Table 5).

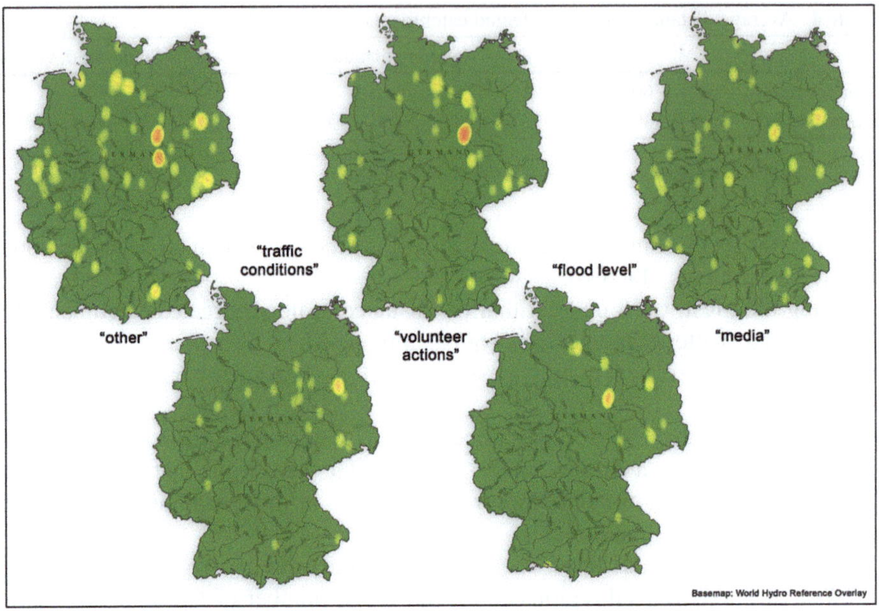

Fig. 5 Spatial distribution of twitter messages depending on content classification

Table 5 Average distances to flood-affected catchments depending on categories

Category	# tweets	Average distance [km]	Standard deviation
VA	113	34	110
FL	32	44	113
TC	30	78	125
O	166	90	114
M	57	150	108

By analysing flood-related tweets based on their content we find that the spatial distribution of tweets differs between the various categories. Especially tweets containing information about VA and FL tend to be concentrated in proximity to flood-affected catchments. Tweets containing information about "media" and "other" show the opposite characteristics.

The application of the post-hoc analysis (LSD) confirms that the differences between categories are statistically significant (Table 6). The average distance of the messages in the categories VA and FL do not differ significantly. In contrast, messages in VA and FL do have significantly different average distances from the messages in O and M.

Table 6 Average differences in distance to flood areas between the categories

	VA	FL	TC	O	M
VA	X	−10 (0.677)	−44 (0.065)	−56 (0.000)*	−116 (0.000)*
FL	10 (0.677)	X	−34 (0.244)	−46 (0.041)*	−107 (0.000)*
TC	44 (0.065)	34 (0.244)	X	−12 (0.617)	−72 (0.006)*
O	56 (0.000)*	46 (0.041)*	12 (0.617)	X	−61 (0.001)*
M	116 (0.000)*	107 (0.000)*	72 (0.006)*	61 (0.001)*	X

6.2.3 RQ3: Is Distance to Flood Phenomena a Useful Parameter to Prioritize Social Media Messages in Order to Improve Situation Awareness?

Analysing the results of the previous research question, we observe that the categories of twitter messages can be divided into three groups as regards to the distance to flood-affected areas:

- Group A: messages in FL and VA are the closest messages to the flooded areas.
- Group B: messages in TC have average distance between the other groups.
- Group C: messages in O and M are more distant to flooded areas.

Applying the criteria we defined for usefulness of social media messages for improving situation awareness (Sect. 5.2), we can conclude that messages in Group A are the most useful ones. Indeed, information about current flood levels is crucial for situation awareness and can complement existing water level measurements, which are only available for determined geographical points where gauging stations are located. Since volunteer actions are increasingly organised via social media, this is a type of information which is very valuable and completely missing from other sources. Hence, our results show that the twitter messages that are closest to the flood-affected areas (Group A) are also the most useful ones. Therefore, we can answer positively our research question, concluding that the distance to flood phenomena is indeed a useful parameter to prioritize twitter messages towards improving situation awareness.

7 Discussion and Conclusion

In this chapter we present a new approach to extract crisis-relevant information from social media platforms like Twitter. Our results show that the spatial distribution of twitter messages referring to the flooding of the river Elbe in Germany in June 2013 is significantly different from the spatial distribution of off-topic messages. We further found that flood-related tweets that contain more useful information for situation awareness (e.g. volunteer actions and flood level) are significantly closer to flood-affected regions than others. This implies that distance to flood phenomena is a useful parameter to prioritize social media messages.

This approach to leverage geographical relations to prioritize social media messages can make a contribution for both research and practice. One potential use of our approach is for enhancing other approaches to classification of social media messages. This could be accomplished by using the prioritization according to geographical relations produced by our approach as weights in the algorithms of existing machine learning techniques. Moreover, the proximity to disaster hotspots could be used for ranking messages to be processed by volunteers in crowdsourcing deployments.

The generality of the results presented here should be investigated by applying our approach in similar analyses of other flooding events. Future work should also concentrate on refining the approach by including information from other social media platforms (e.g. Instagram or Flickr). The integration of other official datasets, e.g. precipitation data or satellite images, is another avenue for future work towards better understanding the relations between social media and crisis phenomena from a geographical perspective. Implementing more detailed hydrological models will additionally extend the validity of our method regarding flood phenomena. Furthermore, our results could be generalised by investigating the value of exploring geographical relations for prioritizing social media messages in other disasters besides floods.

Acknowledgments The authors would like to thank Matthias Stork for the ideas and discussions on hydrological models, as well as the German Federal Waterways and Shipping Administration and the German Federal Institute for Hydrology for providing the water level data. João Porto de Albuquerque would like to express gratitude to FAPESP (grant 2012/18675-1) and CAPES (grant 12065-13-7) for funding.

References

Chatfield AT, Brajawidagda U (2013) Twitter early tsunami warning system: a case study in Indonesia's natural disaster management. In: System sciences (HICSS), 2013 46th Hawaii international conference on (pp. 2050–2060). IEEE

Endsley MR (1995) Toward a theory of situation awareness in dynamic systems. Hum Factors 37(1):32–64

Gao H, Barbier G, Goolsby R (2011) Harnessing the crowdsourcing power of social media for disaster relief. Intell Syst IEEE 26(3):10–14

Goodchild MF, Glennon JA (2010) Crowdsourcing geographic information for disaster response: a research frontier. Int J Digital Earth 3(3):231–241

Goolsby R (2010) Social media as crisis platform: the future of community maps/crisis maps. ACM Trans Intell Syst Technol (TIST) 1(1):7

Graham M, Poorthuis A, Zook M (2012) Digital trails of the UK floods—how well do tweets match observations? The Guardian Datablog. http://www.guardian.co.uk/news/datablog/2012/nov/28/data-shadows-twitter-uk-floods-mapped. Accessed 20 June 2013

Hiltz SR, Plotnick L (2013) Dealing with information overload when using social media for emergency management: emerging solutions. In: Proceedings of the 10th international ISCRAM conference, pp 823–827

Hudson-Smith A, Crooks A, Gibin M, Milton R, Batty M (2009) NeoGeography and web 2.0: concepts, tools and applications. J Location Based Serv 3(2):118–145

Imran M, Elbassuoni SM, Castillo C, Diaz F, Meier P (2013) Extracting information nuggets from disaster-related messages in social media. In: ISCRAM, Baden-Baden, Germany

Kaewkitipong L, Chen C, Ractham P (2012) Lessons learned from the use of social media in combating a crisis: a case study of 2011 thailand flooding disaster

Kongthon A, Haruechaiyasak C, Pailai J, Kongyoung S (2012) The role of Twitter during a natural disaster: case study of 2011 Thai flood. In: Technology management for emerging technologies (PICMET), 2012 proceedings of PICMET'12: 2227–2232

Lehner B, Verdin K, Jarvis A (2008) New global hydrography derived from spaceborne elevation data. EOS, Trans Am Geophys Union 89(10):93–94

Lofi C, Selke J, Balke WT (2012) Information extraction meets crowdsourcing: a promising couple. Datenbank-Spektrum 12(2):109–120

MacEachren AM, Robinson AC, Jaiswal A, Pezanowski S, Savelyev A, Blanford J, Mitra P (2011) Geo-twitter analytics: applications in crisis management. In: Proceedings of the 25th international cartographic conference, Paris, France

Meier P (2012) Crisis mapping in action: how open source software and global volunteer networks are changing the world, one map at a time. J Map Geogr Libr 8(2):89–100

Okolloh O (2009) Ushahidi or "testimony": web 2.0 tools for crowdsourcing crisis information. Participatory Learn Action 59(1): 65–70

Palen L (2008) Online social media in crisis events. Educause Q 31(3):12

Rogstadius J, Kostakos V, Laredo J, Vukovic M (2011) Towards real-time emergency response using crowd supported analysis of social media. In: Proceedings of CHI workshop on crowdsourcing and human computation, systems, studies and platforms

Sächsisches Landesamt für Umwelt, Landwirtschaft und Geologie (2013): Gewässerkundlicher Monatsbericht mit vorläufiger Auswertung des Hochwassers Juni 2013. http://www.umwelt.sachsen.de/umwelt/wasser/707.htm. Accessed 1 September 2013

Sakaki T, Okazaki M, Matsuo Y (2010) Earthquake shakes Twitter users: real-time event detection by social sensors. In: Proceedings of the 19th international conference on world wide web, pp 851–860

Terpstra T, de Vries A, Stronkman R, Paradies GL (2012) Towards a realtime Twitter analysis during crises for operational crisis management. In: Proceedings of the 9th international ISCRAM conference, Vancouver, Canada

Tobler WR (1970) A computer movie simulating urban growth in the Detroit region. Econ Geogr 46:234–240

Triglav-Čekada M, Radovan D (2013) Using volunteered geographical information to map the November 2012 floods in Slovenia. Nat Hazards Earth Syst Sci Discuss 1(3):2859–2881

Vieweg S, Hughes AL, Starbird K, Palen L (2010) Microblogging during two natural hazards events: what twitter may contribute to situation awareness. In: Proceedings of the SIGCHI conference on human factors in computing systems, pp 1079–1088

Yin J, Lampert A, Cameron M, Robinson B, Power R (2012) Using social media to enhance emergency situation awareness. IEEE Intell Syst 27(6):52–59

Event Identification from Georeferenced Images

Yeran Sun and Hongchao Fan

Abstract Geotagged images (e.g., Flickr) could indicate some social events (e.g., festival, parade, protest, sports, etc.) with spatial, temporal and semantic information. Previous researches relied much on tag frequency, thus some images which do not have clearly tag indicating the occurrence of social event would be missing in this case. One potential way to address this problem or enhance the event identification is to make more use of the spatial and temporal information. In this chapter, we take into consideration the underlying spatio-temporal pattern of social events. Particularly, the influence of urban land use and road on the occurrence of event is considered. Specifically, with a spatio-temporal cluster detection method, we firstly detected spatio-temporal clusters composed of geotagged images. Among these detected S-T clusters, we furthermore attempted to identify social events in terms of a classification model. Specifically, land use and road were considered to generate new kinds of spatial characteristics used as dependent variables incorporated into the classification model. In addition to this, user characteristics (i.e., the number of images and the number of users), spatial and temporal range of images, and the heterogeneity of temporal distribution of images were considered as the other dependent variables for the classification model. Consequently, with a binary logistic regression (BLR) method, we estimated the categories (i.e., 'event' or 'non-event' one) of the S-T clusters (cases). Experimental results demonstrated the good performance of the method with a total accuracy of 71 %. With the variable selection process of the BLR method, empirical result also indicates that (1) some characteristics (e.g., the distance to the road and the heterogeneity of temporal distribution of images) do not have considerable influence on the occurrence of 'event'; and (2) compared to the other urban land categories (i.e., residential and recreational land), commercial land has a relatively high influence on the occurrence of 'event'.

Y. Sun (✉) · H. Fan
GIScience Research Group, Institute of Geography, University of Heidelberg, Heidelberg, Germany
e-mail: yeran.sun@geog.uni-heidelberg.de

J. Huerta et al. (eds.), *Connecting a Digital Europe Through Location and Place*,
Lecture Notes in Geoinformation and Cartography, DOI: 10.1007/978-3-319-03611-3_5,

Keywords Volunteered geographic information · Event detection · Flickr · Classification model · Spatial environment

1 Introduction

Currently, volunteered geographic information (VGI) has been widely used in various applications. Some social events are more or less involved in VGI. e.g., geotagged images of Flickr indicate social events, such as parade, festival, concert, etc. Spatial, temporal, and textual information of geotagged images can be used to indicate the location, time and the related topic of the social event. Similarly, geotagged posts of Twitter can indicate the occurrence of events, including social events and natural disasters, in a more real-time way. Geotagged images and posts have the metadata containing longitude, latitude, time, tags, and other semantic information, etc. Geotagged images, in addition, have the imagery information. Such information has the potential to be used in event detection, in which spatial, temporal and semantic information can be used separately or together. Some researchers have already proposed approaches to explore the possibility (e.g., Rattenbury et al. 2007; Earle et al. 2011; Chae et al. 2012). As the most frequently used VGI data sources in social analysis, Flickr georeferenced images and Twitter georeferenced posts have higher potential to event detection with the growing popularity in social media (e.g., Rattenbury et al. 2007; Jankowski et al. 2010; Sakaki et al. 2010; MacEachren et al. 2011; Ruocco and Ramampiaro 2012).

Social event usually occurs at a place during a certain time period. For instance, parade usually occurs at the important avenue and plaza of the city in the day time instead of the night. A music concert held at a pub in the evening has a spatial area no larger than this pub and a temporal period less than 12 h simultaneously, if there are some geotagged images indicating that this event should be geolocated within the building representing this pub and be taken during the evening and night time of a certain day. At the same time, there are still some social events, e.g., festival and sports game, occurring at couples of places in a city or lasting several consecutive days. Even some music concert might occur at a large open air area, e.g., stadium or plaza, whose area is much larger than normal building. In spite of the occurrence location is within a building, some conferences last nearly one week. Despite existence of such social events that occur at relatively large scale area or last a long time, for the other ones which are more, basically, the place where and the time period when they occur are both not large (e.g., within a region with an area below 20×20 m and within on day). User-generated georeferenced images reflecting the same social events of limited spatial and temporal space can probably constitute a spatio-temporal cluster. Therefore, some researchers attempted to detect spatio-temporal clusters to identify events (e.g., Rattenbury et al. 2007; Chen and Roy 2009; Zhang et al. 2012). However, these researches relied mainly on the tag of the image. Since a huge amount of images which do not have the tag indicating the occurrence of event clearly, some events would be missing when only such tag-based methods are applied.

Previous researches relied much on tag frequency, thus some images which do not have clearly tag indicating the occurrence of social event would be missing in this case. One potential way to address this problem or enhance the event identification is to make more use of the spatial and temporal information. In this chapter, we take into consideration the underlying spatio-temporal pattern of social events. Particularly, the influence of urban land use and road on the occurrence of event is considered. Specifically, with a spatio-temporal cluster detection method, we firstly detected spatio-temporal clusters composed of geotagged images. Among these detected S-T clusters, we furthermore attempted to identify social events in terms of a classification model incorporating user, spatial and temporal characteristics. The remainder of this chapter is organized as follows. Section 2 introduces the related works made by the other researchers. Section 3 shows the methodology of this chapter. In Sect. 4 the experimental results were shown. Finally, this chapter made a conclusion with future works.

2 Related Works

To detect social event using VGI, several researches proposed methods in which spatial, temporal, and semantic attributes of images are exploited. Here we briefly introduced some of them.

Rattenbury et al. (2007) firstly used geo-tagged images to detect social event. A method called Scale-structure Identification is proposed and compared with two well-known burst-analysis methods (i.e., Naive Scan Method and Spatial Scan Method). Semantics of tags were extracted and then identified as place or event with the three methods. To detect social events and retrieve associated images from Flickr, Brenner and Izquierdo (2012) proposed a classification model combining locational, temporal, textual and visual features of geo-tagged images. Additionally, external information from datasets and online web services were incorporated to further improve the detection performance. Liu et al. (2011) made use of geotagged images and videos to detect events. They proposed an approach dedicated to the application of image's tag, description and location to event occurrence detection in nine venues across the globe. The result shows the ability of the approach to not only detect events with high accuracy but also identify events that have not been published in popular event directories. In addition to the textual identification of events, they show how to build visual summaries of past events providing viewers with a more compelling feeling of the event's atmosphere.

Wang et al. (2012) incorporated online social interaction features in the detection of social events. In addition to the image's metadata including time, location, tags and description, the "social affinity" of two images was utilized to help event detection. An Incremental clustering method was used to group images to clusters representing events. To find out the occurrence of local events such as local festivals, Lee and Sumiya (2010) made use of geographical regularities deduced from the usual behavior patterns of crowds with geo-referenced posts of Twitter. By comparing these regularities with the estimated ones, they decided whether there were any unusual events happening in the monitored geographical area. Based on Flickr image, Becker et al. (2010) presented several techniques for identifying events and their associated social media documents, by combining multiple context features of the document in a variety of disciplined ways. They proposed a general framework for identifying events in social media documents via clustering, and used similarity metric learning approaches in this framework, to produce high quality clustering results. Pan and Mitra (2011) proposed two event detection approaches aiming at document collection of news article. One is to combine the popular Latent Dirichlet Allocation (LDA) model with temporal segmentation and spatial clustering. The other is to adapt an image segmentation model, SLDA, for spatial-temporal event detection on text. Using Flickr images to detect event, Chen and Roy (2009) employed a wavelet transform to suppress noise. Subsequently, like Rattenbury et al. (2007) they identified tags related with events, and further distinguished between tags of aperiodic events and those of periodic events.

Considering each Twitter user as a sensor, Sakaki et al. (2010) proposed a probabilistic spatiotemporal model for the target event that can estimate the centers of earthquakes and the trajectories of typhoons. To exploit the Twitter data for more details, Watanabe et al. (2011) proposed an automatic geotagging method that identifies the location of non-geotagged document. Empirical results demonstrate that their system can detect local events that are difficult to identify using existing event detection methods. Chae et al. (2012) presented a visual analytics system that provides users with scalable and interactive social media data analysis and visualization. With this system, analyst can explore and examine abnormal topics and events within various social media data sources, such as Twitter, Flickr and YouTube. Latent Dirichlet Allocation method was used to rank extracted topics, whereas seasonal trend decomposition together with traditional control chart methods was applied to find unusual peaks and outliers within topic time series.

3 Methodology

3.1 Definition of Event in the Context of Flickr Image

Becker et al. (2010) defined that an 'event' is something that occurs in a certain place at a certain time. Moreover, Becker et al. (2011) gave a definition of 'event' in the context of twitter post. In addition to this, some studies attempted to categorize the event based on some characteristics (e.g., periodicity) (Chen and Roy 2009). Following the definition form Becker et al. (2010), here we need to take the scale of space and time into consideration. Specifically, there are some events (e.g., a festival or a celebration) occur in a large place (a park or a few landmarks) at a long time (e.g., one week or one month). In contrast, more events (e.g., a music concert or a party) occur in a small place (e.g., a building or a plaza) at a short time (e.g., one

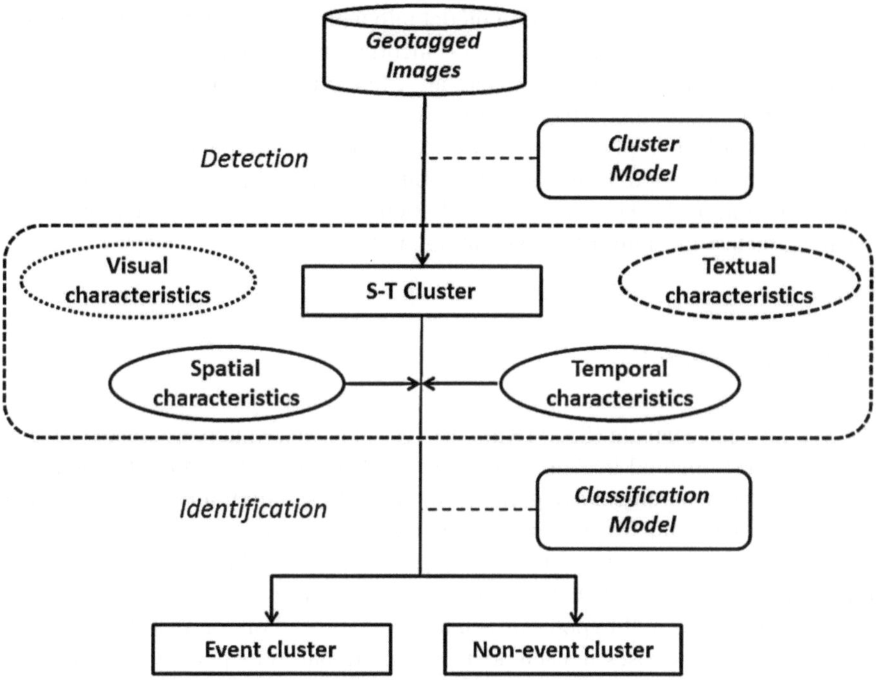

Fig. 1 A prototype of event detection using georeferenced images

day or one night). Both of these two kinds of events are considered in our study. In this study, we attempt to identify the latter type of event with a small spatio-temporal range.

Basically, such event could be indicated by a spatio-temporal (S-T) cluster composed of a number of geoferenced images. For instance, a number of images recording the same event could constitute a cluster since they are highly spatially and temporally close to each other. For instance, if a number of images recording a music concert, these images tend be less than 100 m and less than 6 h far away from each other in spatial and temporal dimensions. In other words, a spatio-temporal cluster of Flickr images might indicate an event. In addition to event, bulk upload as well as landmark can lead to a spatio-temporal cluster of geoferenced images.

3.2 Overview of the Proposed Approach

Figure 1 shows a prototype of event detection using georeferenced images. Initially, using cluster model (e.g., clustering method, cluster detection method, etc.), a number of spatio-temporal clusters are detected from geotagged image data set. Subsequently, spatial, temporal, textual and visual characteristics of the clusters are extracted. Finally, by a classification model into which the spatial, temporal, textual and visual characteristics of clusters are incorporated, 'event clusters' are distinguished from 'non-event clusters'. Here an 'event cluster' consists of images most of which (e.g., more than 50 %) record or imply an 'event'. To determine whether an image record or imply an event, we exploit the tag frequency and sometimes visual content of image. In this study, we take spatial and temporal characteristics into consideration. In the future, we will attempt to incorporate textual as well as visual characteristics into the classification model.

3.3 Cluster Model

The Spatial Scan Statistic (SSS) (Kulldorff 1997) is chosen here as clustering algorithm. The SSS method is able to detect local clusters. Furthermore, the analyst is able to determine parameters (e.g. numbers of clusters, minimum numbers of points in a cluster, the radius of the circle) to detect cluster with specific purposes. The SSS method uses a circular window to identify clusters. This study uses the Poisson model to detect significant local spatial cluster of observed cases compared to expected cases. The null hypothesis is that the cases are Poisson-distributed and the expected number of cases in each area is proportional to its population size. Thus, no spatial clusters exist. The likelihood function for the Poisson model is defined as:

$$A = \left(\frac{c}{E(c)}\right)^{c} \left(\frac{C-c}{C-E(c)}\right)^{C-c} I() \tag{1}$$

where C is the total number of cases, c is the observed number of cases within the window and $E[c]$ is the covariate adjusted expected number of cases within the window under the null-hypothesis. Within the window the expected number of cases is proportional to the population size. Note that since the analysis is conditioned on the total number of cases observed, $C - E[c]$ is the expected number of cases outside the window. $I()$ is an indicator function. When SaTScan is set to scan only for clusters with high rates, $I()$ is equal to 1 when the window has more cases than expected under the null-hypothesis, and 0 otherwise. The opposite is true when SaTScan is set to scan only for clusters with low rates. When the program scans for clusters with either high or low rates, then $I() = 1$ for all windows.

3.4 Spatio-Temporal Characteristics of Cluster

Table 1 shows the user, spatial and temporal characteristics of S-T cluster considered in the event identification. Moreover, some of these attributes will be explained in more detail.

Table 1 User, spatial and temporal characteristics of the S-T cluster for the event identification

Characteristic type	Feature variables	Meaning
User characteristics	X_1	The number of images
	X_2	The number of users
Spatial characteristics	X_3	The radius of gyration (ROG) of the images (The spatial range of cluster)
	X_4	The percentage of images within *commercial land*
	X_5	The percentage of images within *residential land*
	X_6	The percentage of images within *recreational land*
	X_7	The average distance of images to the nearest *primary road*
	X_8	The average distance of images to the nearest *secondary road*
	X_9	The average distance of images to the nearest *local road*
Temporal characteristics	X_{10}	The standard deviation (SD) of the time of the images (The temporal range of cluster)
	X_{11}	The entropy of time of the images (the temporal heterogeneity of the distribution of images)

First of all, user characteristics are basically considered in the event identification. The number of images (X_1) and the number of users (X_2) in an S-T cluster are chosen here as the user characteristics. The main reason to choose X_1 and X_2 is to eliminate the influence of bulk upload. Like 'event' and landmark, bulk upload can generate S-T cluster. A user sometimes uploaded an entire photoset that are georeferenced with one location and one date, thus the images of this photoset are spatial and temporally close to each other. For instance, compared the cluster resulted from 'event', the one resulted from bulk upload seems to have a smaller number of users.

Subsequently, we take into consideration the spatial and temporal range of the images in a cluster. The *radius of gyration* (ROG) of the images (X_3) is to measure the spatial range of images in an S-T cluster. ROG is expressed as

$$ROG\,(C) = \sqrt{\frac{1}{n}\sum_{i=1}^{n}(x_i - x_c)^2 + (y_i - y_c)^2} \tag{2}$$

$$x_c = \frac{1}{n}\sum_{i=1}^{n}x_i \tag{3}$$

$$y_c = \frac{1}{n}\sum_{i=1}^{n}y_i \tag{4}$$

where $P_i(x_i, y_i)$ are the location of the ith image in an S-T cluster C, and (x_c, y_c) are the center of the mass of the cluster C.

Similarly, the standard deviation (SD) of the time of the images (X_{10}) is to measure the temporal range of images in an S-T cluster. It is expressed as

$$SD\_of\_time\ (C) = \sqrt{\frac{1}{n}\sum_{i=1}^{n}(t_i - t_c)^2} \tag{5}$$

$$t_c = \frac{1}{n}\sum_{i=1}^{n} t_i \tag{6}$$

where t_i is the time of the ith image in an S-T cluster C, and t_c is mean time of the images of the cluster C.

In addition to the temporal range of the images, the entropy of time (X_{11}) is to characterize the heterogeneity of the temporal distribution of the images in an S-T cluster. Intuitively, a landmark can attract persons to take pictures in the whole day since persons can visit it in the morning, afternoon or even evening; whereas an 'event' tend to attract persons to take pictures within a relatively limited time period (e.g., 2 or 3 h) since persons would not do at all once it is already over or disappears. In this case, since they are within a relatively short time period (e.g., 2 or 3 h), the images recording an 'event' seems to be more heterogeneously temporally distributed than the images recording a landmark. The entropy is widely used to measure the heterogeneity of phenomena in various disciplines. In this study we thus make use of the entropy to characterize the temporal heterogeneity of the distribution of the images in an S-T cluster. This entropy is defined as

$$Entropy\_of\_time(C) = -\sum_{i=1}^{K}\frac{Num_i^T\ (C)}{Num^T\ (C)} log_2\left(\frac{Num_i^T\ (C)}{Num^T\ (C)}\right) \tag{7}$$

$$Num^T\ (C) = \sum_{i=1}^{K} Num_i^T\ (C) \tag{8}$$

where $\mathrm{Num}_\mathrm{i}^\mathrm{T}$ (C) represents the number of images within the ith (i = 1, 2, ..., K(K = 24)) hour on one day. The lower the Entropy_of_time (C) is, the images in the cluster C are more heterogeneously temporally distributed.

X_{10} and X_{11} can partially eliminate the influence of the inconsistency of time zone between the local region where user are at present and the original region where user came before. Since the time when the picture was taken depends on the time that is recorded on the device with which the photos are taken. For instance, a tourist visits to a region whose time zone is different with that of his or her hometown. If she forgets to change the time zone of her camera from the one of her hometown to the one of this region, the images taken by her in this region would be uploaded with the time computed based on the time zone of users' hometown instead of the time computed based on the time zone of this local region. In this situation, such recorded time is not the really local time when the picture is taken. X_{10} and X_{11} can eliminate much influence when the number of users is small or difference of time zone between these two regions is small in a cluster.

X_3, X_{10} and X_{11} can eliminate the influence of bulk upload. Due to a lower spatial and temporal range, the cluster resulted from bulk upload seems to have a lower X_3, X_{10} and X_{11} than the once resulted from 'event'.

Furthermore, to get a better understanding of the influence of the spatial environment on the occurrence of 'event', we attempt to explore if land use as well as road affect the occurrence of 'event' in the context of Flickr. Specifically, three typical urban land use types (i.e., commercial land, residential land and recreational land) and three main road types (i.e., prime road, secondary road and local road) are chosen. For an S-T cluster, X_4, X_5 and X_6 represent the percentage of images geolocated within commercial land, residential land and recreational land respectively; X_7, X_8 and X_9 represent the average distance of images to the nearest prime road, secondary road and local road respectively.

3.5 Classification Model

The binary logistic regression (BLR) method is able to measure the relationship between the binary dependent variable and one or several independent variable(s), by computing the probabilities of the outputs for the different category. BLR allows assessing the contribution of individual variables to the classification and subsequently choosing the important variables to build a new model, leading to an enhancement of the classification. The BLR model is represented as

$$P(y) = \frac{1}{1 + e^{-(a_0 + a_1 x_1 + a_2 x_2 + \cdots + a_i x_i)}} \quad (9)$$

where $y \in \{1, 0\}$ is the dependent variable, and $(x_1, x_2, \ldots, x_i)$ are the independent variables. a_0 is the intercept, and $(a_1, a_2, \ldots, a_i)$ are the coefficients of the independent variables $(x_1, x_2, \ldots, x_i)$. In this classification problem, the targeted classification category (1 or 0) and the 11 feature variables in Table 1 correspond to the dependent variable Y and the initially independent variables in Eq. (9) respectively.

3.6 'Event' and 'Non-event' S-T Cluster

In Eq. (9), there are two targeted classification categories, i.e., 1 and 0.1 means that an S-T cluster is an 'event' one whereas 0 means that an S-T cluster is a 'non-event' one. Specifically, we firstly exploit the tag frequency and sometimes visual content of image to determine whether an image record or imply an event. Subsequently, if more than 50 % of images record or imply an event, the cluster is considered as an 'event' one; otherwise, it is considered as a 'non-event' one.

4 Experimental Results

4.1 Case Study

The proposed algorithm is implemented and tested with Flickr photos in Munich, Germany. All the images were taken during the year 2010 and 2011. Totally, 45, 950 images were acquired via Flickr API (http://www.flickr.com/services/api/). From the metadata of the image, location (longitude and latitude), taken time (date and time) as well as tags of each image were acquired.

4.2 Data Preprocessing

Since there are positional and temporal errors as well as bulk uploads in the Flickr images (e.g., Senaratne et al. 2011; Zielstra and Hochmair 2013), we need some strategies and steps to eliminate or decrease the influences of them. First, we discarded the images whose recorded accuracy level of the location information is below the city level. Secondly, we discarded the images with a taken time of 00:00:00 since for the vast majority of images 00:00:00 means a wrong time.

The inconsistency of time zone is difficult to be resolved completely. The temporal characteristics (i.e., the SD of the time of the images and the entropy of time of the images) can partially eliminate the influence of the inconsistency of time zone between the local region where user are at present and the original region where user came before (see Sect. 3.4). Furthermore, some characteristics of S-T cluster (e.g., the number of users, the ROG of images, etc.) are defined to decrease the influence of the bulk upload (see Sect. 3.4).

4.3 Cluster Detection

Location and date of the image is inputted into the software SaTScan (http://www.satscan.org/) enabling the implementation of SSS method. We set the temporal size of the scanning window as one day and the spatial size of the scanning window as 100 m. Both the spatial and temporal size of the scanning window is relatively small since the majority of the events in which this chapter is interested have a relatively

Table 2 Clusters detected with SSS method

Cluster	Radius (m)	Date	LLR	Obs. Num	Exp. Num	ODE
1	100	2011/7/6	173.817	533	385.302	1.383
2	100	2011/4/3	61.763	190	137.350	1.383
–	–	–	–	–	–	–
212	0	2010/7/15	9.413	29	20.964	1.383

Note Obs. Num is the observed number, *Exp.* Num is the expected number, *ODE* is the ratio of Obs. *Num and Exp. Num, and LLR* is log-likelihood ratio

Table 3 The dependent and independent variables of the S-T clusters (cases)

Cluster (case)	Independent variable											Dependent variable
	X_1	X_2	X_3	X_4	X_5	X_6	X_7	X_8	X_9	X_{10}	X_{11}	Y
1	533	2	1	100%	0%	0%	545	301	8	84	2.17	1
2	190	1	1	100%	0%	0%	44	28	81	68	1.75	1
–	–	–	–	–	–	–	–	–	–	–	–	–
212	17	2	9	10%	90%	0%	179	215	20	38	0.32	0

Note the units of X_3, X_7, X_8 and X_9 are meter, and the unit of X_{10} is minute.

small spatial and temporal range. With SSS method, 212 S-T clusters were finally detected as shown in Table 2.

4.4 Event Identification

In this section, we introduced the process and the result of event identification as follows.

1. Calculating the characteristics of clusters (i.e., the variables of the classification model)

First of all, we need to extract or calculate the characteristics (i.e., user, spatial and temporal characteristics) of S-T cluster mentioned in Sect. 3.4 and determine the category (i.e., 'event' cluster and 'non-event' cluster) of S-T cluster in terms of the rules in Sect. 3.6. Consequently, we acquired the characteristics and the category for each one of the 212 S-T clusters. For the BLR model, the dependent and independent variables of the S-T clusters (cases) are as shown in Table 3. The dependent and independent variables are already explained in Table 1.

2. Selecting the training and test data set

Among the detected S-T clusters (cases), 1/5 and 4/5 of them were chosen as test and training data set respectively. To acquire the actual result of the identification, we made use of the tag frequency plus the visual content of image. The most frequent

Table 4 The best fitting model learnt from the training data set

	Variable (X)	Meaning	Coefficient of the variable (A)	p-value
Variables in the model	X_1	The number of images	0.010	0.036
	X_2	The number of users	0.878	0.016
	X_3	The ROG of the images (The spatial range of cluster)	−0.020	0.097
	X_4	The percentage of images within *commercial land*	0.828	0.041
	X_{10}	The SD of the time of the images (The temporal range of cluster)	0.004	0.160
	A_0	Intercept	−2.975	0.000
Variables not in the model	X_5	The percentage of images within *residential land*	–	0.321
	X_6	The percentage of images within *recreational land*	–	0.477
	X_7	The average distance of images to the nearest *primary road*	–	0.725
	X_8	The average distance of images to the nearest *secondary road*	–	0.998
	X_9	The average distance of images to the nearest *local road*	–	0.998
	X_{11}	The entropy of time of the images (the temporal heterogeneity of the distribution of images)	–	0.750

event-indicated tag (e.g., parade, festival, concert, skiing, etc.) within the cluster was used to determine the category of the event.

3. Determining important variables for classification model

In BLR method, it is allowed to discard trivial variables to get the best fitting model containing important variables. We used the backward elimination selection procedure to ascertain the best fitting model based on statistical criteria of the change in log-likelihood. With the SPSS software, the backward elimination begins with initially entering all variables into the model. Consequently, five variables with higher statistical significances (i.e., p-values are less than 0.2) were kept in the model whereas the others were discarded. Table 4 shows the best fitting model acquired in terms of the training data set. We thus showed this BLR model as

$$P(y) = \frac{1}{1 + e^{-(-2.975+0.010\times x_1+0.878\times x_2-0.020\times x_3+0.828\times x_4+0.004\times x_{10})}}$$

$$y = \begin{cases} 1, & \mathit{if}\ P(y) > 0.5 \\ 0, & \mathit{otherwise} \end{cases} \tag{10}$$

Table 5 Classification accuracy of event and non-event S-T clusters

	Estimation			Accuracy (%)
	Class	1	0	
Observation	1	6	3	67
	0	9	24	73
	Total			71

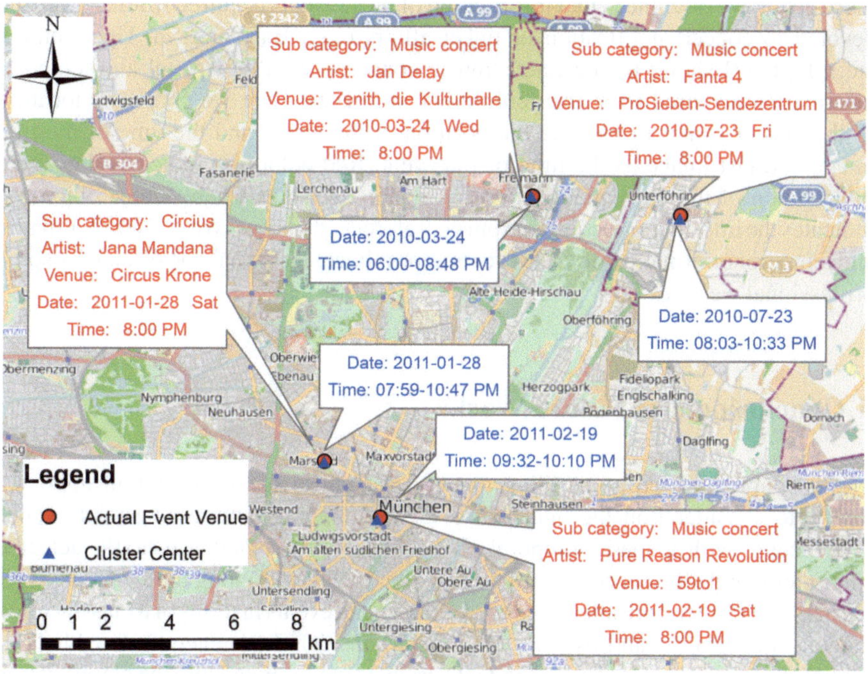

Fig. 2 Some detected music concerts and corresponding information from other website sources (basemap: OpenStreetMap)

The empirical results of variable selection indicate that (1) user characteristics (i.e., the number of images and the number of user) make relatively large contributions to the event identification; (2) spatial and temporal range of images (i.e, ROG of the images and The SD of the time of the images) make relatively large contributions as well; (3) compared to the other urban land categories (i.e., residential and recreational land), commercial land has a relatively high influence on the occurrence of 'event'; (4) the distance to the road tend to make no contribution since the p-values are close to 1; and (5) compared to the temporal range, the heterogeneity of temporal distribution of images makes a relatively low contribution.

4. the result of the classification (event identification)

With the BLR model in Eq. 10, among 42 test cases (clusters), 30 ones were estimated correctly, thus the total accuracy was 71 %. Table 5 shows the results of estimation with details. Class *1* and *0* mean *event* and *non-event* S-T cluster respectively.

5. Visualizing the identified 'events'

Besides, we geo-visualized some detected social events. To evaluate the detection in more detail, we offer the actual information of these events acquired from the website devoted to concert notification or ticket affairs (e.g., Bandsintown, Upcomming, etc.). Figure 2 shows the detected 'music concerts' with the spatial and temporal information acquired from the georeferenced images. Here we chose the location of the cluster center to indicate the position of the event, and the time of the first image and last image as the time of the event. Red color text indicates the actual information of these events acquired from some websites whereas blue color text indicates the information acquired from georeferenced images. We can see that spatial and temporal information of these events approximate the actual one offered by some professional websites.

5 Conclusion and Future Works

In this chapter, we exploited Flickr images to identify social event. Our approach took into consideration the underlying spatio-temporal patterns of social events. Specifically, with a spatio-temporal cluster detection method, we firstly detected spatio-temporal clusters composed of geotagged images. Among these detected S-T clusters, we furthermore attempted to identify social events in terms of a classification model. Specifically, land use and road were considered to generate new kinds of spatial characteristics used as dependent variables incorporated into the classification model. In addition to this, user characteristics (i.e., the number of images and the number of users), spatial and temporal range of clusters, and the heterogeneity of temporal distribution of images were considered as the other dependent variables for the classification model. Consequently, with a BLR method, we estimated the categories (i.e., 'event' or 'non-event' one) of the S-T clusters (cases). Experimental results demonstrated the good performance of the method with a total accuracy of 71 %. With the variable selection process of the BLR method, empirical result also indicates that (1) some characteristics (e.g., the distance to the road and the heterogeneity of temporal distribution of images) do not have considerable influence on the occurrence of 'event'; and (2) compared to the other urban land categories (i.e., residential and recreational land), commercial land has a relatively high influence on the occurrence of 'event'.

There are some aspects needed to be considered in the future works to enhance this approach. Firstly, as the most important characteristics, textual and visual characteristics should be combined with the spatial and temporal characteristics used in this chapter to enhance the event identification. Secondly, different types of events have different levels of spatio-temporal range. For instance, a festival or sport game tends to last longer (e.g., a week) within a larger spatial area (e.g., a large park or green space) than a music concert which might take place at a building and be over within one day. Therefore, it will be interesting to take the level of spatio-temporal range into account since distinct types of events might correspond to distinct level of spatio-temporal range. Accordingly, the spatial and temporal size of the scanning window for the cluster detection method would be changed.

References

Rattenbury T, Good N, Naaman M (2007) Towards automatic extraction of event and place semantics from Flickr tags. In: Proceedings of the 30th ACM international conference on research and development in information retrieval, Amsterdam, Netherlands, 23–27 July 2007

Earle PS, Bowden DC, Guy M (2011) Twitter earthquake detection: earthquake monitoring in a social world. Ann Geophys 54(6):708–715

Chae J, Thom D, Bosch H, Jang Y (2012) Spatiotemporal social media analytics for abnormal event detection and examination using seasonal-trend decomposition. In: Proceedings of the 2012 IEEE conference on visual analytics science and technology (VAST), Seattle, WA, 14–19 Oct 2012

Chen L, Roy A (2009) Event detection from flickr data through wavelet-based spatial analysis. In: Proceeding of the 18th ACM conference on information and knowledge management, Hong Kong, China, 02–06 Nov 2009

Brenner M, Izquierdo E (2012) Social event detection and retrieval in collaborative photo collections. In: Proceedings of the 2nd ACM international conference on multimedia retrieval, Hong Kong, China, 05–08 June 2012

Liu X, Troncy R, Huet B (2011) Using social media to identify events. In: Proceedings of the 3rd ACM SIGMM international workshop on Social media, Scottsdale, Arizona, USA, 30–30 Nov 2011

Wang Y, Sundaram H, Xie L (2012) Social event detection with interaction graph modeling. In: Proceedings of the 20th ACM international conference on multimedia, Nara, Japan, 29 Oct–02 Nov 2012

Lee R, Sumiya K (2010) Measuring geographical regularities of crowd behaviors for Twitter-based geo-social event detection. In: Proceedings of the 2nd ACM SIGSPATIAL international workshop on location based social networks, San Jose, California, 02–02 Nov 2010

Becker H, Naaman M, Gravano L (2010) Learning similarity metrics for event identification in social media. In: Proceedings of the third ACM international conference on web search and data mining, New York, USA, 04–06 Feb 2010

Pan C, Mitra P (2011) Event detection with spatial latent Dirichlet allocation. In: Proceedings of the 11th annual international ACM/IEEE joint conference on digital libraries, Ottawa, Ontario, Canada, 13–17 June 2011

Sakaki T, Okazaki M, Matsuo Y (2010) Earthquake shakes Twitter users: real-time event detection by social sensors. In: Proceedings of the 19th international conference on world wide web, Raleigh, North Carolina, USA, 26–30 April 2010

Watanabe K, Ochi M, Okabe M, Onai R (2011) Jasmine: a real-time local-event detection system based on geolocation information propagated to microblogs. In: Proceedings of the 20th ACM international conference on Information and knowledge management, Glasgow, Scotland, UK, 24–28 Oct 2011

Kulldorff M (1997) A spatial scan statistic. Commun Stat Theory Methods 1997(26):1481–1496

Becker H, Naaman M, Gravano L (2011) Beyond trending topics: real-world event identification on twitter. In: Proceedings of the fifth international AAAI conference on Weblogs and social, media (ICWSM'11)

Zielstra D, Hochmair HH (2013) Positional accuracy analysis of Flickr and Panoramio images for selected world regions. J Spat Sci 58(2):251–273

Senaratne H, Bröring A, Schreck T (2011) Assessing the credibility of VGI contributors based on metadata and reverse viewshed analysis: an experiment with geotagged Flickr images. Comput Graph Forum 30(3):871–880

MacEachren AM, Robinson AC, Jaiswal A, Pezanowski S, Savelyev A, Blanford J, Mitra P (2011) Geo-twitter analytics: applications in crisis management. In: Proceedings of the 25th international cartographic conference, Paris, France

Jankowski P, Andrienko N, Andrienko G, Kisilevich S (2010) Discovering landmark preferences and movement patterns from photo postings. Trans GIS 14(6):833–852

Ruocco M, Ramampiaro H (2012) A scalable algorithm for extraction and clustering of event-related pictures. Multimedia Tools Appl J doi:10.1007/s11042-012-1087-z

Zhang H, Korayem M, You E, Crandall DJ (2012) Beyond co-occurrence: discovering and visualizing tag relationships from geo-spatial and temporal similarities. In: Proceedings of the fifth ACM international conference on web search and data mining, Seattle, Washington, USA, 8–12 Feb 2012

Part II
Trajectory Analysis

A Recursive Bayesian Filter for Anomalous Behavior Detection in Trajectory Data

Hai Huang, Lijuan Zhang and Monika Sester

Abstract This chapter presents an original approach to anomalous behavior analysis in trajectory data by means of a recursive Bayesian filter. The anomalous pattern detection is of great interest in the areas of navigation, driver assistant system, surveillance and emergency management. In this work we focus on the GPS trajectories finding where the driver is encountering navigation problems, i.e., taking a wrong turn, performing a detour or tending to lose his way. To extract the related features, i.e., turns and their density, degree of detour and route repetition, a long-term perspective is required to observe data sequences instead of individual data points. We therefore employ high-order Markov chain to remodel the trajectory integrating these long-term features. A recursive Bayesian filter is conducted to process the Markov model and deliver an optimal probability distribution of the potential anomalous driving behaviors dynamically over time. The proposed filter performs unsupervised detection in single trajectory with solely the local features. No training process is required to characterize the anomalous behaviors. Based on the results of individual trajectories collective behaviors can be analyzed as well to indicate some traffic issues, e.g., turn restriction, blind alley, temporary road-block, etc. Experiments are performed on the trajectory data in urban areas demonstrating the potential of this approach.

H. Huang (✉)
Institute of Applied Computer Science, Bundeswehr University Munich,
85577 Neubiberg, Germany
e-mail: hai.huang@unibw.de

L. Zhang · M. Sester
Institute of Cartography and Geoinformatics, Leibniz University Hannover, Appelstr. 9a,
30167 Hannover, Germany
e-mail: Lijuan.Zhang@ikg.uni-hannover.de

M. Sester
e-mail: Monika.Sester@ikg.uni-hannover.de

J. Huerta et al. (eds.), *Connecting a Digital Europe Through Location and Place*,
Lecture Notes in Geoinformation and Cartography, DOI: 10.1007/978-3-319-03611-3_6,

1 Introduction

Anomalous behavior detection refers to the problem of finding patterns in data that do not conform to expected behaviors. It is of great interest for the applications of navigation/driver assistant system, surveillance and emergency management.

The techniques employed for anomalous pattern detection in the last years are summarized in (Chandola et al. 2009) with following classes: classification based techniques, parametric or non-parametric statistical techniques, nearest neighbor based techniques, clustering based techniques, spectral techniques and information theoretic techniques. A significant number of works related to automated anomaly detection in trajectory data involve trajectory learning, i.e., cluster models of trajectories corresponding to normal cases are learned from historical trajectories and new trajectories are typically assigned an anomaly score based on the distance to the closest cluster model or likelihood of the most probable cluster model (Morris and Trivedi 2008). Hu et al. (2006) propose an algorithm for automatic learning of motion patterns and use these patterns for anomaly detection and behavior prediction. Trajectories are clustered hierarchically using spatial and temporal information and then use a chain of Gaussian distributions to present each motion pattern. Based on the learned motion patterns, statistical methods are used to detect anomalies and predict behaviors. Besides the cluster based trajectories learning method, Piciarelli et al. (2008) propose a trajectory learning and anomaly detection algorithm based on one-class Support Vector Machine. The algorithm can automatically detect and remove anomalies in the training data. They first evenly sample points from the raw trajectory and then model each trajectory with a fixed-dimensional feature. Bu et al. (2009) build local clusters using continuity characteristics of trajectories and monitor anomalous behavior via efficient pruning strategies. Ma (2009) presents a method of real-time anomaly detection for users following normal routes. Trajectories are modeled as a discrete-time series of axis-parallel constraints ("boxes") and then incrementally compared with a weighted trajectory collected from N norms.

Current approaches include (Kim et al. 2011), in which Gaussian process regression is used for the recognition of motions and activities (also anomalous events given already learned normal patterns) of objects in video sequences. Pang et al. (2011, 2013) adapt likelihood ratio test statistic to learn traffic patterns and detect anomalous behavior from taxis trajectories to monitor the emergence of unexpected behavior in the Beijing metropolitan area.

Recursive Bayesian estimation (or Bayes filter) (Masreliez and Martin 1977), e.g., the Kalman filter (Kalman 1960) for linear and normally distributed variables, are widely used in the areas of signal processing, navigation and robot/vehicle control. A main character of Bayes filters is the dynamic updating (actually two steps: prediction and updating) the estimation of the underlying variable(s) based only on the most recently acquired measurement data. Kalman filter and its extension have been proved appropriate for trajectory analysis. Recent works include (Prevost et al. 2007), which presents an extended Kalman filter to predict the trajectory of a moving object with the measurement data from a moving sensor—an unmanned aerial system (UAS). An Unscented Kalman filter is used in (Sun et al. 2012) for the trajectory tracking based on the satellite data with weak observability and inherent large initial error.

This chapter presents an original approach to anomalous behavior analysis of GPS trajectory data of vehicles. A variant of recursive Bayesian filter is proposed for a dynamic inference process. One of the original ideas is to find where the driver is meeting navigation problems, i.e., taking a wrong turn, performing a detour or tending to lose his way. Differing from most of the previous approaches:

1. The filtering is conducted for a high-level feature, i.e., the belief of behavior character, instead of the vehicle state, i.e., position and orientation.
2. The pattern detection is performed on single trajectory and no previous learning process is required to distinguish normal and anomalous behaviors.

For this purpose high-level features, i.e., (1) turns, their combination and density, (2) the degree of detour and (3) the route repetition, are required and a long-term perspective is taken to extract them from the original data. We use an extended high-order Markov chain to remodel the trajectory integrating these long-term features. The proposed recursive Bayesian estimator processes the Markov model and deliver an optimal probability distribution of the potential anomalous drive behaviors over time.

The chapter is organized as follows. In Sect. 2 we introduce the anomalous behaviors in the trajectories and the features we employed to recognize them. Section 3 presents the Markov model adapted to the trajectory data and the recursive belief filter. Experiments and results are demonstrated in Sect. 4. The chapter ends up with conclusions in Sect. 5.

2 Anomalous Behaviors and Features

In this work, we focus on the anomalous patterns in driving. In contrast to a normal drive from the start spot to the predetermined destination, anomalous behaviors may happen in many various situations, e.g., taking a wrong turn, getting lost, road-block, temporary stopover, etc. Please note that we, just like other works, also work based on the basic GPS data, i.e., time stamps and the positions. Unlike the conventional measurements of position and velocity, however, the anomalous patterns can usually not be observed at a single time point. High-level as well as long-term measurements, i.e., behavior features, are required as the "observations" of the underlying state.

2.1 Turns Combination and Density

Turning is one of the most basic movements in the trajectory data. Although a single turn is not indicating any anomalous behavior, their combination and density can

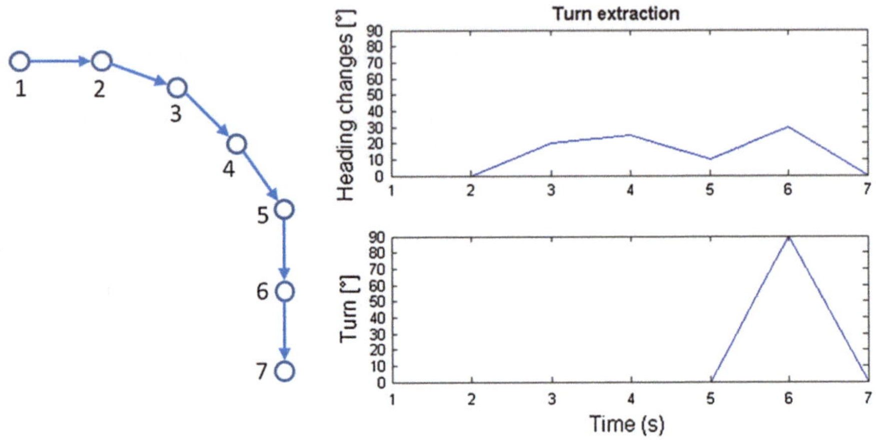

Fig. 1 Turn extraction: the accumulated heading changes are extracted as a turn movement at the last state

deliver some anomalous patterns like forming a detour/loop and unusually intensive turns. In the GPS data a turn is normally finished within several time stamps (several seconds). As shown in Fig. 1, at the first state where a turn starts the values of heading changes are counted and the turn is "marked" at the last state when the turn has been finished. We use the total absolute heading change of 40° as the threshold to determine a turn. Turns right to the previous direction is defined as positive.

In comparison with the detection of a single turn, a perspective with even longer term is taken to observe and evaluate the combination and density of multiple turns. Turns inside a given time interval (a "memory" of normally 1 to 2 min at urban driving speed) are recorded with the directions of turning. Intensive sequential same turns, e.g., double or triple left turns, have more impact on the belief of anomalous behaviors than the sequential different turns because they are implying a potential detour (see below) or the tendency of looping.

2.2 Detour Factor

Detour often happens when the driver meets traffic issues, e.g., road-blocking and traffic jam, or fails to find the correct or best way to the destination. A detour factor is conducted to quantify the degree of detour as an anomalous feature. From a start point, if the trajectory tends to go backwards, or in other words, the heading change is about 180°, it will be treated as a detour and the detour factor will be calculated for all the points in the backward segment. As shown in Fig. 2, the detour factor of an individual point is the ratio of the length of the trajectory from the start point (solid blue) and the direct distance (green) between the start point and the current position.

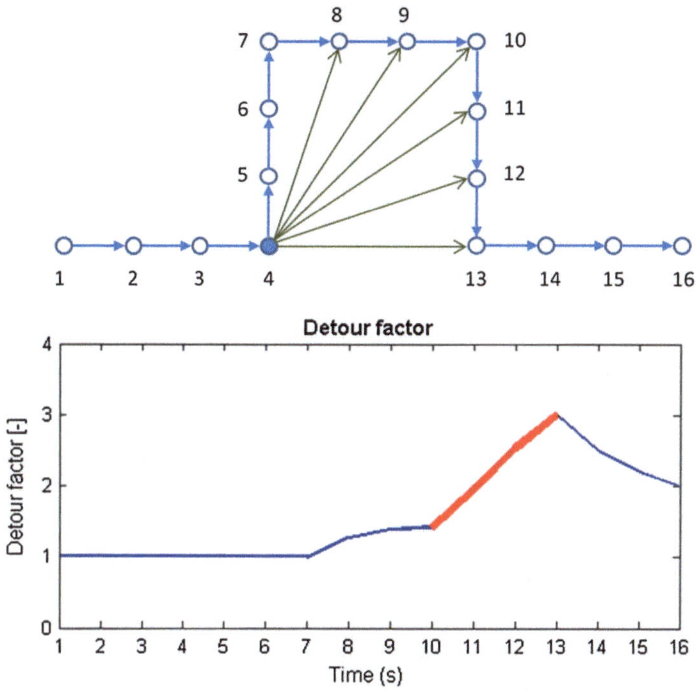

Fig. 2 Detour factor: calculated from the first turn of turn-combination of U-form. The detour factor values are only given to a certain segment (*bold*, *red*) after the heading change is accumulated to about 180° until the value decreases. A value of 1 is then given to the rest states meaning no detour

2.3 Route Repetition

The most prominent feature in a one-way trajectory is the route repetition, i.e., the driver goes back to the same road part, from either same or opposite direction, on his way to the destination. Route repetition with the opposite direction is mostly the result of performing an U-turn while that with same direction often happens after driving a loop. The current trajectory segment, i.e., between the current and the last steps, is repeating a former route when any prior trajectory segment(s) fall inside the buffer of the current segment and approximately parallel to it.

3 A Recursive Belief Filter

A Bayes filter is conducted to estimate an unobserved state, i.e., the belief of anomalous behaviors, recursively over time. The extracted features mentioned above can be considered as the measurements/observed states in the Hidden Markov Model (HMM).

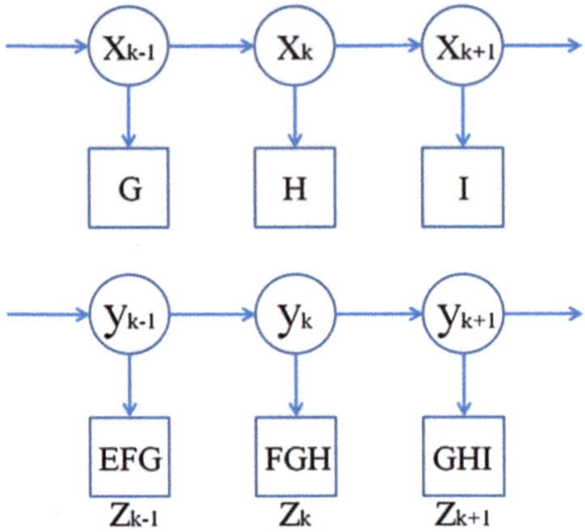

Fig. 3 An example of remodel the high-order Markov chain with the "memory" $m=3$. The last 3 observations of each state in X is integrated into new observations Z for the new chain Y

3.1 The High-Order Markov Chain

Markov chains are commonly used in the modeling of state changes over time sequence. The first order Markov chain is the basis of most Bayes filter variants, e.g., Kalman filter, and can be easily considered as an appropriate model for trajectory data. In this work we use an extended high-order Markov chain to integrating the long-term features mentioned above. Let x be the unobserved state (here the probability of anomaly) and z the measurement from long-term observations, the HMM as the process model is presented in Fig. 3.

The proposed high-order Markov chain $X = \{x_0, ..., x_n\}$ still follows the Markov assumption, i.e., the probability of the current state given a limited number of previous ones is conditionally independent of the other earlier states:

$$p(x_k|x_{k-1}, x_{k-2}, ..., x_{k-m}, ..., x_0) = p(x_k|x_{k-1}, x_{k-2}, ..., x_{k-m}) \tag{1}$$

with $m < k$. The measurement $Z = \{z_0, ..., z_n\}$ at each state is dependent not only on the corresponding state, but also several previous states:

$$p(z_k|x_k, x_{k-1}, ..., x_{k-m}, ..., x_0) = p(z_k|x_k, x_{k-1}, ..., x_{k-m+1}) \tag{2}$$

The higher order implies also, however, (1) the number of to be solved parameters grows exponentially with the order $O(|x|^m)$ (with $|x|$ the number of possible states of x and m the order) and (2) the reliability of parameter estimation decreases. We then remodel the trajectory by constructing a new chain $Y = \{y_0, ..., y_n\}$ with a m-tuple of x states:

$$y_k = (x_k, x_{k-1}, ..., x_{k-m+1}) \tag{3}$$

so that the new chain Y over the m-tuple is equivalent to a first order Markov chain keeping the conventional Markov property

$$p(y_k | y_{k-1}, y_{k-2}, \ldots, y_0) = p(y_k | y_{k-1}) \tag{4}$$

with a "memory" of m, and

$$p(z_k | y_k, y_{k-1}, \ldots, y_0) = p(z_k | y_k) \tag{5}$$

as shown in Fig. 3 (bottom).

3.2 The Belief Filter

The proposed filter is a simple variant of recursive Bayesian estimator keeping the dynamic property and the prediction/updating scheme. Although normally the prediction and updating steps work alternately and provide required inputs for each other, either of them has also the probability to be skipped. In this work both of these cases will happen:

- We are using multiple measurements (behavior features) for the updating. Sometimes more than one feature can be extracted at the same time stamp. With the assumption that the features are independent to each other (simplified assumption) the updating step is performed multiple times before the next prediction.
- These long-term features, however, cannot be continuously observed. In the interval of the given observations the prediction will be performed solely for multiple times.

Prediction

The prediction step calculates the total probability, i.e., the integral of the products of the transition probability $p(x_k | x_{k-1})$ and the probability of the previous state $p(x_{k-1} | z_{k-1})$ over all possible x_{k-1}. In this case we have only one variable, i.e., the belief, to be estimated and in principle it cannot be predicted based on any current measurements. We assume that the anomalous behaviors are rather "transitory" than the normal drive and, therefore, use a simple exponential decay to predict the belief of the next state:

$$x_k = x_{k|x_{k-1}} = F \cdot x_{k-1} + w_k \tag{6}$$

where

$$F = e^{-s \cdot k};\ w_k \sim \mathcal{N}(0, \sigma^2). \tag{7}$$

F simulates the decay given to the belief of anomaly along with the driving. A Gaussian noise is added by w_k. k is used to count the number of previous state(s) without new anomalous feature(s) being reported. The accumulation of k makes sure that the belief decays rapidly after the driver performs normally. The decay tendency can actually be tuned by the factor s. Generically we give no weight to k for each step, i.e., $s = 1$, in the urban area. Fine tuning of s can adapt the filter to:

- the driving in suburban areas with higher speed and sparse street crossings, in which case two potential anomalous behaviors may have longer time interval and the belief decay should be set slower to keep the pattern being found, and
- the pedestrian trajectory in dense urban areas, where the decay speed may need to be further enhanced to avoid continuous accumulation of the belief.

The predicted state estimate is taken as prior estimate for the current state.

Updating
The update step uses Bayes rules. The prior estimate is refined with the observation on the current state and deliver a posterior estimate.

$$p(x_k|z_k) = \frac{p(z_k|x_k) \cdot p(x_k|z_{k-1})}{p(z_k|z_{k-1})} \tag{8}$$

where the prior distribution is actually

$$p(x_k|z_{k-1}) = \prod_{i \in \mathcal{V}} p(z_{i,k}|x_k) \cdot p(x_k|x_{k-1}) \tag{9}$$

with multiple measurements. $p(x_k|x_{k-1})$ is the initial distribution after prediction. z_i with $i \in \mathcal{V}$ the observation(s) that have been previously integrated ($\mathcal{V}$) in the current updating phase.

3.3 Belief Inference

We employ two simple typical cases: detour and wrong turn with simulated data to present the inference process using the proposed belief filter. Besides the information of turns these two cases have their particular features that another one does not have i.e., the detour case has only detour factor and no repeated route while the wrong turn case has the latter only. So that the influence of the individual features can be well demonstrated. Figure 4 presents a simple simulated trajectory with detour (left) and the extracted high-level features plotted over time (right). The bold red line shows the inferred belief of anomalous patterns over time. The possibility values are also presented in the trajectory with scaled colors. Please note that the green circle and red asterisk are used to mark the start and end positions of the trajectory, respectively. The value/color of each line segment is determined by its start point. We use these two examples to demonstrate some typical situations in the inference process.

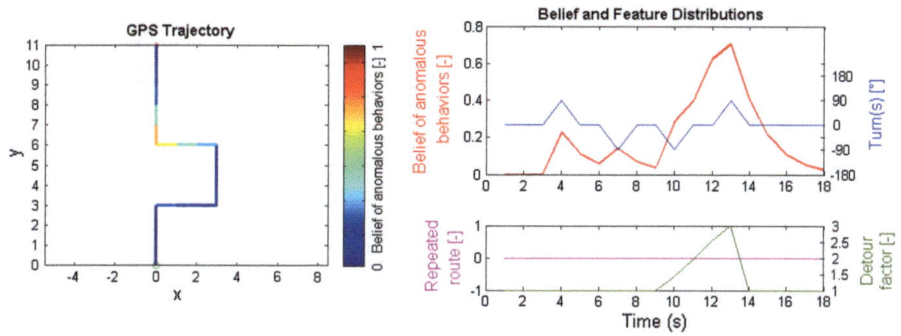

Fig. 4 Simulated trajectory of detour (*left*) with start position (*green circle*), end position (*red asterisk*) and the belief of anomalous behavior shown with scaled color. Three high-level features: Turns (*blue*), repeated route (*magenta*) and detour factor (*green*) are plotted together with the belief of anomaly over time (*right*)

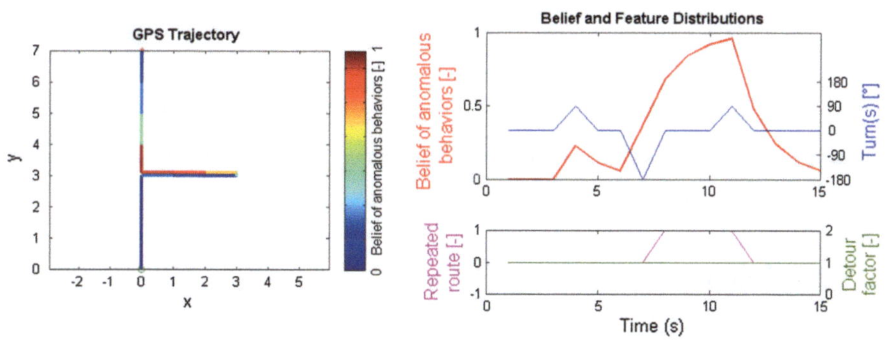

Fig. 5 Simulated trajectory of detour: trajectory with colors indicating the belief of anomaly (*left*) and the distributions of the belief and behavior features over time (*right*)

- Double different turns is considered normal. If the current turn has the different direction to the previous one, less probability is given to guarantee the continuous decay of the belief of anomaly.
- Double same turns, in the contrast, mean potential detour or even looping. Probability gain is added when the second turn happens.
- Detour factor increases and reach the maximum value when the detour is finished. The belief of anomaly has the peak value at this time as well.
- After the detour the belief of anomaly has a fast decay.

Another typical case of wrong turn is shown in Fig. 5. All the states of repeated route have the same feature value of 1. The belief increases as long as the vehicle stays in the wrong way and reaches the peak value at the spot where the wrong turn started. The belief decays to normal value when the vehicle goes back to the previous road.

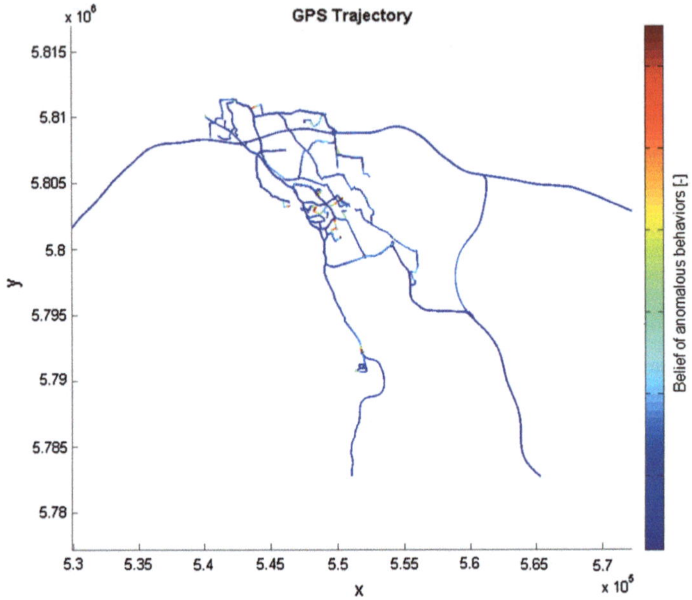

Fig. 6 Experiment on a VGI dataset with 100 Trajectories. Colors of the trajectory indicate the belief of anomaly

4 Experiments

Experiments are performed on the volunteered geographic information (VGI) data, an open trajectory dataset as well as the trajectories from self-acquisition. Figure 6 shows an experiment on a VGI dataset, which has been gathered by one business commuter in two years inside the city of Hannover, Germany. 100 trajectories are randomly selected to test the presented algorithm. Anomalous behaviors, with a probability of anomaly over 85 %, are detected in 12 trajectories (12 % of the total number). 368 out of total 68136 (0.54 %) GPS nodes are labeled presenting potential anomaly. Detailed analyses on several individual trajectories can be found in the follow-up figures.

In comparison with the two simulated cases (cf. Figs. 4 and 5) Figs. 7 and 8 show the detour and wrong turn detections on the actual VGI trajectories.

A more complicated case with turns, loop and an incomplete detour is given in Fig. 9 (left). As shown in the belief and feature distributions (Fig. 9, right), the influence of the first two turns decrease rapidly along the driving and present actually a normal segment. Forming a loop is in contrast a prominent anomalous behavior, which consists of three sequential turns and route repetition when the vehicle goes back to the former road. A small segment containing turns combination is detected afterwards and part of it is recognized as a detour (second peak of the green line) with the proposed method.

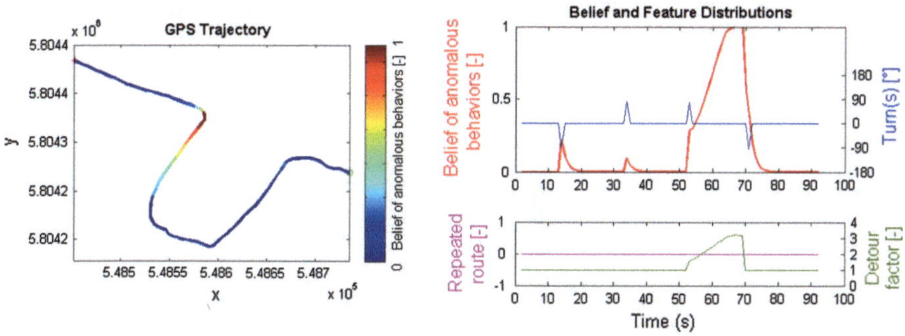

Fig. 7 An example of detour: trajectory with colors indicating the belief of anomaly (*left*) and the distributions of the belief and behavior features over time (*right*)

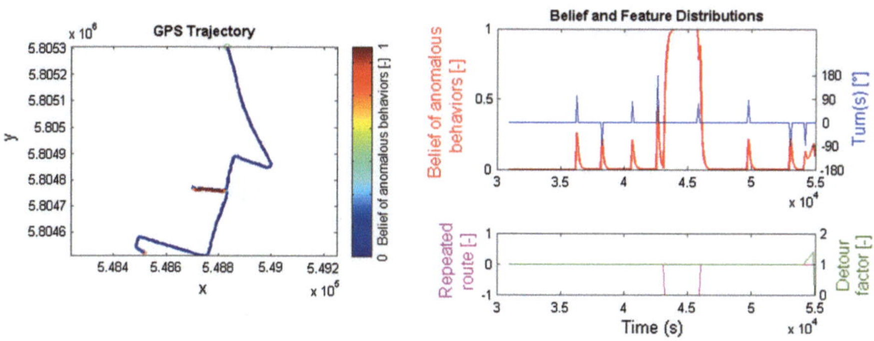

Fig. 8 An example of wrong turn: trajectory with colors indicating the belief of anomaly (*left*) and the distributions of the belief and behavior features over time (*right*)

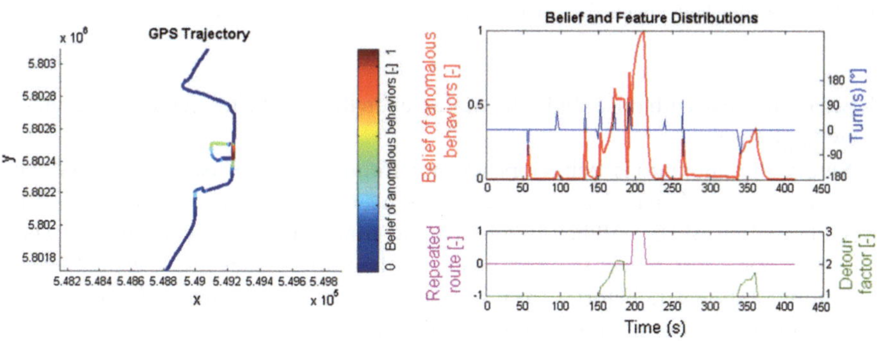

Fig. 9 A single trajectory with turns, loop and detour: trajectory with colors indicating the belief of anomaly (*left*) and the distributions of the belief and behavior features over time (*right*)

Figure 10 demonstrates a trajectory which is correctly recognized as normal driving, i.e., no obvious anomalous pattern are found. The belief distribution function

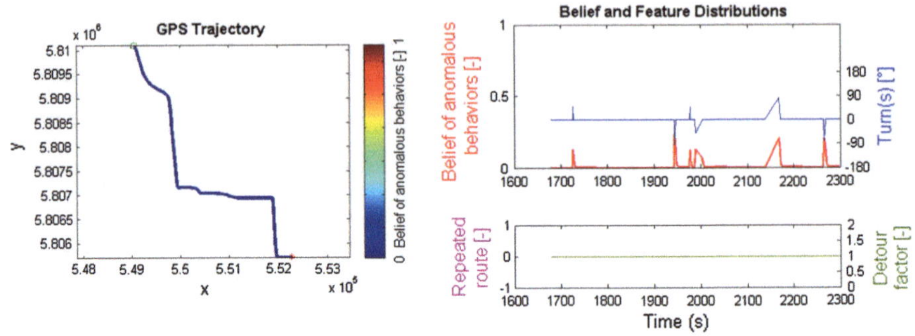

Fig. 10 An example of normal driving: trajectory with colors indicating the belief of anomaly (*left*) and the distributions of the belief and behavior features over time (*right*)

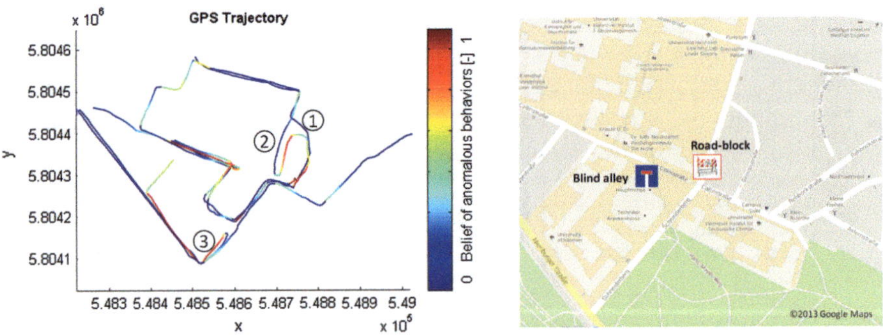

Fig. 11 Collective behaviors in the cases of road-blocking and blind alley: a collection of trajectories in the same area and similar time period (*left*) and the street map with the locations of the road-block and the blind alley being manually labeled (*right*)

shows robustness with only slight fluctuates, even though the trajectory also contains a few large turns.

Surely the anomalous behaviors are not frequent in the usual trajectory data. In some cases of urban traffic, e.g., road-blocking, blind alley or turn restriction, however, anomalous traces will be often found and concentrated in a certain area. These collective behaviors reflect to a certain extent the traffic issues mentioned above.

Figure 11 shows trajectories from self-acquisition (with known traffic conditions and driver behaviors) in an urban area, where a temporary road-block as well as a blind alley nearby (right) near a road crossing can be found. Anomalous patterns are found at the end of the blind alley and from multiple sides of the road-block. As shown in the trajectories (left), driver 1 from the north saw the sign of road-block and made a detour, driver 2 missed the warning sign of road-block, had to make a U-turn right before the road-block and then performed a detour to go on with the same direction. Driver 3 from the south turned around even earlier because of the warning sign and observing a traffic jam before the crossing. Although the blind

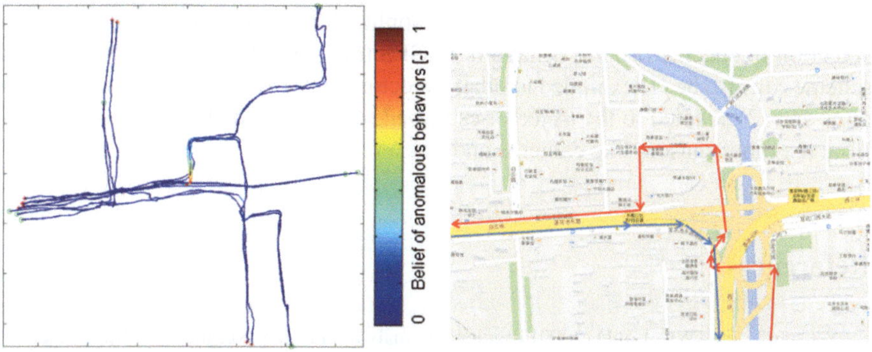

Fig. 12 Collective behaviors in the case of turn restriction: trajectories (*left*) passing the road intersection shows no anomaly except that from the bottom to the left and the street map (*right*) with the routes from bottom to left (*red*) and reversed (*blue*) indicating the turn restriction

alley on the west side is not a temporary setup, it causes U-turns sometimes for the drivers who are not familiar with this area.

Figure 12 presents an experiment on the open trajectory dataset: "GeoLife GPS Trajectories" (Zheng et al. 2008, 2009, 2010) of Beijing, China. Inside the shown segment trajectories from the bottom to the left show coincident detour while the reversed (from the left to the bottom) trajectories have no anomaly. We assume that such phenomena may indicate a potential left turn restriction, which is proven by the street map shown in Fig. 12 (right), i.e., no left turn is possible here because of the cloverleaf junction and the direction restrictions in the streets.

5 Conclusion

This chapter presents an original approach to anomalous behavior detection in the trajectory data by means of a recursive Bayesian filter. The main contributions of this work can be summarized as follows:

- A recursive belief filter is conducted for the dynamic detection of anomalous patterns.
- Long-term behavior features are integrated using high-order Markov model.
- Unsupervised detection in single trajectory with local features.

By these means the belief of anomalous behaviors can be inferred dynamically over time. In single trajectory, the result indicates where the driver is likely meeting navigation problem and an assistant is needed. Furthermore, a potential of reflecting traffic issues, e.g., turn restrictions, unexpected blind alleys and temporary road-blocks, is shown as well by analyzing the collective behaviors of multiple trajectories. We are, however, actually aware that the geometric features only are still limited for a plausible anomaly detection. Further semantic information and background knowledge might be helpful to estimate the anomalous behaviors more precisely.

In this chapter we demonstrate one application of the proposed filter—the detection of specific driving behaviors in GPS trajectory data. We assume that this scheme can also be (1) extended to other trajectory patterns given corresponding features and (2) adapted to the trajectories of pedestrian or animals, which are derived from the other sensors like cameras and trackers.

References

Bu Y, Chen L, Fu AWC, Liu D (2009) Efficient anomaly monitoring over moving object trajectory streams. In: Proceedings of the 15th ACM SIGKDD international conference on Knowledge discovery and data mining, KDD '09. ACM, New York, NY, USA, pp 159–168. doi:10.1145/1557019.1557043

Chandola V, Banerjee A, Kumar V (2009) Anomaly detection: a survey. ACM Comput Surv 41(3):15:1–15:58. doi:10.1145/1541880.1541882

Hu W, Xiao X, Fu Z, Xie D, Tan T, Maybank S (2006) A system for learning statistical motion patterns. IEEE Trans Pattern Anal Mac Intell 28(9):1450–1464. doi:10.1109/TPAMI.2006.176

Kalman RE (1960) A new approach to linear filtering and prediction problems. Trans ASME-J Basic Eng 82(Series D):35–45

Kim K, Lee D, Essa I (2011) Gaussian process regression flow for analysis of motion trajectories. In: Proceedings of IEEE international conference on computer vision (ICCV). IEEE computer society

Ma TS (2009) Real-time anomaly detection for traveling individuals. In: Proceedings of the 11th international ACM SIGACCESS conference on computers and accessibility, Assets '09. ACM, New York, NY, USA, pp 273–274. doi:10.1145/1639642.1639712

Masreliez C, Martin R (1977) Robust bayesian estimation for the linear model and robustifying the kalman filter. IEEE Trans Autom Control 22(3):361–371. doi:10.1109/TAC.1977.1101538

Morris B, Trivedi M (2008) A survey of vision-based trajectory learning and analysis for surveillance. IEEE Trans Circuits Syst Video Technol 18(8):1114–1127. doi:10.1109/TCSVT.2008.927109

Pang LX, Chawla S, Liu W, Zheng Y (2011) On mining anomalous patterns in road traffic streams. In: Proceedings of the 7th international conference on advanced data mining and applications–volume Part II, ADMA'11. Springer, Berlin, Heidelberg, pp 237–251

Pang LX, Chawla S, Liu W, Zheng Y (2013) On detection of emerging anomalous traffic patterns using gps data. Data Knowl Eng 87:357–373. http://www.sciencedirect.com/science/article/pii/S0169023X13000475

Piciarelli C, Micheloni C, Foresti G (2008) Trajectory-based anomalous event detection. IEEE Trans Circuits Syst Video Technol 18(11):1544–1554. doi:10.1109/TCSVT.2008.2005599

Prevost C, Desbiens A, Gagnon E (2007) Extended kalman filter for state estimation and trajectory prediction of a moving object detected by an unmanned aerial vehicle. In: American control conference, ACC '07, pp 1805–1810. doi:10.1109/ACC.2007.4282823

Sun L, Li D, Yi D, Liu J (2012) Trajectory tracking based on iterated unscented kalman filter of boost phase. In: 2012 IEEE International conference on service operations and logistics, and informatics (SOLI). pp 232–235. doi:10.1109/SOLI.2012.6273537

Zheng Y, Li Q, Chen Y, Xie X, Ma WY (2008) Understanding mobility based on gps data. In: ACM conference on ubiquitous computing (UbiComp 2008). Korea, ACM Press, Seoul, pp 312–321

Zheng Y, Zhang L, Xie X, Ma WY (2009) Mining interesting locations and travel sequences from gps trajectories. In: International conference on World Wild Web (WWW 2009). ACM Press, Madrid Spain, pp 791–800

Zheng Y, Xie X, Ma WY (2010) Geolife: a collaborative social networking service among user, location and trajectory. IEEE Data Eng Bull 33(2):32–40

Using GPS Logs to Identify Agronomical Activities

Armanda Rodrigues, Carlos Damásio and José Emanuel Cunha

Abstract The chapter presents an approach for collecting and identifying the daily rounds of agronomists working in the field for a farming products company. Besides recognizing their daily movements, the approach enables the collection of data about the shape and size of land parcels belonging to the company's clients. The work developed involved the design of spatial movement patterns for data collection through GPS logs, with minimal disruption of the agronomists' activities. The extracting of these patterns involved place and activity extraction, with specific algorithms proposed for marking and unmarking exploration parcels. These algorithms were evaluated by field testing with very positive results.

1 Introduction

The use of mobile sensor data for identifying spatially mapped activities is, currently, a research field under steady development (Ashbrook and Starner 2003; Hecker et al. 2011; Ye et al. 2009). Mostly, researchers aim to identify frequent locations that users go to, trends in their movement, so that commercial companies can propose/recommend products or services that they provide. Some generic algorithms have been developed, using probabilistic methods (Lee and Cho 2011;

A. Rodrigues (✉)
CITI/DI, Faculdade de Ciências e Tecnologia, Universidade Nova de Lisboa,
2829-516 Caparica, Portugal
e-mail: a.rodrigues@fct.unl.pt

C. Damásio · J. E. Cunha
CENTRIA/DI, Faculdade de Ciências e Tecnologia, Universidade Nova de Lisboa,
2829-516 Caparica, Portugal
e-mail: cd@fct.unl.pt

J. E. Cunha
e-mail: j.cunha@campus.fct.unl.pt

J. Huerta et al. (eds.), *Connecting a Digital Europe Through Location and Place*,
Lecture Notes in Geoinformation and Cartography, DOI: 10.1007/978-3-319-03611-3_7,

Liao et al. 2005, 2007), to extract the form of spatially mapped activities from logs generated by GPS devices, when the patterns underlying these activities have not been identified. These algorithms, although very powerful and having reported very high precision, involve a level of complexity, parameterization and training which hampers their immediate use in different domains. Moreover, a comprehensive approach towards identifying unknown patterns may prove unnecessary, when the aim of logging/tracing your steps is, in fact, collecting data for well specified operational activities.

In this chapter, we present an applied research work, which was developed in cooperation with a Portuguese farming products company (Borrego Leonor e Irmão S.A.). The aim of this work was to develop and support an inexpensive/self-providing method for collecting data about the shape and size of land parcels belonging to this company's more than 500 clients, distributed over a large geographical area. It was also of importance to be able to collect and identify the daily rounds of the company's agronomists. The work developed thus involved the design of spatial movement patterns for collecting data about the agronomists' activities, which included walking and/or driving around land parcels, as well as their daily movements to and from the clients' farms. The design of these patterns enabled the informed collection of data, through GPS logs, about the location, shape and size of clients' land parcels with minimal disruption of the agronomists' activities. This information is fundamental for the company, since it simplifies the evaluation of each client's needs. The amount of fertilizer or pesticides, needed in a particular situation, depends on the specificity of the crop, soil, area and problem to address. This capability can thus improve productivity for the company, as well as for the farmers, namely by planning ahead their stocks of products, like pesticides and fertilizers, in order to guarantee the fulfillment of each client's possible needs at the right time. The use of this method to collect this type of data is motivated by the changing rate of the exploration parcels shape and size, as crops and farmers frequently rotate in the considered region, occasionally several times per year.

1.1 User Story

The daily life of most people follows some general regular patterns, in fact recurrent patterns in time and space. This insight is the basis of our approach to extract activities from GPS logs, captured by the user's mobile device. In the case of our domain of application, a typical user story can be found in Table 1. It is particularly striking that driving periods provide immediately a segmentation of the GPS log, allowing to reduce the search space and to focus in the relevant periods. Detection of the sequences of GPS trackpoints, where the user starts and returns to the same point, provide the remaining clues to determine what she is doing. In particular, inside the non-driving periods, and by looking at these sequences, we can find, in the morning, the sequences 9–13 (visit to the field), 10–11 (marking of parcel), 11–12 (observation in the parcel), and 10–12 (visit to the parcel), and in the afternoon 16–17 (at company),

Table 1 User story

No.	Driving?	Time	Place	Description
1		7:00	Home	Turns on the smartphone, checks email and verifies the agenda for the day
2		7:15	Home	Goes to the car and drives to the favourite coffee house to have an expresso
3	Driving	7:20	Coffee house	Stops the car, walks to the coffee house, and drinks the coffee. Meanwhile, receives a call
4		7:35	Coffee house	Enters the car, and drives to company offices to get some products to deliver to clients
5	Driving	8:05	Company	Parks the car, loads the car with the goods to deliver, and takes care of the necessary paper work. The boss sets a meeting with a client for lunch time
6		8:24	Company	Leaves by car to a new parcel of a client to check the status of the crops
7	Driving	9:30	Client's office	Parks the car, meets the client, talks a while, makes some business
8		10:00	Client's office	Head to the new parcel in the client's car
9	Driving	10:20	Near parcel	The client stops her car, and they walk to the parcel
10		10:35	Parcel	The agronomist marks the parcel, checking the border plants, and returns to the beginning of the parcel
11		11:45	Parcel	Marking ends, and then goes inside the parcel to check some problems
12		12:15	Parcel	Returns to the starting point, and then goes to the car
13		12:17	Near parcel	They enter the car and return to the client's office
14	Driving	12:34	Client's office	They arrive at client's office, say goodbye
15		12:40	Client's office	Starts the car and drives in a hurry to the office
16	Driving	13:05	Company	Parks the car, and boss is already waiting
17		13:15	Company	They start walking to a nearby restaurant
18		13:25	Restaurant	Reach the restaurant, and have a business lunch
19		15:35	Restaurant	Leave the restaurant and walk back to the company's office
20		15:45	Company's office	Enter the building, and stay inside for a meeting
21		17:55	Company	Leaves the office, and enters the car
22	Driving	18:03	Gas station	No gas. Stops to fill-in
23		18:17	Gas station	Drives home
24	Driving	18:30	Home	Stops the car and turns-off the phone

18–19 (at restaurant), 20–21 (meeting), as well as some non-interesting sequences like 16–20, or 16–21. Notice that we are interested in the innermost sequences, since the other, longer sequences are usually aggregations of activities or bad pairing, but we do not know which.

As a side remark, the previous observations also pave the way to a construction of a hierarchy of activities with interesting potential applications (for instance, the sequence 1–24 is the working day of the user, while 7–15 is the complete visit to the client).

This motivates our approach, that we first overview in Sect. 3, and further detail in the remaining parts of the chapter. In Sect. 2, we relate the issues underlying the developed work with existing related work. Section 4 describes the only pattern "taught" to the agronomists in order to mark/unmark parcels, while the more important and novel algorithms are presented in Sect. 4. Section 5 provides the preliminary evaluation of the system and, finally, the conclusions appear in Sect. 7.

2 Related Work

Extraction of data related to the shape of land parcels is mostly realized offline through automatic extraction from maps or photos (Clementini and Ippoliti 2013; Pitarch et al. 2011). This does not work if you know where an exploration parcel will be in the future, but no crop has been created yet. Mostly, drawing the shape of the parcel on a google map interface may also not work, as you may need to be in the field to recognize the local references that define this shape. Drawing on a mobile device, on location, has its limitations, coming from problems with the screen, dust and light. The methodology used in this work is motivated by these reasons. We decided to explore the agronomists current work habits, which involve surrounding exploration parcels in the field, and drawing on chapter maps, by adding data collecting through GPS.

Li et al. (2008), Ye et al. (2009) and Zheng et al. (2009) use GPS trajectories generated by users to find similarities between them, based on the sequences of places they visited. Ye et al. (2009) define the concept of Staypoint, used in this chapter, to represent a physical location where the user remains for a period of time. Li et al. (2008) describe an algorithm for staypoint detection.

The collection of data, captured through GPS, is subject to error and showing evidence of staying in one place for a while may be difficult. Clustering is thus used in this context, specifically density clustering, which allows for irregular formed clusters (Zhou et al. 2007). This technic was used to cluster staypoints, specifically the DJ-Cluster algorithm described in Zhou et al. (2004).

Concerning activity extraction from GPS tracking, Lee and Cho (2011) propose a system in which, using contextual data produced by smartphones, it is possible to extract information about the activities realized by the users. The system is supported by *Hierarchical Bayesian Networks*. Another method for classification is used by Liao et al. (2005), using a framework based on *Relational Markov Networks(RMN)*, which can extract information about a user day-to-day activities. RMN are an extension of *Condition Random Fields*, graph-based models which become very effective for classifying activities. These techniques have been analyzed and produce good results. However, the activities to be identified in this chapter are very well defined, with specifically designed patterns. To use methods this elaborate would add, not needed, complexity to the system. We thus decided to address the problem by designing specific algorithms.

3 Approach

The approach used in the project, with the aim of capturing an agronomist's daily activities, involves five stages: capturing GPS logs; extracting well–known places; activities extraction; results visualization and correction, when needed.

Capturing GPS logs: The starting point for the flow is the capturing of GPS logs, using the mobile device. The main point considered in this stage was the fact that the agronomist's daily schedule should not be deranged by the use of the mobile device for data capture. This, associated to the difficulty of viewing data on a smartphone display, under the sun, led to the development of a very simple user interface for manipulating the mobile device, which simply involved starting the app, in the morning, before starting daily rounds. All data input came from the agronomist's movements, which were recorded by the device. Prior to this, the technician was briefed on the patterns to be used for collecting parcel shape. The design of these patterns will be described in Sect. 4. The mobile part of this process ends at this stage, as the rest is executed from the desktop, using the logs generated by the agronomist, as input. Data extraction starts at this point and involves the following stages.

Extracting well-known places: Once the capture of GPS logs is finished, extraction of well-known places visited by the technician, during capture, is realized. The thousands of captured points are restricted using the *Staypoints* technique (Li et al. 2008; Ye et al. 2009; Zheng et al. 2009). This result is improved with clustering, through the *DJ-Clustering* algorithm (Zhou et al. 2007). The resulting staypoints' coordinates are then submitted to a Geocoding service, to add information to the visited places, and provide context to the agronomist's activities. Previously visited places are stored in the database supporting the applications, with a count of the number of visits.

Activities extraction: The next stage of the approach involves identifying activities performed by the agronomist, during her daily work schedule. Several activities were collected and analyzed but this chapter mainly focuses on "Parcel marking/unmarking" and "Driving". The basis for the methodology used was the fact that the agronomists' movements can be classified as in *driving mode* or in *walking mode*. When in driving mode, the technician will be going towards a specific place. As she reaches her aim, she will then step out of the car and perform additional activities in walking mode. When in walking mode, the activity of marking/unmarking parcels should be identified, from the designed patterns movements performed. To instantiate this methodology from the GPS logs, an analysis of the velocity of the movement between staypoints is firstly performed, enabling the evaluation of whether the technician is walking or driving. Due to the highly error-prone captures, it was not possible to determine the type of movement directly from the original logs. We thus resort to using a fixed size window of GPS points to reduce error, designated by Movement Window, which will be detailed in Sect. 5.1. In fact, the introduction of this concept enabled the measuring of velocity between staypoints to be based on the average coordinates extracted for each Movement Window. This analysis resulted in a classification of the logs in terms of walking segments and

Fig. 1 Application interface. The solution used google maps as the basis for overlaying captured logs

driving segments. The refinement of this methodology enabled us to isolate, inside walking segments, those where movement velocity was nearly zero, where parcel marking and unmarking patterns could be extracted, as this activity was performed on foot. The extraction of these patterns will be described in detail in Sect. 5.2.

Results visualization and correction: A preliminary interface for visualizing the results was developed. In a cartographic window it is possible to overlay the resulting segments, classified according to the performed activities. The aim of this interface is to enable the agronomists to evaluate results and insert corrections if needed (Fig. 1).

In the next sections we will describe the design of the patterns used in parcel marking and unmarking, as well as the implementation of the activity extraction algorithms.

4 Parcel Marking Pattern

Although the design of the patterns was developed in the context of the agronomist activities, taking into account their information needs, these patterns can be applied to other types of activities, such as the definition of security perimeters or any type of activity that involves the need of completely surrounding a two-dimensional region. The circumscription of the parcel is an activity actually performed by agronomists to mark parcels, typically with very precise GPS equipment. So, our pattern reflects this behavior and is natural to users.

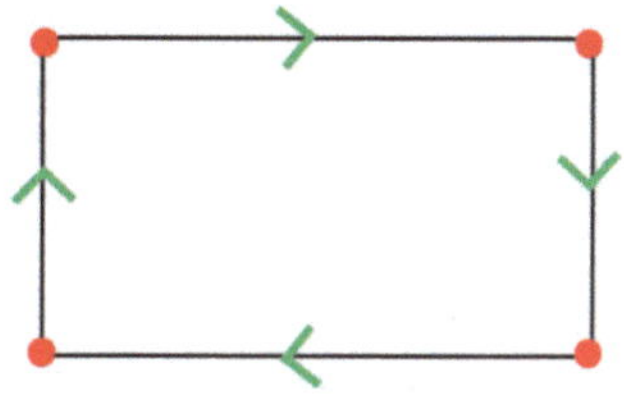

Fig. 2 Agronomist marking pattern

The agronomists were given instructions for data collection when marking or observing exploration parcels, so that these activities could be later recognized through the patterns involved, when analyzing the generated records.

The parcel marking activity thus follows the general technique for capturing polygons in Geographic Information Systems, with the inside of the parcel kept to the right of the marker (direction is clockwise). Marking should always start and end on the same vertex of the parcel. Accordingly, it is possible to remove a region from the parcel (for example, a small lake) in the same fashion, while moving in the opposite direction, that is, counterclockwise (with the area to be removed on the left side of the marker). An example of a marking pattern is shown in Fig. 2.

These patterns are identified inside a more general pattern which aims to isolate sequences of movements where the agronomist drives to a recognized staypoint, walks in several periods and finally goes back to the car location and drives again.

5 Activity Extraction: Implementation

Activity extraction from the initial capture is performed in three steps, and the result of each step is used as input for the following one. In this way, the information used is increasingly restricted until it becomes tailored to what is necessary to extract the relevant activities. The sequence of steps can be described as:

1. Analysis and identification of the most frequent type of movement performed between staypoints and classified as: *Driving, Walking or AlmostStoppedOr Stopped*;
2. Relevant sequences extraction: this step involves the extraction of sequences, from the logs, which will be relevant for the identification of the activities;
3. Activity extraction, from the sequences generated in step 2.

5.1 Most Frequent Type of Movement

The aim of this step is to identify the type of movement performed between staypoints identified in the logs. The main cue for the type of activity underlying the agronomist's movements comes from the velocity of movement, which will enable the separation between segments associated with the *driving* activity and those which may involve

Fig. 3 Example of movement window: the data associated with each window is shared between consecutive windows. In a new window, the oldest trackpoint is lost and the most recent one is added. In this way, the mean of the coordinates evolves smoothly and is less prone to errors of data capture

parcel marking/unmarking. The GPS logs used, mainly collected in a rural region, were very error-prone, leading to the use of a subterfuge for evaluating velocity, which involved the use of *Movement Windows*. As shown in Fig. 3, a movement window includes a sequence of consecutive trackpoints with a fixed length (60 trackpoints—one trackpoint every second). Each window is represented by one coordinate, the mean of all the coordinates included in it. The velocity measured between staypoints is not determined directly from the trackpoints obtained through GPS tracking but as the mean velocity in the movement window. The calculus of the velocity in this way enables the classification of movement periods between staypoints in terms of the most frequent type of movement, namely driving, walking or stopped.

The process to determine the type of movement of each window is somewhat elaborated in order to handle speed variations by the user. The basic problem here was to identify periods where the agronomist was genuinely walking and (for example) not simply stopped in a traffic light. The algorithm conceived for this purpose evaluates the state of the moving device by identifying moments of change in velocity (slow to fast and fast to slow) which enables the recognition of a situation where the technician steps out of the car to some walking activity.

Accordingly, each movement window will be automatically classified as Driving whenever the window velocity is greater than or equal to the *drivingVelocity* parameter, set in our implementation to 2.75 m/s corresponding to approximately 10 Km/h (a running pace). However, if the velocity drops below the *drivingVelocity* limit then we still classify the period as driving until *numMinWindows* (30) have continuously stayed below the driving velocity. If more than *numMinWindows* are below the driving velocity, then the algorithm switches to a state detecting low speed movement or absence of movement.

In this state, an additional algorithm *determineIfWalkingOrAlmostStopped* was developed for separating situations where the technician is Walking from AlmostStoppedOrStopped (see Algorithm 1). With this methodology, it is possible to isolate periods when movement velocity was zero or nearly zero (maybe due to noise) from genuine walking periods. The algorithm uses a simple statistical test to determine if the user is walking or not (i.e. AlmostStoppedOrStopped).

It computes the average of the movement windows velocity, and its standard sample deviation, and calculates a threshold that separates walking periods from AlmostStoppedOrStopped periods:

Algorithm 1: Window Movement Identification

```
1  Input: windows[]: array of Window
2  Output: List of pairs <action,List of Windows> (resultList)
3
4  //initialization
5  lowSpeed:=false, merged:=false, create new empty List<Window> resultList
6
7  //classify the first window
8  lowSpeed := ( windows[0].avgVelocity ≤ drivingVelocity)
9
10 create new empty List<Window> currList
11 currList.add(windows[0])
12
13 //iterate over all the remaining windows except the first (already treated) one
14 for j:= 1 to windows.size-1 do
15   // Classify the previously seen windows as Driving
16   if(windows[j].avgVelocity <= drivingVelocity && !lowSpeed) then
17     lowSpeed:=true
18     if(!merged) then
19       add new pair <driving,currList> to resultList
20     else
21       resultList.last().append(currList)
22     end
23     create new empty currList
24   end
25
26   // Classify previously seen windows as Walking/AlmostStoppedOrStopped
27   if(windows[j].avgVelocity > drivingVelocity && lowSpeed) then
28     lowSpeed:=false
29
30     //condition that verifies if it was not just a momentary reduction of speed
31     if(currList.size ≥ numMinWindows || resultList.size == 0) then
32       //Obtain List of pairs <action,list of windows> newListOfPairs from currList
33       newListOfPairs = determineIfWalkingOrAlmostStopped(currList)
34       //add all the results
35       for p ∈ newListOfPairs do
36         resultList.add(p)
37       end
38       create new empty currList
39     else
40         merged:=true
41     end
42   end
43   currList.add(windows[j])
44 end
45
46 treat last currList using the code for driving (17-24), or non-driving (line 34)
47
48 return resultList
```

$$threshold := |(average - (1.96 \times standardDeviation)| \quad (1)$$

If the window velocity is below the above threshold then it is classified as AlmostStoppedOrStopped, otherwise it is a Walking period. If the movement window velocity follows a normal distribution then it is expected that only 2.5 % of the values will be below the threshold. The absolute value is used when the expression inside becomes negative due to a low average, favoring AlmostStoppedOrStopped classifications.

The result of this analysis is the classification of each window occurring between staypoints. The classification assigned to the whole period is determined from the majority classification of each window in the period, resulting in a final classification of the segment as Walking, Driving or AlmostStoppedOrStopped. This result is the starting point for the next step, with the aim of isolating sequences to be used in activity extraction. The worst-case complexity of this classification process is linear on the number of trackpoints.

5.2 Relevant Sequences Extraction

Given that the most relevant activities to be extracted are non-driving activities, and that for these to be performed, it is necessary to drive to a staypoint, the aim of this step is to identify sequences which do not include driving periods, but are placed inside driving periods. The results from the previous step are thus used to focus the processing in the interesting non-driving periods. Several algorithms were developed to achieve this aim and which are used in sequence, taking the output from the previous algorithm described in Sect. 5.1 as input:

1. Algorithm for identifying possible activity periods;
2. Algorithm for identifying activity occurring sequences;
3. Algorithm for super-sequence removal.

Identifying possible activity periods: This algorithm starts from a list of pairs (*staypoint* , *typeOfMovement*) which associate to each staypoint, the most frequent type of movement practiced on the way to the staypoint. From this dataset, the algorithm extracts a list of sequences of trackpoints that are included inside two external driving movements, for which at least one of the corresponding staypoints is located outside of localities (settlements). The rationale is that the relevant non-driving activities for agronomists are performed outside community settlements.

Identifying activity occurring sequences: The next algorithm uses a sequence of trackpoints at a time, resulting from the previous step, and applies a variant of the movement window algorithm described in Sect. 5.1. A different parameterization is used, to determine the locations where the user really stops, namely by reducing the window size to 10, and numWindows to 5. Sequences are considered if they involve starting and ending in the same location (creating the shape of a parcel), by finding two windows where user has stopped (i.e. AlmostStoppedOrStopped), are closely located to each other (in radius of 25 m, a reasonable GPS error), and are separated by at least 90 s. It is also assumed that activities starting and ending in the same place are separated by at least 5 min (the approximate time to walk a square parcel with 10,000 m^2 of area at 5 Km/h, i.e. a small parcel in the region at a good walking pace). Sequences are also filtered in terms of their minimum area (a low parametric value of 10 m^2). The details can be found in Algorithm 2. Anyway, if the data involves several moments where the technician goes back to the same position, several sequences may be extracted, with repeating sub-sequences. The aim is to identify the smallest sequence which involves movement around the same location, as seen in the next algorithm, below. This step is quadratic in the number of AlmostStoppedOrStopped windows.

Algorithm 2: Relevant Sequences

```
Input: Window stoppedWindows[]
Output: List of Lists <Trackpoint> (resultList)

//iterate over all coordinates that represent almostStopped/Stopped periods
for i:=0 to stoppedWindows.size do
  //start iterating from the next one
  for j:=i+1 to stoppedWindows.size do

    //check if both coordinates are near each other
    if distance(stoppedWindows[i], stoppedWindows[j]) <= radiusNearPlaces then

      //verify if already has elapsed minimumTimeToComeback
      if interval(stoppedWindows[i], stoppedWindows[j] >= minTimeToComeback then
        create List<Trackpoint> currList
        currList := get all trackpoints between beginning of window
            stoppedWindows[i] and ending of window stoppedWindows[j]

        //verify if the area is big enough
        if calculateArea(currList) >= minSizeArea then
          add currList to resultList

          //optimization that moves the index of the external cycle to force
          //a difference of at least separationTime between the sequences
          a:=i+1
          while(a<stoppedWindows.size) do
              if interval(stoppedWindows[i], stoppedWindows[a]) > separationTime then
                i=a-1
                break
              else
                a++
              end
          end
          break
        end
      end
    end
  end
end

return removeSuperSequences(resultList)
```

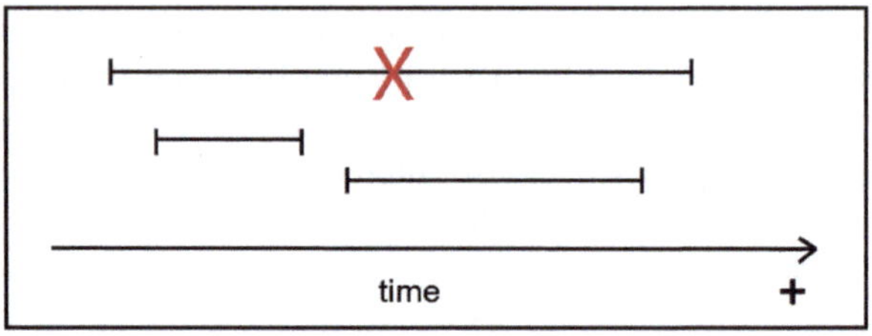

Fig. 4 Super-sequence removal example

Super-sequence removal: The last algorithm in this step receives all the relevant sequences generated by the previous one and removes super-sequences, which include data repeated in smaller sequences, as shown in Fig. 4.

This image represents a situation where the agronomist leaves the car once he arrives in the property, marks two parcels and returns to the car. The first sequence is a super-sequence that includes the following two, and is thus removed from the list of relevant sequences, since it corresponds to a higher-level activity which we are not interested in detecting (in this simple case, "visit to the client" as in the user story). The algorithm iterates over the sequences resulting from the previous step and verifies, for the current sequence under analysis, if it starts and ends after the next one (temporally). If these conditions are true, the sequence is added to the result set. If not, the algorithm re-iterates all the sequences in the result set and removes any that temporally terminates after the current one. Once it finishes, the algorithm returns all the minimum sequences which represent the time periods during which relevant activities might have occurred. This is a worst-case quadratic algorithm in the number of sequences.

The results of the algorithms presented in this section filter the sequences to the ones where parcel marking and unmarking may be found, and the calculation of the area of the parcels underlying these sequences. However, it is still necessary to verify whether the aim of the technician was to mark or unmark the parcel, that is, whether the areas should be added to or removed from the final set. This will be addressed in the next section.

5.3 Activity Extraction

Segments of the initial log classified as driving are immediately assigned the driving activity. From the staypoints, visits to usual places can also be extracted, since the number of times at that place is recorded, and the corresponding usual activity. The remaining relevant sequences, extracted in the previous step, are used to identify the performed activities in the field. In this section, and due to the limitations of the space provided for the writing of the chapter, we simply describe the work developed with the aim of identifying marking or unmarking of farming parcels. The method

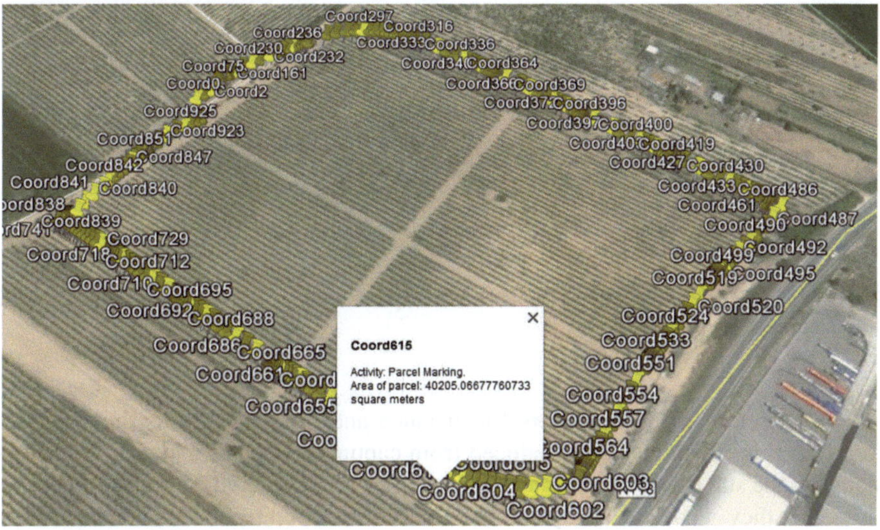

Fig. 5 Parcel marking example

used for this was detecting the direction of movement, while walking around the parcel (clockwise or counter-clockwise).

Detection of direction was achieved through the use of the PostGIS operator, *isCounterClockWise,* which takes a set of coordinates as an input parameter, and evaluates it according to direction. An example of parcel marking can be seen in Fig. 5. Parcel unmarking corresponds to the removal of part or of the total of a parcel area.

6 Parameterization and Evaluation

The performance of the approach was evaluated through 5 capture sessions provided by two agronomists. Capture was performed with the *OSMTracker* app.[1] and the resulting data was verified by the technicians involved in the evaluation. The logs resulting from the first two sessions were used to parameterize the algorithms developed and once the results from these were acceptable, they were re-submitted to the system, in order to verify that the solution involved no data loss. After this, 3 additional capture sessions were performed and submitted to the system, with the objective of evaluating the solution's performance and results, based on the current parameterization. All the results generated by the system were confirmed with the data collector.

---

[1] http://wiki.openstreetmap.org/wiki/OSMtracker

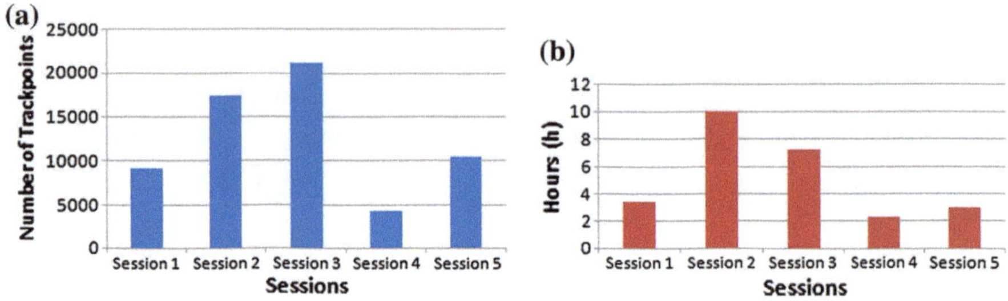

Fig. 6 **a** Number of trackpoints collected in each capture session; **b** Total time taken to generate the data in each capture session

The currently used parameters for distance and time, as well as those needed for extracting staypoints were deduced from capture sessions 1 and 2. In these sessions, all the staypoints' coordinates extracted were considered correct by the users. However, they were not always associated with the right point of interest (POI). The identification of the POI associated with a staypoint was achieved in two ways: using a publicly available Geocoding service, and using information directly inserted by the agronomists. The second method is used when the identified coordinate has already been recognized in the context of the application. The geocoding service is used when a new staypoint is identified. This method has presented some limitations, as the suggested POI for a coordinate is biased by the importance of POIs in the used system, which means that larger, most popular places may take precedence over more accurate locations. The reason for this is the fact that GPS logs are notoriously error-prone, particularly inside buildings that can alter the reference coordinate used.

Figure 6 presents the number of trackpoints and the time needed to collect them, for each capture session performed. We can see the relationship between these two variables, although it is not always this direct (for example, when the device cannot detect enough satellites for positioning).

Tables 2 and 3 summarize the results of the evaluation performed, when applying the developed approach to the 5 capture sessions. Table 2 presents the results of the performance of the system for staypoint identification and Point of Interest identification. Mostly, staypoints are correctly identified and positioned with few exceptions. Points of Interest are also mainly well identified. Some exceptions are noted: in session 1, positioning is correct but geocoding delivered one mistake in identifying the right POI. In session 2, one identified staypoint was verified to be noise, while one POI, from verified staypoints, was wrongly classified in geocoding. Session 4 was exceptionally developed in an urban area and the only staypoint and POI collected was wrongly placed in the ocean.

Activity identification is summarized in Table 3 There were only two wrongly identified activities. In session 2, we concluded that parcel marking was not correct, as the technician had not followed the provided instructions. In session 5, it was concluded that the parcel to be marked was very narrow, which may have led to a lesser result. This conclusion requires additional verification.

Table 2 Staypoint and point of interest identification

Session	Staypoint identification			Point of interest identification		
	Total	Right	Right %	Total	Right	Right %
1	3	3	100	3	2	66.7
2	6	5	83.3	5	4	80
3	14	14	100	14	14	100
4	1	0	0	N/A	N/A	N/A
5	4	4	100	4	4	100
Total	28	26	92.9	26	24	92.3

Table 3 Activity identification

Session	Activity identification		
	Total	Right	Right %
1	6	6	100
2	8	0	0
3	21	20	95.2
4	3	3	100
5	4	3	75
Total	42	32	76.2

The evaluation performed involves a small set of tests. However, it does show very good results, which need to be verified in large scale use of the solution. One down point of the implementation is battery use, which needs to be addressed. Mostly, the developed work has been considered as a contribution to the identification of agronomical activities, with users considering the approach a success. In fact, most activities in the logs are identified, including driving, having lunch, pumping gas, executing technical visit and marking/unmarking of parcels.

7 Conclusions and Further Work

In this chapter, a solution for identifying and extracting information about agronomic activities from GPS logs is presented. The use of a smartphone GPS device enables the capture of an agronomist's daily movements and, at the end of the day, the creation of a complete report of the technicians activities. This was achieved by identifying the places where she spent most time, through the use of the staypoints technique, associated with density clustering. The resulting places were complemented with data from a geocoding service, when needed.

Because existing techniques for activity extraction were considered too generic for the focus of the work, the problem was addressed through specifically designed solutions. The identification of activities is thus supported by previously obtained staypoints and by isolating sequences delimited by driving periods.

The results obtained from this approach are good and promising. Although the number of tests was limited, feedback was very positive, with places being accurately identified, and a few misses in relating these with geocoding results. Activity identification is a success, with minor problems in marking narrow parcels, which needs to be address.

This thread of research, particularly what concerns the identification of driving/walking activities, is currently receiving major attention (Hemminki et al. 2013) which motivates the extension of the work presented in this chapter. The problems with the current solution will be addressed, as well as other developments which can positively enhance the current approach. Mainly, we aim to design patterns for marking the existence of specific crops and equipment in the field, which will enable the development of algorithms for identifying these patterns. Moreover, the design of the application must be subjected to usability evaluation, which has not be addressed until now, as the focus of the work was on functionality development and testing.

The used approach of finding recurrent patterns in GPS log data is also envisaged as a technique to be employed to extract and hierarchically structure the activities of users, in particular for aggregating activities into more complex ones. This is a promising avenue of research that we intend to explore in this area of application.

Acknowledgments This work was supported by FCT—project PTDC/AAC-AMB/120702/2010.

References

Ashbrook D, Starner T (2003) Using gps to learn significant locations and predict movement across multiple users. Pers Ubiquit Comput 7(5):275–286

Clementini E, Ippoliti E (2013) Automatic extraction of complex objects from land cover maps. In: Vandenbroucke D, Bucher B, Crompvoets J (eds) Geographic information science at the heart of Europe. Lecture notes in geoinformation and cartography, Springer International Publishing, Switzerland, pp 75–93

Hemminki S, Nurmi P, Tarkoma S (2013) Accelerometer-based transportation mode detection on smartphones. In: Proceedings of SenSys'13, Roma, Italy, 11–15 Nov 2013

Hecker D, Körner C, Stange H et al (2011) Modeling micro-movement variability in mobility studies. In: Geertman S, Reinhardt W, Toppen F (eds) Advancing geoinformation science for a changing world. Lecture notes in geoinformation and cartography, Springer, Berlin, pp 121–140

Lee Y, Cho S (2011) Human activity inference using hierarchical bayesian network in mobile contexts. In: Lu B-L, Zhang L, Kwok J (eds) Neural information processing. Lecture notes in computer science, vol 7062. Springer, Berlin, pp 38–45

Li Q, Zheng Y, Xie X et al (2008) Mining user similarity based on location history, In: Proceedings of the 16th ACM SIGSPATIAL international conference on advances in geographic information systems, Irvine, CA, USA, ACM, New York, 5–7 Nov 2008

Liao L, Fox D, Kautz H (2005) Location-based activity recognition using relational markov networks. In: Proceedings of the 19th international joint conference on artificial intelligence. Morgan Kaufmann Publishers Inc., San Francisco, pp 773–778

Liao L, Fox D, Kautz H (2007) Extracting places and activities from gps traces using hierarchical conditional random fields. Int J Rob Res 26(1):119–134

Pitarch Y, Vintrou E, Badra F et al (2011) Mining sequential patterns from MODIS time series for cultivated area mapping. In: Reinhardt W, Toppen F, Geertman S (eds). Advancing geoinformation science for a changing world, Lecture notes in geoinformation and cartography, Springer, Berlin, pp 45–61

Ye Y, Zheng Y, Chen Y et al (2009) Mining individual life pattern based on location history. In: Mobile data management: systems, services and middleware, IEEE MDM'09, Taipei, Taiwan, 18–20 May 2009, pp 1–10

Zheng Y, Zhang L, Xie X et al. (2009) Mining interesting locations and travel sequences from gps trajectories. In: Proceedings of the 18th international conference on world wide web, Madrid, Spain, ACM, 20–24 April 2009, pp 791–800

Zheng Y, Chen Y, Li Q et al (2010) Understanding transportation modes based on GPS data for web applications. ACM Trans Web 4(1):1–36

Zhou C, Bhatnagar N, Shekhar S et al (2007) Mining personally important places from gps tracks. In: Data engineering workshop, 2007 IEEE 23rd international conference on data engineering, Istanbul, Turkey, 17–20 April 2007, pp 517–526

Zhou C, Frankowski D, Ludford P et al (2004) Discovering personal gazetteers: an interactive clustering approach, In: Proceedings of the 12th annual ACM international workshop on geographic information systems, Washington, DC, USA, ACM, 8–13 Nov 2004, pp 266–273

Assessing the Influence of Preprocessing Methods on Raw GPS-Data for Automated Change Point Detection

Tomas Thalmann and Amin Abdalla

Abstract The automated recognition of transport modes from GPS data is a problem that has received a lot of attention from academia and industry. There is a comprehensive body of literature discussing algorithms and methods to find the right segments using mainly velocity-, acceleration- and accuracy-values. Less work is dedicated to the derivation of those variables. The goal of this chapter is to identify the most efficient way to preprocess GPS trajectory data for automated change-point (i.e., the points indicating a change in transportation mode) detection. Therefore the influence of different kernel based smoothing methods as well as an alternative velocity derivation method on the overall segmentation process is analyzed and assessed.

1 Introduction

GPS sensors have become a standard feature of modern mobile devices, such as smart phones or tablet computers. Consequently, a plethora of tracking applications have been developed to monitor user movement. The output is mostly a list of points with time-stamps, and therefore not very conclusive. Thus, deducing transportation modes from raw GPS-tracks, i.e., the determination of walking, biking, driving (car) or public transport can add considerable value to it. It can, for example, improve the availability of data for transport studies, where such information is still mainly gathered through hand written diaries. The idea of automated recognition of transport modes is not new, and has been a topic of interest in various research communities (Li et al. 2010; Reddy et al. 2010; Stopher et al. 2008).

T. Thalmann (✉) · A. Abdalla
Research Group Geoinformation, Department of Geodesy and Geoinformation,
Vienna University of Technology, Vienna, Austria
e-mail: tthalmann@gmail.com

A. Abdalla
e-mail: abdalla@geoinfo.tuwien.ac.at

J. Huerta et al. (eds.), *Connecting a Digital Europe Through Location and Place*,
Lecture Notes in Geoinformation and Cartography, DOI: 10.1007/978-3-319-03611-3_8,

A great part of the literature, though, is dedicated to the development and assessment of algorithms and methods to detect change points, i.e., those points in the list that represent changes in transportation means. Interestingly, the values used for the segmentation process remain unquestioned. Neither the derivation of speed from the raw GPS data, nor the potential smoothing of the data to reduce the influence of positional errors were subject of discussion.

The remainder of the work is structured as follows: In Sect. 2 we discuss relevant literature and methods used to deduct change points. Section 3 gives a detailed account of methods used. Sections 4 and 5 present the result of the analysis and the last section concludes with a set of recommendations derived from the findings.

2 Background and Related Literature

Liao et al. (2007) proposed a method to categorize and predict the movement and transportation behavior of individuals using GPS data. Their methodology is based on machine learning techniques and therefore require training data-sets that are specific to a user. This work will focus its attention on methods that are user independent, hence no user history is required for the process of segmentation. Such methods usually rely on a mix of GPS, acceleration, GSM and WIFI data. The main task is to assign a transportation mode for each point in a given GPS trajectory. This classification process consists of three steps:

1. Select and extract sensor measurements and metrics (features or descriptors) from the data.
2. Select a classifier and calculate the required thresholds and parameters.
3. Run tests to qualitatively evaluate the classification result.

As pointed out by Li et al. (2010), mobile phones have become multisensor-systems with great potential. Of course the features/observations from point 1 heavily depend on the chosen sensor(s). By using GPS and Accelerator data, for example, Reddy et al. (2010) achieved a 93 % classification accuracy. Unfortunately, they did not distinguish between the various modes of motorized transportation. The research of Chen and Bierlaire (2013) has focused on the usage of all sensors available on a Smart-phone and relies on a mix of GPS, Accelerator, Blueooth, GSM and WIFI data.

Ogle et al. (2002) draw attention to the fact that GPS-data is subject to systematic and random errors that coarsely depend on the geometric constellation of the receiver and the satellites in the field of view, as well as on atmospheric and clock influences (Hoffmann-Wellenhof et al. 2001). One have to take GPS accuracy into account when calculating and selecting features for transportation mode classification and derivation of other motion patterns.

Laube et al. (2007) identify four levels of analysis, respectively scale of feature calculation: *instantaneous* (local), *interval* (focal), *episodal* (zonal) and *total* (global). In consideration of the fact, that every track-point-location can contain error-components from GPS-measurements, it is assumed that interval-based feature calculation is able to heavily reduce such error influences compared to instantaneous calculation. Therefore researcher have proposed to use a moving interval to calculate basic features such as velocity or acceleration (Gómez and Valdés 2011; Cimon and Wisdom 2004).

Other approaches use statistical smoothing methods to reduce GPS error influences. E.g. Giremus et al. (2007) propose a particle filter to detect and reduce multipath errors and Wann and Chen (2002) investigate Kalman filtering with an additional post-Kalman-smoothing for velocity. Jun et al. (2006) compare Kalman filtering, least squares spline approximation, kernel-based smoothing and an adopted version of Kalman filtering in respect of velocity and acceleration profiles.

Dodge et al. (2008) categorizes the features derived from a generic trajectory as *primitive parameters* (position, time-stamp or interval), *primary derivatives* (distance, direction, duration or velocity) or *secondary derivatives* (change of direction, sinuosity or acceleration).

According to Zheng et al. (2010) the overall goal of an untrained method is to find features, which are not affected by differing traffic conditions, thus better reflect the transportation mode of a user. According to Zheng et al. (2010) these are Heading-Change-Rate, Stop-Rate and Velocity-Change-Rate. A classification algorithm then takes the descriptors (or features) and assigns one of the predefined classes.

As mentioned before the classification systems need parameters and thresholds. They can be determined either by supervised learning or by empirical examination. The current work chose an untrained method proposed by Zheng et al. (2010) that uses GPS data only. They state:

- Our method is independent of other sensor data like GSM signal and heart rate, and map information, for example, road networks and bus stops, etc. Thus, it is generic to be deployed in a broad range of Web applications.
- The model learned from the data-set of some users can be applied to infer GPS data from others; that is, it is not a user-specific model.

Similar to Stopher et al. (2008), their approach is a clear and descriptive two-step algorithm: (1) a change-point based track segmentation; (2) determination of transportation mode per segment.

Interestingly, all of the investigated methods proposed in literature start with existent velocity and acceleration data. The deriving process of those variables from a raw GPS track, or the effect of different methods for calculating speed and acceleration is not discussed. On the other hand approaches for trajectory analysis rarely deal with real data and the problems of uncertainty and sampling that come along with real data (Laube and Purves 2011). The influence of different analysis on practical applications like transportation mode for instance has not been extensively addressed in literature. This work shows that preprocessing of GPS-track data has a considerable effect on the overall performance of common classification methods. The assumption is that the results of such segmentation and classification processes can be improved by appropriate preprocessing.

3 Preprocessing of Raw GPS Data

This section discusses three approaches to preprocess raw GPS data, to reduce positional error influence on velocity estimations: (1) GPS-Point Accuracy (2) Smoothing Filters and (3) Velocity derivation.

It has to be stated, that most of the GPS-capable mobile phones (iOS and Android) already use built-in, model-based filters such as Kalman filters. So in fact the data used in this work already have been partially smoothed.

3.1 *GPS Accuracy*

The accuracy of the GPS position measurement mainly depends on the geometric constellation of the receiver and the available satellites. This is described by the Dilution of Precision (DOP), especially the horizontal DOP (HDOP) values. Widely spread satellites cause a more accurate position fix and a smaller DOP value. Since 4 satellites are necessary to fix a position it is considered to be advantageous to sort out the track points that have less than 4 satellites and a DOP higher than 3 (Stopher et al. 2005). The biggest advantage of this method is its simplicity, but it comes with drawbacks, such as the reduction of points in the GPS trajectory (see Sect. 4.2 for further discussion of this issue).

3.2 *Velocity and Acceleration Derivation*

If GPS data does not contain velocity values from Doppler-measurements the values have to be calculated somehow. A naïve approach, as done by Zheng et al. (2010) for example, is to simply calculate the velocity using distance and time difference from two consecutive points. Considering GPS-errors, the distance d_{AB} can contain large error components resulting from the positional errors of point A and B. The influence of such errors is smaller, the greater the distances get. To reduce the estimation error, Gómez and Valdés (2011) proposed the use of the n-th track point TP_{n+1} instead of the immediate follower $TP_{(n)}$ to calculate velocity value v_n (see Eq. 1).

$$v_n = \frac{d\left(TP_{(n+i)}, TP_n\right)}{\delta t\left(TP_{(n+i)}, TP_n\right)} \tag{1}$$

We will call this a point-based interval of length i. This approach works fine, as long the track follows a relatively straight line, but in reality the trajectory rarely is. Using longer intervals leads to an underestimation of the distance, because the air-line distance could be considerably shorter than the trajectory, especially in urban areas. This issue is known and one possible solution would be to matching GPS trajectories to external map data (Li et al. 2013).

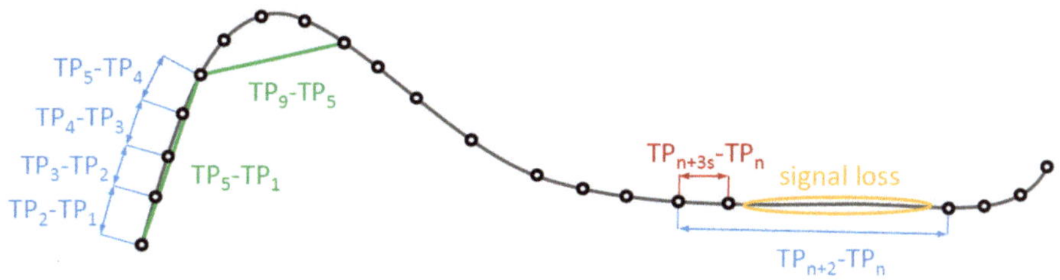

Fig. 1 Trajectory and different intervals for calculating average velocity; point based intervals (Eq. 1) in *blue* and *green*, time based interval (Eq. 2) in *red*

In addition, location observations are not equidistant, as it is assumed in Fig. 1. Observations are only delivered by the sensor in case a GPS signal is available and a position fix obtainable. So time-periods between location-fixes vary. Even if a position fix is delivered constantly every second, the distance between two track-points depends on the current speed. The spatial density of observations strongly varies, and so does the length of the temporal interval. Furthermore, the periods of signal loss can become very long when traveling on the subway. It is not desirable that points at the end of a subway segment skew the velocity derived by them and those before the segment. The use of wide point based-intervals increases the number of erroneous velocity calculation, e.g., if the interval length is set to 20 points, the velocity at 19 points before the gap will be skewed by the error. To avoid such problems, we propose to use time-fixed intervals rather than point-based ones. Eq. 1 becomes:

$$v_t = \frac{d\left(TP_{(t+\delta t)}, TP_t\right)}{\delta t\left(TP_{(t+\delta t)}, TP_t\right)} \tag{2}$$

With Eq. 2 we are able to retain the benefits of the point based-intervals and address the problem of data gaps to reduce the negative impact of signal loss. This can lead to considerable improvement for the analysis of GPS-tracks, especially in cities with a subway-system. The result of Eq. 2 is illustrated in Fig. 2 by the gray curve. It can be seen that the maximum of this curve moved to the left, so that it is now at the same time-stamp as the acceleration began in the original data in blue. By using the adjusted Eq. 3 unwanted time-shifts caused by Eq. 2 are eliminated.

$$v_t = \frac{d\left(TP_{(t-\delta t/2)}, TP_{(t+\delta t/2)}\right)}{\delta t\left(TP_{(t-\delta t/2)}, TP_{(t+\delta t/2)}\right)}. \tag{3}$$

3.3 Smoothing Filters

The third possibility to reduce the influence of positional errors on a velocity profile is statistical smoothing. The goal is to smooth a signal, respectively applying a low-pass-filter. The operation applied is a kernel-weighted average on a sliding

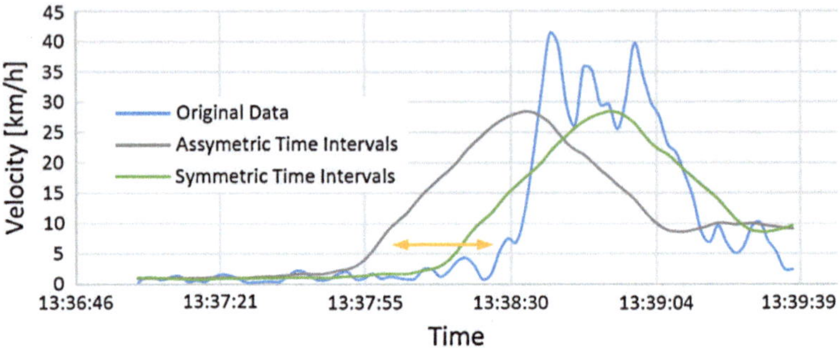

Fig. 2 Asymmetric time intervals of Eq. 2 show an unwanted time-shift, which can be eliminated with symmetric time intervals of Eq. 3

window of size h. Following a similar consideration which lead to time-based intervals in Sect. 3.2, we take a constant bandwidth, e.g. $h = 3\,\text{s}$, instead of a K-Nearest-Neighbor-Bandwidth.

The Nadaraya-Watson kernel smoothing algorithm defined in Eqs. 4 and 5 (Hastie et al. 2009) is used:

$$\hat{f}(x_0) = \frac{\sum_{i=1}^{N} K_h(x_0, x_i) y_i}{\sum_{i=1}^{N} K_h(x_0, x_i)} \tag{4}$$

$$K_h(x_0, x) = D\left(\frac{|x - x_0|}{h(x_0)}\right) \tag{5}$$

$D(t)$ determines the shape of the weighting function. The most popular kernels can be seen in Fig. 3. After visual investigation of differing weighting functions a Tricube-kernel seemed to bear the best results. It is the kernel with the steepest shape except from the Uniform-kernel, and would therefore maintain walk-stop-changes or generally short-time changes of velocity more precisely.

It is defined by:

$$D(t) = \begin{cases} 70/81\left(1 - |t|^3\right)^3, & \text{if } t \leq 1 \\ 0, & \text{otherwise} \end{cases} \tag{6}$$

There are mainly two possibilities to apply a kernel smoothing method on GPS-Tracks, namely on the velocity values itself, or on the positions prior to the velocity calculation. Preceding experiments showed that the velocity-smoothing does not bring any notable benefit, so in the coming experiments only position smoothing is applied to the data. Thus, the kernel smoothing is applied on the locations only, respectively on the Latitude/Longitude pairs of the track, which leads to a translation of every track point depending on the surrounding points. Afterwards the velocity is calculated as described in Sect. 3.2.

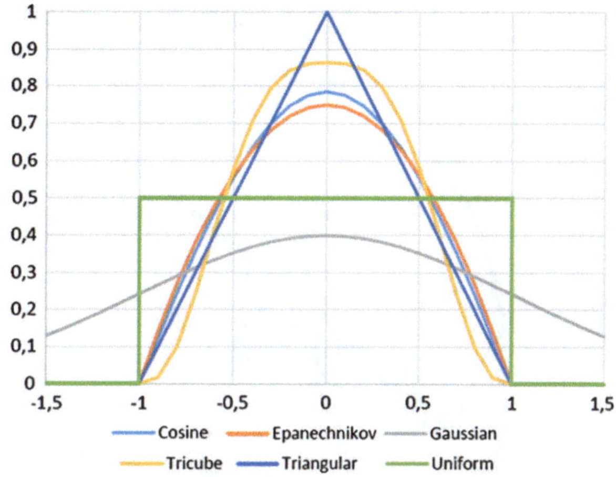

Fig. 3 Most popular kernels

4 Assessing the Preprocessing Methods

In this section a visual analysis of the effect of the three discussed preprocessing methods is used to discuss their effectiveness. It shows that presumably a combination of two will achieve the best results.

4.1 Test Data

The experiments and results in this chapter are based on a data set consisting of 14,260 track points that were tracked over 8 h and 26 min. A total of 40 Change-Points, i.e., the points indicating a change in transportation mode, were manually determined and the track was subsequently segmented into transport types of Walk, Bike, Tram, Bus, Car, Train and Subway. The tracks have been recorded by 4 persons with 3 different Android phones (Sony Xperia Z, HTC One X and Motorola Defy) and the free App GPSLogger.[1] The tracking took place between 7 am and 11 pm on varying days in June 2013 and are spread over the urban area of the city of Vienna, Austria (Fig. 4).

The App delivers .GPX-Files of track points containing Locations, time-stamp, HDOP, Number of satellites and Velocity from Doppler-measurements. The files were imported to a SQL-database for the experiments corresponding to the formal definition of a geo-spatial lifeline from Hornsby and Egenhofer (2002): A list of ordered GPS-track point corresponds to a list of space-time observations of the form `<ID, Location, Time>`, where `ID` is a unique identifier, `Location` is a spatial coordinate pair and `Time` is a sequential time-stamp. Other measurements like HDOP or Doppler-velocity is linked to a track point via the `ID`.

[1] https://play.google.com/store/apps/details?id=com.mendhak.gpslogger

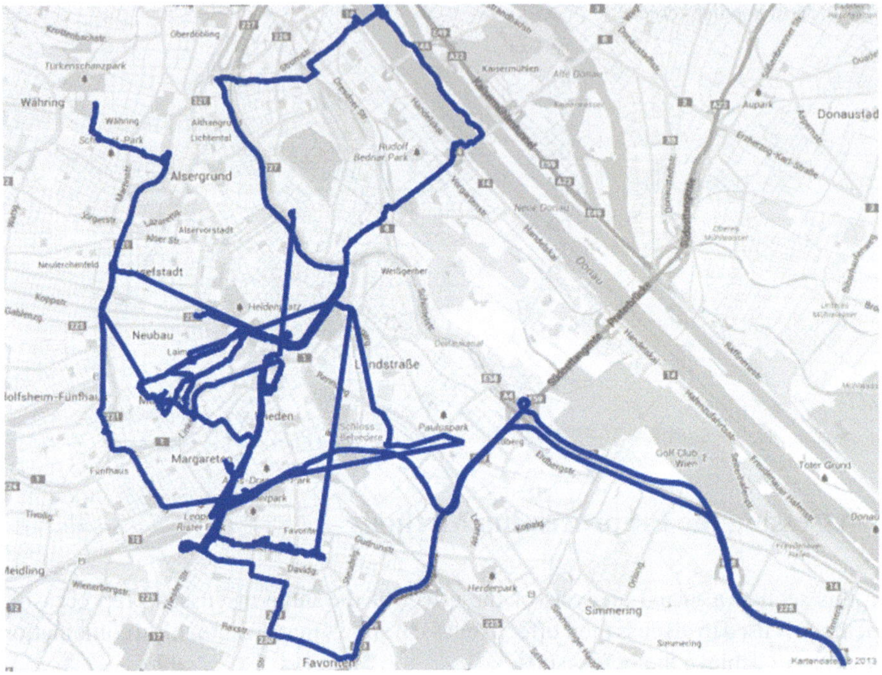

Fig. 4 Distribution of the testdata in the urban area of Vienna

4.2 Visual Investigation

To investigate the effect of the methods, we apply each of it to the raw GPS data separately, from which we then derive velocity based on point-based intervals from Eq. 2. First, we look at filtering out points based on the HDOP and satellite count, as discussed in Sect. 3.1. Figure 5 shows that some of the peaks in velocity, caused by inaccurate GPS-fixes were sorted out. On the other hand it expands some of the periods without positional fixes. Such gaps in the track can make change point-detection more difficult or in worst case impossible.

The effect of the smoothing filter (Sect. 3.3) can be seen in Fig. 6. The velocity is calculated from the smoothed positions with an interval of 1 point. The reader should note that the unfiltered data is still very noisy ($i = 1\text{point},\ h = 0\,\text{s}$). With higher smoothing parameters the data gets smoother. Subsequently, the difference between walk and stop segment becomes less recognizable. On the other hand the peak at around 13:36:05, a result of GPS-position-error, is partly removed from the graph by using a 20 s smoothing parameter. While removing the mentioned peak is desirable, it is also crucial to maintain the distinguishing properties of walk and stop segments. Finally, Fig. 7 illustrates the improvement that can be achieved by using the temporal interval based velocity derivation in Eq. 3.

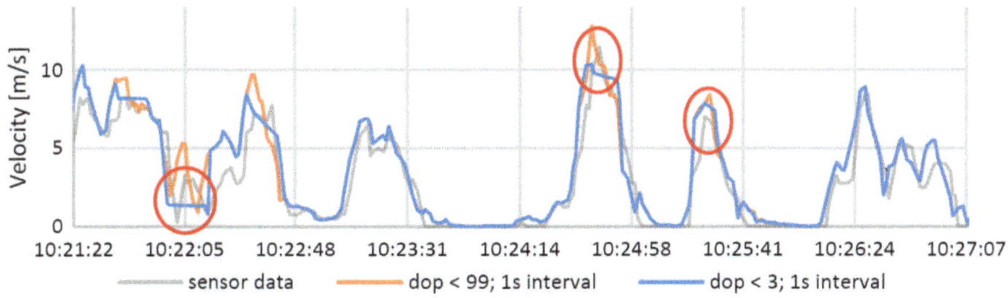

Fig. 5 Effects of a DOP-filter on the *velocity curve*

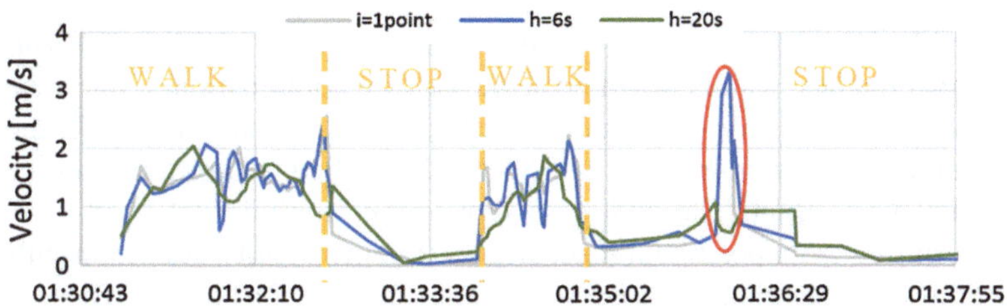

Fig. 6 Different smoothing parameters *h* of kernel smoothing on positions

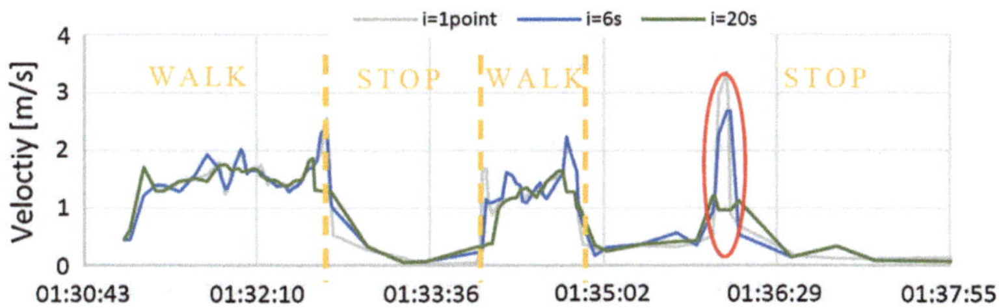

Fig. 7 Different temporal intervals *i*

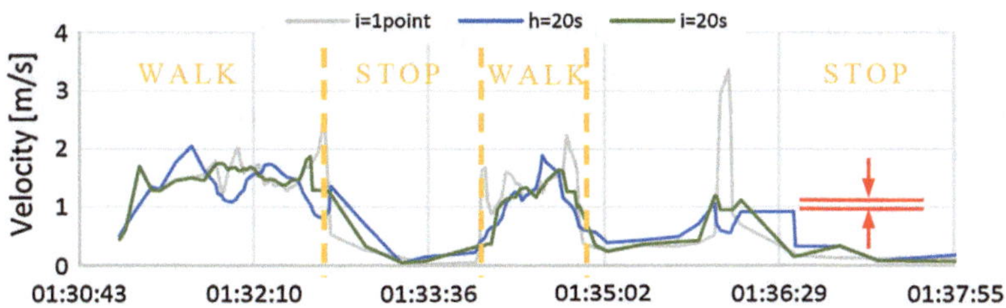

Fig. 8 Comparison of standalone temporal intervals and standalone kernel smoothing

The influence of a longer interval i is similar to that of a higher smoothing parameter h. With the interval-method the difference between walk and stop segment is better preserved than with the kernel smoothing. On the other hand the GPS-error is not considerably lowered (see Fig. 8).

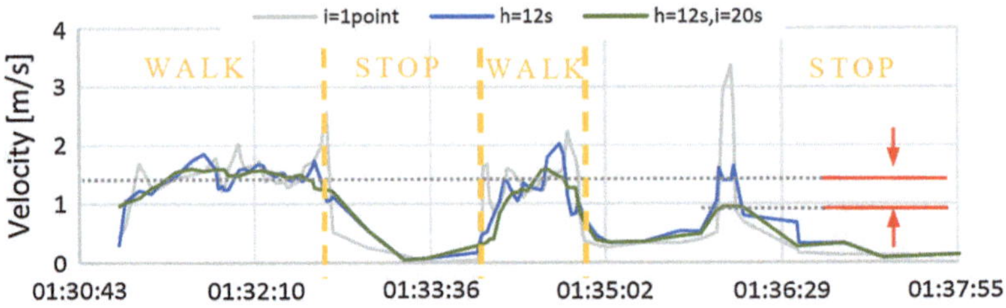

Fig. 9 Reduction of GPS-errors through the combined preprocessing methods

Each of the introduced methods has its pros and cons, but based on visual analysis, we infer that the best results might be achieved by a combination of more than one. Thus, in the following we investigate the effect of using combinations of methods, i.e., the temporal interval-method (see Sect. 3.3) applied to position-smoothed data using a TriCube-kernel. The goal is to reduce the amplitude of short-time peaks usually resulting from positional inaccuracies. In other words, sudden or unfeasible increases in speed should be filtered, such that a classification algorithm does not split a segment at that point. With a combined approach of both methods the maximum of the short-time peak of the GPS error can be downshifted below the average velocity of walk segments. This is necessary for the algorithm to be able to correctly detect walk segments and ignore errors (see Fig. 9).

In conclusion, it is assumed that the best performance can be achieved by a combination of a temporal interval-based-calculation and an applied kernel filter.

5 Change Point Detection and Segmentation Assessment

This section will test the performance of a well-known segmentation-algorithm proposed by Zheng et al. (2010) on various preprocessed GPS data sets. We will show that, as asserted in the previous section, a combination of temporal interval based velocity derivation and a kernel filter results in improved performance of the exemplary algorithm.

5.1 Changepoint Detection

The algorithm used for this study is taken from Zheng et al. (2010) and searches for track points, at which the testimonial changed transportation mode. Such points are called change-points and partition the track into segments. The classification of transportation mode per segment is conducted in a second step.

The algorithm is based on the fact that around 99 % of transportation mode changes happen with an interjacent walk segment. So at first the track is split up into alternating walk and non-walk segments, which greatly reduces the complexity of the segmentation.

- *Step 1*: Using a loose upper bound of velocity (Vt) and that of acceleration (at) to distinguish all possible Walk Points from Non-Walk Points.
- *Step 2*: If the distance or time span of a segment composed by consecutive Walk Points or Non-Walk Points less than a threshold, merge the segment into its backward segment.
- *Step 3*: If the length of a segment exceeds a certain threshold, the segment is regarded as a certain segment. Otherwise it is deemed as an uncertain segment. If the number of consecutive uncertain segments exceeds a certain threshold, these uncertain segments will be merged into one Non-Walk segment.
- *Step 4*: The start point and end point of each Walk segment are potential change points, which are used to partition a trip.

5.2 Assessment Methodology

As mentioned in the last section, the algorithm requires a velocity upper bound u, an acceleration upper bound a, a minimal segment length d and a minimal certain segment length (in this work fixed as 1.5-times minimal segment length) as input. We call such a set of input parameters (the configuration of the algorithm): $p_i \in P = U \times A \times D$.

For our assessment we first apply the differing preprocessing methods described in Sect. 3 (with varying values for time interval $i\ elem\ I$, smoothing parameter $h \in H$, and upper DOP-bound $o \in O$). We then have a set of outputs defined by their input $v \in V$ with $V = I \times H \times O$. Then the differing parameter configurations $p_1 \ldots p_n \in P$ are applied to every output produced by the preprocessing parameters $v \in V$. Finally, for each of the following quality measures a matrix was generated:

- Correct CPs
- Missed CPs
- Wrong detected CPs

A CP is deemed *correct*, if it lies within a distance-threshold of 100 m; *missed* if a real-world-CP is not detected and *wrong* if a detected CP does not exist in real world.

To calculate an overall measure for CP recognition for each (p_i, v_j) pair, the number of correct and missed CPs have to be taken into account as well as the number of wrong CPs. For this purpose the concepts of recall and precision, as done in other work (Olson and Delen 2008), are facilitated:

$$Recall = \frac{(correct\ CPs)}{(total\ real\ world\ CPs)} \tag{7}$$

$$Precision = \frac{(correct\ CPs)}{(correct\ CPs + wrong\ CPs)} \tag{8}$$

For the evaluation of change-point detection, respectively segmentation the recall is viewed to be more important since false positive (wrong) CPs can be corrected

Table 1 Effect of an HDOP filter

	Recall	Precision
$o = 999$	0.89	0.240
$o = 3$	0.875	0.288

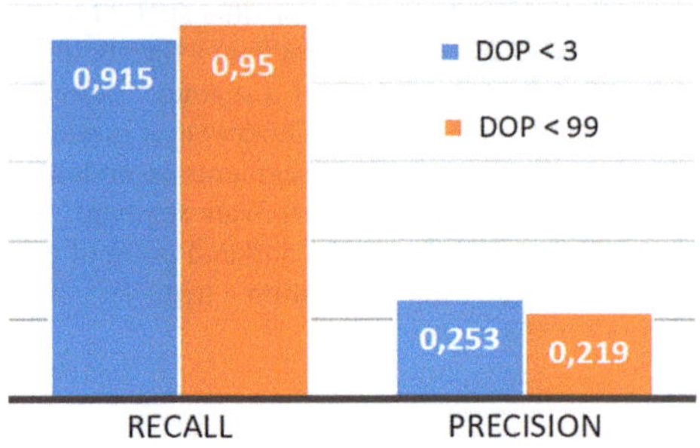

Fig. 10 Overall results of an HDOP filter

afterwards, i.e., if the two adjacent segments are of the same transportation mode. A output-version is calculated by a pair of temporal interval length i and a smoothing parameter h for a given parameter setting p out of the parameter set P. So, the value of a cell is the maximum recall of $v \in V$, which has been achieved by the specified $p \in P$. A matrix is generated for recall and precision.

5.3 Comparison and Assessment

At first we have applied our set P on the original data calculated from the point based-interval of length 1 (v_0) and searched for the parameter $p_{(0,optimal)} = (u = 2.3\,\text{m/s}, a = 0.2\,\text{m/s}^2, d = 15\,\text{m})$ achieving the best result.

The same parameter configuration $p_{(0,optimal)}$ is then applied on a reduced data set, where all track points with a HDOP higher than 3 are sorted out ($o = 3$), using the point based velocity derivation of interval 1 (Table 1):

The HDOP filter causes a reduction in terms of recall, while the precision improved. Figure 10 shows the average recall and precision for versions with HDOP filter and the average recall and precision for versions without HDOP filter. It suggests that HDOP based filtering decreases recall ($-3.5\,\%$) but positively affects the precision ($+3.5\,\%$) of change-point detection. We suppose that this effect is caused by the increasing number and length of gaps in the tracks. Since we are interested in high recalls the set of preprocessing, parameter versions V degrades to $V = I \times H$, in other words we do not filter out points of high uncertainty. Table 2 shows the recalls of the configuration $p_{(0,optimal)}$ for varying point count-based intervals compared to the recalls of $p_{(0,optimal)}$ for varying duration-based interval data (The interval length is in points for the point based intervals and in seconds for the temporal intervals).

Table 2 Recalls from point-count intervals and temporal intervals with varying interval length

Length	Point-count	Temporal
$i = 1$	0.89	0.85
$i = 6$	0.89	0.90
$i = 10$	0.84	0.925
$i = 16$	0.815	0.875
$i = 22$	0.815	0.90

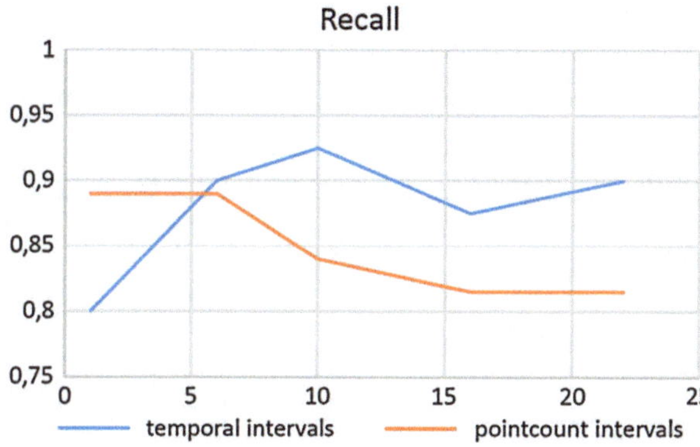

Fig. 11 Recalls from point-count intervals and temporal intervals with varying interval length

Table 3 Recalls of the combined preprocessing methods with the parameter set $p_{(0,optimal)}$

Recall	$h = 0\,\text{s}$	$h = 5\,\text{s}$	$h = 8\,\text{s}$	$h = 11\,\text{s}$	$h = 14\,\text{s}$	
$i = 1$	0.85	0.85	0.875	0.925	0.9	0.88
$i = 6$	0.9	0.875	0.925	0.925	0.9	0.905
$i = 10$	0.925	0.875	0.95	0.925	0.9	0.915
$i = 16$	0.875	0.85	0.925	0.9	0.9	0.89
$i = 22$	0.9	0.9	0.85	0.875	0.875	0.88
	0.89	0.87	0.905	0.91	0.895	Avg

The point-based interval recall values decline with increasing interval length, due to the lack of appropriate signal loss handling. The algorithm is unable to detect subway-segments because of the effect described in Sect. 3.2. On the other hand the temporal intervals are able to handle those subway segments and reduce the influence of GPS errors. Thus the results improve with increasing interval length (see Fig. 11).

The results prove that the proposed method of duration-based intervals improve change-point detection of tracks where there are periods of sustained signal loss. As stated earlier, the assumption is that a combined method of position smoothing and temporal intervals will perform best. To test the assumption, the various versions of preprocessed data, along with the optimal configuration $p_{(0,optimal)}$ found for the raw data set, is used as an input for the algorithm. The results are shown in Table 3.

Table 4 Recalls of the combined preprocessing methods with the optimized parameter set $p_{(i,optimal)}$

Recall	$h = 0$ s	$h = 5$ s	$h = 8$ s	$h = 11$ s	$h = 14$ s	
$i = 1$	0.95	0.975	1.0	1.0	1.0	0.985
$i = 6$	0.975	0.975	1.0	1.0	1.0	0.99
$i = 10$	0.95	0.975	1.0	1.0	1.0	0.985
$i = 16$	0.95	0.95	0.95	0.925	0.925	0.94
$i = 22$	0.925	0.9	0.95	0.9	0.875	0.91
	0.95	0.955	0.98	0.965	0.96	Avg

Table 5 Precisions of the combined preprocessing methods with the optimized parameter set $p_{(i,optimal)}$

Precision	$h = 0$ s	$h = 5$ s	$h = 8$ s	$h = 11$ s	$h = 14$ s	
$i = 1$	0.257	0.201	0.204	0.200	0.206	0.214
$i = 6$	0.192	0.202	0.225	0.247	0.211	0.215
$i = 10$	0.307	0.227	0.25	0.263	0.244	0.258
$i = 16$	0.288	0.222	0.221	0.242	0.266	0.248
$i = 22$	0.291	0.356	0.240	0.234	0.313	0.287
	0.267	0.242	0.228	0.237	0.248	Avg

The first column ($h = 0$ s) contains the standalone interval results, hence no position smoothing filter was applied beforehand. It can be observed, that the additional application of a kernel smoother leads to improvement in recall. Comparing the results to the point based recall values presented in Table 3, a maximum increase in recall by 6 % can be found ($i = 10$ s, $h = 8$ s).

So far we have tested all the versions with the parameter set $p_{(0,optimal)}$ found from the original point based interval data. Since the velocity curves from the different preprocessing versions V have altered characteristics, it seems likely that the optimal parameter set $p_{(i,optimal)}$ of a version v_i is not the same as $p_{(0,optimal)}$. Therefore the whole set P has been applied on all produced version using the preprocessing parameters V, to find the optimal algorithm parameter setting for the preprocessed data. Tables 4 and 5 show the results for recall and precision with the optimal $p_{(i,optimal)}$ for the preprocessed data.

The observable trend in precision shows an improvement with longer intervals and decrease with the additional kernel smoothing. The recall of the combined method shows improvements over the stand-alone interval method in column 1 ($h = 0$ s) as well as over the stand-alone kernel-smoothing method in row 1 ($i = 1$ s). The system was able to correctly detect all change-points. By using the same parameter set $p_{(0,optimal)}$ the original results from the point based-intervals were enhanced by the proposed preprocessing methods by 6 % (Table 6).

Table 6 Improvement by preprocessing with the same parameter set $p_{(0,optimal)}$

	$(v_0, p_{(0,optimal)})$	$(v_{13}, p_{(0,optimal)})$	Improvement
Recall	0.89	0.95	+6 %
Precision	0.240	0.316	+7.6 %

Table 7 Improvement by preprocessing with the optimal parameter set $p_{(i,optimal)}$

	$(v_0, p_{(0,optimal)})$	$(v_{14}, p_{(14,optimal)})$	Improvement
Recall	0.89	1.0	+11 %
Precision	0.240	0.263	+2.3%

Optimizing the parameter setting for various preprocessed versions of the data set resulted in a maximum of 11 % increase in recall, effectively resulting in recall values of 100 % for some versions (Table 7).

6 Conclusion and Recommendations

Our results showed that the preprocessing of raw GPS-data to obtain velocity data plays a considerable role in the process of automated transport mode derivation. Literature about algorithms and methods has, to our knowledge, not addressed the issue sufficiently. The chapter introduces and discusses various methods that can potentially improve the velocity derivation process from raw GPS-data. It proved their efficiency by comparing results of a well-known untrained method applied to a data set before and after it was preprocessed by the proposed methods.

To quantify the enhancement the recall of change-points, i.e., the points that mark a change in transportation mode was used as a measure. With the application of kernel smoothing on track point positions we were able to increase the recall of the investigated algorithm by a few percent. By using temporal interval based velocity derivation in addition to the position smoothing, an improvement of 6 % in recall was achieved. By optimizing the algorithm's parameters a maximum of 11 % improvement was reached, effectively resulting in a 100 % recall value in some cases. Thus, the application of each of the methods alone is not sufficient; rather a combination of them is necessary to achieve optimal results. The applied methods, on the other hand, caused a decrease in precision, i.e., increase of false change-points in addition to the correct ones. Since false positive change-points can be filtered out afterwards, e.g., if the two adjacent segments are of the same transportation mode, we deemed the recall to be a more important quality measure.

The chapter showed that the input parameters and the interval lengths of the smoothing window and velocity derivation, i.e., preprocessing, plays a role for the results of the algorithm. Further optimization of the algorithm's input parameters was only possible because of the available ground-truth data set, hence the model was fostered to that data set. Future research will need to investigate whether the parameters found to work best in this model are generally applicable for other data sets or, if not, how they can be estimated.

In general, we assume that the proposed preprocessing methods will have similar enhancement effects on other algorithms that make use of velocity as a primary classification determinant. While the work has focused on change-point detection there was no thorough testing of the overall improvement for the actual transportation mode classification, something open to future analysis.

References

Chen J, Bierlaire M, Flötteröd G (2011) Probabilistic multi-modal map matching with rich smartphone data. In: Proceedings of the Swiss Transport Research Conference (STRC), Switzerland, 11–13 May 2011

Cimon NJ Wisdom MJ (2004) Accurate velocity estimates from inaccurate GPS data. In: Proceedings of the tenth forest service remote sensing applications conference

Dodge S, Weibel R, Lautenschütz AK (2008) Towards a taxonomy of movement patterns. Inf Visual 7(3–4):240–252

Giremus A, Tourneret JY, Calmettes V (2007) A particle filtering approach for joint detection/estimation of multipath effects on gps measurements. IEEE Trans Signal Proc 55(4):1275–1285

Gómez-Torres NR, Valdés-Díaz DM, (2011) GPS capable mobile phones to gather traffic data. In: Ninth LACCEI Latin American and Caribbean conference (LACCEI, (2011) engineering for a smart planet, innovation, information technology and computational tools for sustainable development, medelln, Colombia

Hastie T, Tibshirani R, Friedman J (2009) The elements of statistical learning. Springer, New York

Hoffmann-Wellenhof B, Lichtenegger H, Collins J (2001) GPS: theory and practice. Springer, New York

Hornsby K, Egenhofer MJ (2002) Modeling moving objects over multiple granularities. Ann Math Artif Intell 36(1–2):177–194

Jun J, Guensler R, Ogle JH (2006) Smoothing methods to minimize impact of global positioning system random error on travel distance, speed, and acceleration profile estimates. Transp Res Rec J Transp Res Board 1972(1):141–150

Laube P, Dennis T, Forer P, Walker M (2007) Movement beyond the snapshot—dynamic analysis of geospatial lifelines. Comput Environ Urban Syst 31(5):481–501

Laube P, Purves RS (2011) How fast is a cow? cross-scale analysis of movement data. Trans GIS 15(3):401–418

Li X, Ortiz PJ, Browne J, Franklin D, Oliver JY, Geyer R, Chong FT (2010) Smartphone evolution and reuse: establishing a more sustainable model. In: IEEE 39th international conference on parallel processing workshops (ICPPW) 2010, pp 476–484

Li Y, Huang Q, Kerber M, Zhang L, Guibas L (2013) Large-scale joint map matching of GPS traces. In: Proceedings of the 21st ACM SIGSPATIAL international conference on advances in geographic information systems, ACM, pp 214–223

Liao L, Patterson DJ, Fox D, Kautz H (2007) Learning and inferring transportation routines. Artif Intell 171(5):311–331

Ogle J, Guensler R, Bachman W, Koutsak M, Wolf J (2002) Accuracy of global positioning system for determining driver performance parameters. Transp Res Rec J Transp Res Board 1818(1): 12–24

Olson DL, Delen D (2008) Advanced data mining techniques [electronic resource]. Springer, Berlin

Reddy S, Mun M, Burke J, Estrin D, Hansen M, Srivastava M (2010) Using mobile phones to determine transportation modes. ACM Trans Sens Netw (TOSN) 6(2):13

Stopher PR, Clifford E, Zhang J, FitzGerald C (2008) Deducing mode and purpose from GPS data. Institute of Transport and Logistics Studies

Stopher PR, Jiang Q, FitzGerald C (2005) Processing GPS data from travel surveys. In: 2nd international colloqium on the behavioural foundations of integrated land-use and transportation models: frameworks, models and applications, Toronto

Wann CD, and Chen YM (2002) Position tracking and velocity estimation for mobile positioning systems. In: The IEEE 5th international symposium on wireless personal multimedia communications, vol 1. pp 310–314

Zheng Y, Chen Y, Li Q, Xie X, Ma WY (2010) Understanding transportation modes based on gps data for web applications. ACM Trans Web (TWEB) 4(1):1

Part III
Data Mining, Fusion and Integration

Mining Frequent Spatio-Temporal Patterns in Wind Speed and Direction

Norhakim Yusof, Raul Zurita-Milla, Menno-Jan Kraak and Bas Retsios

Abstract Wind is a dynamic geographic phenomenon that is often characterized by its speed and by the direction from which it blows. The cycle's effect of heating and cooling on the Earth's surface causes the wind speed and direction to change throughout the day. Understanding the changeability of wind speed and direction simultaneously in long term time series of wind measurements is a challenging task. Discovering such pattern highlights the recurring of speed together with direction that can be extracted in specific chronological order of time. In this chapter, we present a novel way to explore wind speed and direction simultaneously using sequential pattern mining approach for detecting frequent patterns in spatio-temporal wind datasets. The Linear time Closed pattern Miner sequence (LCMseq) algorithm is constructed to search for significant sequential patterns of wind speed and direction simultaneously. Then, the extracted patterns were explored using visual representation called *TileVis* and 3D wind rose in order to reveal any valuable trends in the occurrences patterns. The applied methods demonstrated an improvement way of understanding of temporal characteristics of wind resources.

Keywords Wind · Speed · Direction · Sequential pattern mining · *TileVis* and 3D wind rose

1 Introduction

Wind is a dynamic and continuous space-time phenomenon that is often characterized by its speed and by the direction from which it blows. From a physical point of view,

N. Yusof (✉) · R. Zurita-Milla · M.-J. Kraak · B. Retsios
University of Twente, Enschede, The Netherlands
e-mail: n.yusof@utwente.nl

N. Yusof
Universiti Teknologi Malaysia, Johor Bahru, Malaysia
e-mail: norhakim@utm.my

J. Huerta et al. (eds.), *Connecting a Digital Europe Through Location and Place*, Lecture Notes in Geoinformation and Cartography, DOI: 10.1007/978-3-319-03611-3_9,

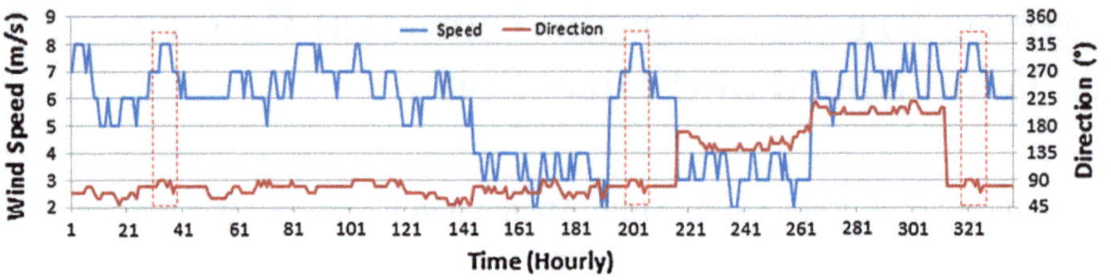

Fig. 1 Hourly wind speed between 2001 and 2010

wind is the air motion caused by the heating and cooling cycles from the Sun on the Earth's surface and the effect of topography (Şahin 2004). This effect causes the wind speed and direction to change throughout the day. Understanding these changes (i.e. wind dynamics) is important for various applications such as renewable energy, fire growth, or pollutant dispersion (Cabello and Orza 2010; Huang et al. 2009). For instance in wind research, wind speed and direction are the main parameters for determining efficient wind energy harvesting (de Prada Gil et al. 2012). Information on wind speed can provide estimation of amount of power that can be generated by the wind turbines. Wind direction is important for setting the wind turbine in the field. This is, to align with the wind direction to get most power output from a given speed (Erdem and Shi 2011). Theoretically, slight changes in wind speed and direction might cause significant changes in the amount of the total power generated from the wind.

Understanding the changeability of wind speed and direction in long term time series of wind measurements is a challenging task. To illustrate this, consider Fig. 1 that shows the changes in wind speed and direction for every hour for station Valkenburg (210) recorded in 14 days. Interesting questions on this dataset is: "Are there any interesting patterns can be observed in this time series dataset?" or "Can we find any interesting frequently recurring patterns in this time series dataset?". Finding such patterns in temporal wind measurements can provide valuable insights that inform the wind characteristics.

In Fig. 1, the line segments in red boxes are the patterns that occurred three times in the dataset. Note that the occurring patterns are similar in both wind speed and direction. However, the occurring patterns were discovered in separate lines which make it difficult to identify the matching patterns from speed and direction simultaneously. It is reasonable to assume that discovering pattern from wind speed in connection with the wind direction will differ from discovering pattern from these attributes independently. Therefore, the challenge is to discover such patterns from multiple attributes simultaneously. Besides, the recurring patterns are also important to be considered in temporal wind dataset.

Discovering patterns in wind speed and direction simultaneously over time could provide new insight to the wind pattern. This type of pattern highlights the recurring of speed together with direction that can be extracted in specific chronological order of time. The importance of these patterns not only able to realize the fully potential of wind characteristics (Jung and Tam 2013) but also provide useful information for further action plan (Buddhakulsomsiri and Zakarian 2009). For instance, detailed characteristics of wind patterns that discovered non-dispatchable wind speed can improve power generation prediction (Erdem and Shi 2011). Frequent sequential pattern mining technique is efficient for discovering such pattern (Floratou et al. 2010). Sequential pattern mining is appearing as a useful alternative to statistical and modelling techniques (Barszcz et al. 2012; Herrera et al. 2005) for providing details about the underlying patterns in large datasets (Grosser et al. 2005).

Despite the fact that data mining has received much attention in various wind applications (Colak et al. 2012), little can be found on how to discover patterns from wind speed and direction simultaneously in temporal wind dataset with the aid of spatio-temporal visualization. Hence, to gain insights of the wind sequential patterns will necessitate the visual representation for visualization purpose. Visualization technique would easily reveal valuable trends for evaluating the occurrences patterns in the wind dataset (Keim 2002).

The aim of this chapter is to present a novel way to explore wind characteristics using sequential pattern mining approach for detecting frequent patterns in spatio-temporal wind datasets. For this, we propose a new method to combine sequential data mining and visualization in order to recognize the wind speed and direction patterns using case study data consisting of 10 years of hourly wind data. The algorithm is constructed to search for significant sequential patterns of wind speed and direction simultaneously in temporal wind datasets. Then, the generated sets of unique sequential patterns are evaluated with the aid of visual representation in order to reveal any valuable trends in the occurrence patterns in space and time context.

2 Related Works

The sequential pattern mining was first introduced by Agrawal and Srikant (1995) using the Apriori algorithm. This algorithm defined the sequential patterns mining as finding the maximal (longest) sequences of items that have a certain user-specified minimum support. Several sequential pattern algorithms have been proposed as improvement over earlier version such as SPADE (Zaki 2001), PrefixSpan (Jian et al. 2004), VOGUE (Zaki et al. 2010) and LCMseq (Nakahara et al. 2010). The differences between these algorithms are mostly related to enhance the computational time by imposing some constraints on the mining process, or in some subtle differences in how they handle the sequence mining process (Buddhakulsomsiri and Zakarian 2009). For instance, Chen et al. (2008) include user-defined constraints in order to discover pattern that meets the user needs, Kuo et al. (2009) and Raïssi and Pei (2011) focus on achieving better computational efficiency by implementing K-mean algorithm and frequent sequence tree respectively, and Li et al. (2012) and Zaki et al. (2010) provided algorithms for mining with gap-constrained in the subsequences to extract frequent sequence patterns with gaps between elements. One important limitation of these mining techniques is the lack of visual representation for providing overviews or selected interesting subsets of pattern whereas such a visualization yields a high degree of satisfaction in understanding the discovered patterns (Han and Kamber 2001).

Even though sequential pattern mining proved to be efficient for extracted patterns, however relying on the patterns itself does not help much in understanding the pattern behavior (Chang 2011). Moreover, extracting and providing numerous patterns to end users makes it difficult for them to analyse these patterns without proper interpretation tools (Sallaberry et al. 2011). Tool such as visualization provide the ability to navigate the underlying patterns through spatio-temporal data and to interactively manipulate them using a variety of visual representations (Schumann and Tominski 2011).

In most visual representation simple graphics such as bar chart, x–y plot and rose diagram are commonly used to support the data mining process. These simple graphics easy-to-use because it shows only highly aggregated data and present small portion of data values (Keim et al. 2002). However, visualizing highly aggregated data as an overview is not sufficient for large volume of data analysis. The detailed information of the data is also required to support the discovery of interesting patterns.

In the previous works, there are various of visualization techniques related to multivariate data and time dependent. Keim et al. (2002) presented a pixel bar chart approach for visualizing multi-attribute which able to reduce losing information by overplotting and aggregation. Samuel et al. (2013) proposed multi-attribute ranking visualization called *LineUp*. This technique used bar charts in various configurations consisting of a serial combination (stacked bars) and parallel combination (bars placed next to each other) to encode the ranked of each attribute. Ho et al. (2011) proposed extended Parallel Coordinates Plot (PCP) by attaching histogram at each axis. The axes are split into equal height bin size and the width of the bin indicates the frequency of temporal indicator data. The most common wind rose diagram typically used to visualize the distribution of wind speed and direction only at single location at a time. The diagram will aggregate the speed and direction into a percentage of amounts of time (Mukulo et al. 2014). This visualization could limit the detail information of when the high wind speed occurred and how these speed distributed in time.

Even though these methods show good visualization results for multivariate data, however, these approaches are less suitable to be used within the temporal wind sequential pattern exploration. Based on our approach, we enable to visualize the temporal trend discovery by overcome these two main limitations. First, systematic integration elements of spatial, attribute of sequential patterns and time are required to reduce complexity or visual cluttering in multidimensional view. This integration will aid users to explain the recurring of wind patterns when involve with long time series dataset. Second, these sequential patterns are extracted in specific chronological order of time (per hour), therefore, new type of visualization is needed to represent the wind speed and direction at any moment in time to reveal every single occurrence patterns. This will require visualization that visualizes the differences between the wind speed and direction (per hour) and comparison between the patterns.

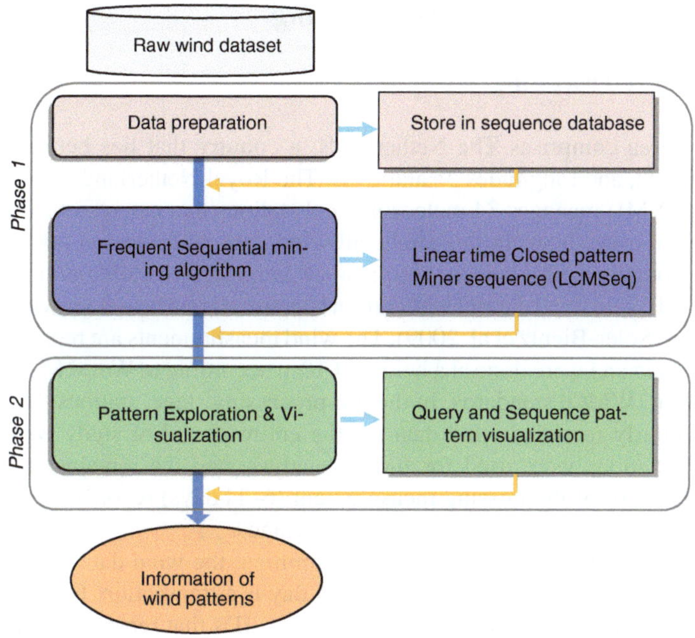

Fig. 2 The overall method for exploring spatio-temporal pattern of wind speed and direction

In this chapter, we proposed a method to extract patterns of wind speed and direction simultaneously using sequential pattern mining approach for detecting sequence patterns in spatio-temporal wind dataset. Next, the extracted patterns will be explored using visual representation called *TileVis* and 3D wind rose techniques to improve the understanding of temporal characteristics of wind resources.

3 Methods

The study is divided into two main phases in order to investigate the temporal patterns of wind speed and direction. In phase one, frequent wind patterns are extracted by using current state-of-art sequential mining algorithms. In the second phase, the behavior of wind according to the extracted patterns was explored utilizing new visualization approaches. The visualization is needed to better understand and assess the relevance of the mined patterns. The overall methodological phases are shown in Fig. 2. The next two subsections describe in detail the data preprocessing as well as the data mining and the visualization approaches used in this chapter.

3.1 Data Preparation and Data Mining

3.1.1 Data Pre-Processing

The study area comprises The Netherlands, a country that lies between latitudes 50° and 54° N, and longitudes 3° and 8° E. The Royal Netherlands Meteorological Institute (KNMI) manages 34 meteorological stations that can measure wind speed and direction every hour. In this study, all wind data in the period 2001–2010 was collected from the KNMI website (http://www.knmi.nl/klimatologie/uurgegevens/). A period of 10 year wind dataset was chosen because it is sufficient for wind resources observation (Soler-Bientz et al. 2009). The wind measurements are representative for an open terrain and were done at a height of 10 m according to World Meteorological Organization (WMO) standards. In the pre-processing stage, stations that consist of complete hourly recorded wind data for the entire period of study were selected. Hence, 29 stations were used for further analysis and the remains stations were removed because of the missing measurement (6–11 days) records. Figure 3 shows the distribution of these stations in The Netherlands.

To be able to perform sequential pattern mining, the wind dataset was stored in a sequence database where the data for each day is a transaction. Every transaction consists of a sequence of 24 elements or itemsets IDs that represent a unique combination of wind speed and direction. In order to get such IDs a lookup table was created to transform the speed and direction attributes and assigned into a unique and discrete ID (refer to Lookup table in Fig. 4).

In this work, the wind speed and direction were respectively grouped into seven and eight difference classes (Table 1). Every unique combination of these classes was assigned a ID. Using this lookup table, a complete list of itemsets' ID for each transaction was produced and used as an input to the sequential patterns mining.

3.1.2 Frequent Sequential Pattern of Wind Dataset

The Linear time Closed pattern Miner sequence (LCMSeq) algorithm (Uno et al. 2005) was used to mine frequent sequential patterns from the wind sequence database. The advantages of using LCMseq compared to other sequential pattern mining algorithms such as SPADE, VOGUE and PrefixSpan (see in paragraph 4 Sect. 1) are as follows.

The LCMseq algorithm uses a technique called prefex preserving extension that can timely enumerate frequent sequential patterns even in huge databases that require larger than the memory size (Nakahara et al. 2010). LCMseq also provide several constraint functions such as it can only extract sequence patterns that appear in a specific window width, restricted gap length and specific size of sequence . (Nakahara et al. 2010). This mining algorithm only needs practical computational knowledge for generating the sequential pattern.

For extracting the sequential wind patterns with LCMseq, two types of constraint were used together with the specification of the minimum support (min_sup) unit

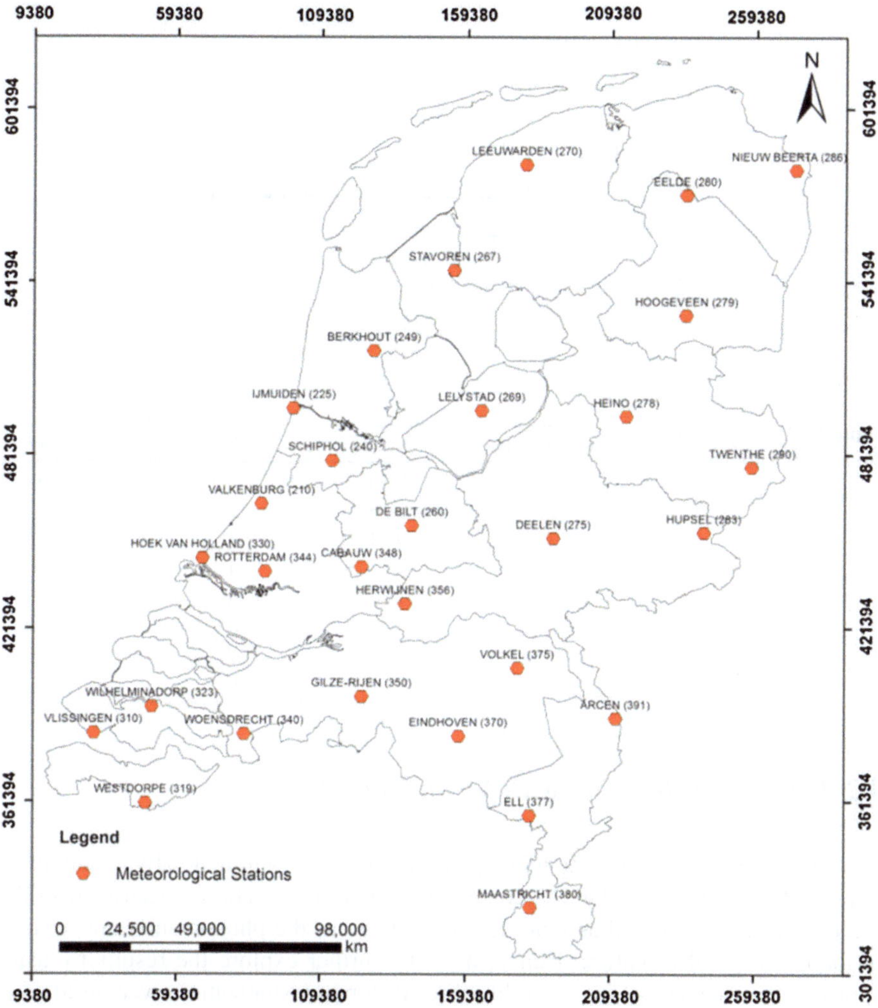

Fig. 3 Distribution of the selected meteorological stations in the study area

of the patterns in the sequence database. The constraint parameters comprise the minimum sequence gap (g) and the length size (l). Let denote a sequence as S, where St is the itemset at the time stamp t. The St values can be grouped into vectors that represent the sequence patterns of the wind for a given time lengths size, l. In this study, we chose to analyse daily wind patterns so t contains 24 discrete time stamps (hourly wind conditions). Since we are interested in continuous patterns we fixed g to 0. In this case, the "free parameter" l can vary between 2 and 24. For example, if l equals 12 then LCMseq will extract wind sequence patterns with at least a time length of 12 h. Notice that the end user can decide upon different ways to discretize the continuous wind sequence. This depends on the target applications such as wind variation assessment or wind power estimation per day.

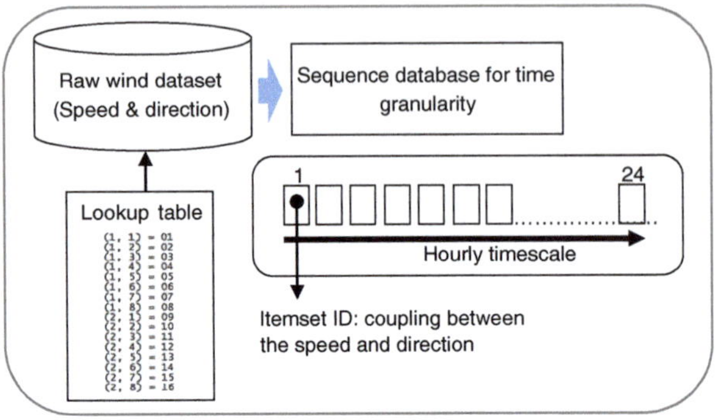

Fig. 4 Sequence database for wind sequential pattern mining

Table 1 Wind speed and direction IDs for each group

ID_Speed	Speed (m/s)	ID_Direction	Degree
1	0–4	1	45
2	5–8	2	90
3	9–12	3	135
4	13–16	4	180
5	17–20	5	225
6	21–24	6	270
7	25–28	7	315
		8	0/360

3.2 Pattern Exploration with Visualization

The obtained set of frequent wind patterns does not automatically result in an improved understanding of the behavior of wind variations. This is because the LCM-seq outputs are presented as plain text files whereas the phenomenon under study is spatio-temporal in nature. Hence, here we further explore the results by using appropriate visual techniques. For this, three different visualizations were used: (1) a *TileVis* representing the overview of wind sequential patterns, (2) a 3D wind rose for visualizing different types of wind sequential pattern, and (3) wind frequent sequential patterns exploration with *TileVis* using specific type of questions related to space, time and attribute.

3.2.1 *TileVis* Based Visualization

In order to visualize the wind sequential patterns across different time granularities and for set of geographical locations, *TileVis* was developed. The *TileVis* uses position in two dimensions grid. The position along the horizontal axis refers to date, while the position on the vertical axis refers to geographic locations. In the *TileVis*, consist several plots is called sub-Map. Each sub-Map used to plot the sequential patterns in

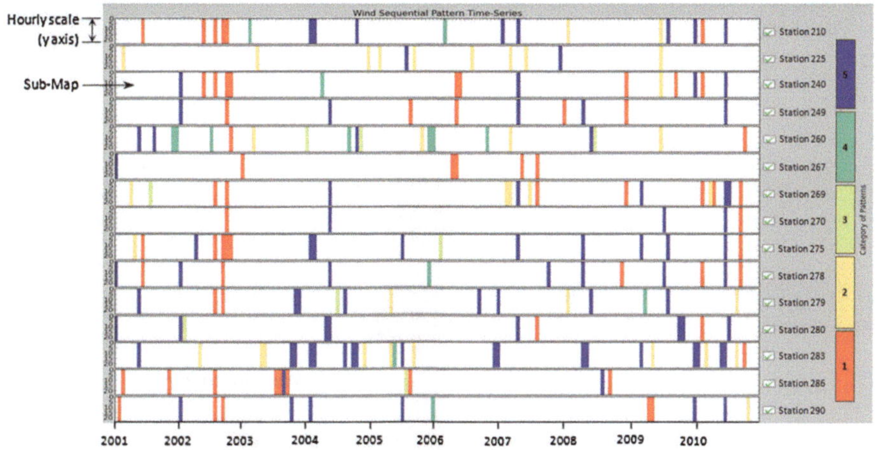

Fig. 5 Wind sequential patterns overview using *TileVis* visualization

hourly time scale in vertical direction. The sub-Maps are arranged based on the ID number of the station starting from station 210 (the first sub-Map) to 391 (the last sub-Map). To visualize the overview of the extracted patterns in the *TileVis*, each pattern is encoded with different colour. Figure 5 shows the example of the main interface of *TileVis* visualizing five classes of wind patterns associated to 15 different stations.

3.2.2 3D Wind Rose

To understand the sequential wind patterns extracted by LCMseq, we visualized them using a three dimensional graphic. Since the extracted wind patterns incorporate time, we need a representation whereby we can visualize the wind speed and direction at any moment in time. The 3D wind rose is developed based on the classic 2D wind rose concept. The length of the rose bar represents the wind speed and the facing direction of the rose bar indicates the wind direction. For an easing interpretation, each rose bar is filled with a colour that is linked to the wind speed. The third dimension of the rose (i.e. the vertical axis) represents the temporal dimension which is based on the length of the sequence patterns. This means that rose bars are stacked one on top of the other (Fig. 6).

3.2.3 Pattern Exploration with *TileVis*

In order to gain insight in the wind characteristics, the extracted patterns need to be interpreted. Therefore, the exploration of these patterns are based on specific types of question according to time, attributes and location elements were applied (Andrienko and Andrienko 2006; MacEachren 2004) (Fig. 7).

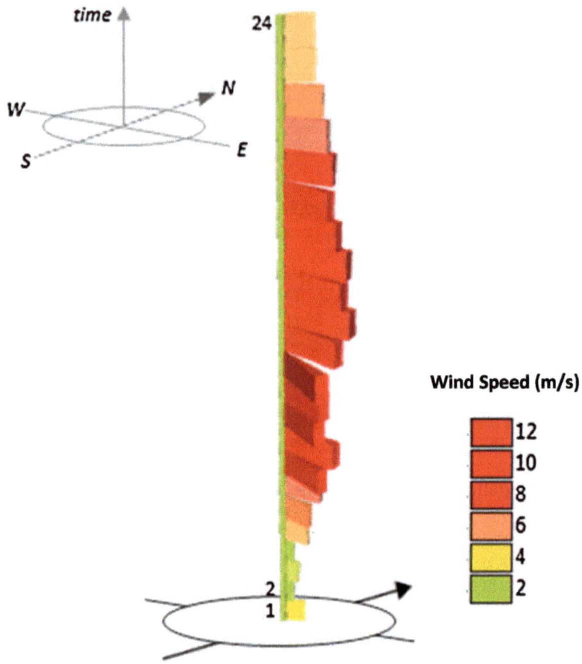

Fig. 6 3D wind rose used for visualizing wind sequence patterns

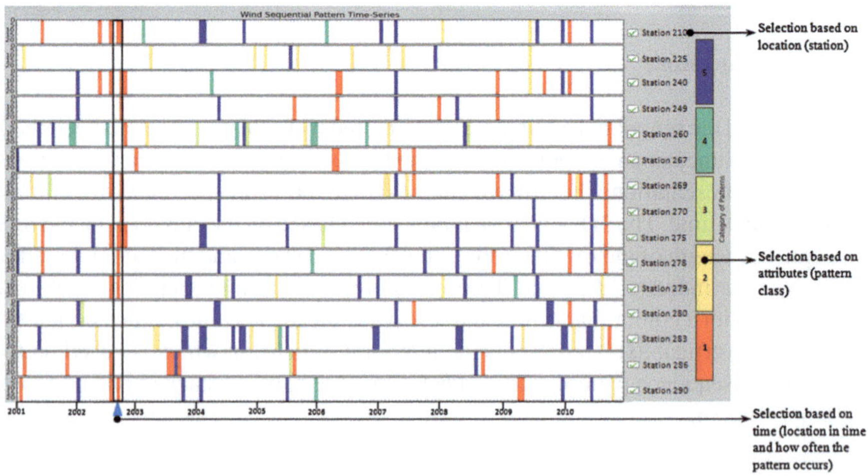

Fig. 7 *TileVis* interface with the interactive functions

Table 2 Wind sequential patterns extracted from the LCMseq

Pattern	Sequence patterns	Frequency
Pattern_1	13 13 13 13 13 13 13 13 13 13 13 13 13 13 13 13 13 13 13 13 13 13 13 13	137
Pattern_2	3 3 3 3 3 3 3 3 3 3 3 3 3 3 3 3 3 3 3 3 3 3 3 3	81
Pattern_3	17 17 17 17 17 17 17 17 17 17 17 17 17 17 17 17 17 17 17 17 17 17 17 17	24
Pattern_4	5 5 5 5 5 5 5 5 5 5 5 5 5 5 5 5 5 5 5 5 5 5 5 5	27
Pattern_5	16 16 16 16 16 16 16 16 16 16 16 16 16 16 16 16 16 16 16 16 16 16 16 16	168

Formulating different types of question depends on which information the user needs to know. The question can be formulated using time, attributes and location elements as follows:

- *Time*: What patterns did occur in December 2002?
- *Attribute*: Which station belongs to pattern "1"?
- *Location*: What patterns are found in station X?
- *Time*: How often does a patterns occur from 9:00 to 17:00 h?

The above questions are basic examples used to explore the wind patterns, however the user can further explore the pattern by increasing the level of complexity (Bertin 1967). In this study, we only focused on these four questions in order to interact with the *TileVis* visualization for exploring the wind sequential patterns.

4 Results and Discussion

4.1 Wind Sequential Pattern Overviews

For demonstrating the LCMseq sequential pattern mining, the parameters l (length size) and g (sequence gap) were set to 24 and 0 respectively. The results from the data mining show there are five different patterns can be extracted from the wind sequence database. These five patterns as well as their frequencies are shown in Table 2.

To provide a better overview of these wind sequential patterns for the entire period of 10 years the *TileVis* is used. Figure 8 shows the distribution of the 5 wind sequential patterns that occurred at all stations in the study area. Each pattern is characterized by a different colour in the *TileVis*. This visualization shows that patterns 1 and 5 are the most dominant ones in the period under study.

4.2 Wind Sequential Pattern with 3D Wind Rose

Next, we can visualize the detailed characteristics of the temporal wind sequential patterns using the 3D wind rose. Each itemset in the sequence was decoded into the speed and direction values (Table 1) in order to visualize it in the 3D wind rose.

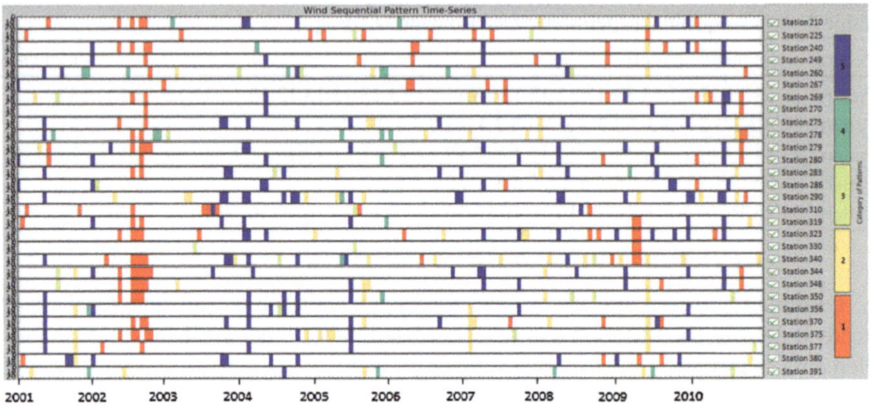

Fig. 8 Wind sequential patterns overview in the *TileVis*

Figure 9 illustrates the sequence of wind speed and direction for every hour (24 h overall) for the five extracted wind sequential patterns using 3D wind rose.

Figure 9 shows that the wind speed and direction for pattern 1 range from 5–8 m/s and 45–90° (Northeast-East), respectively. The differences of color depict the wind speed; the darker red color indicates the highest wind speed (8 m/s) while the darker green color indicates the lowest speed (5 m/s). For pattern 2, the speed and direction are 2–4 m/s and 45–90° (Northeast-East), respectively. The highest wind speed is 4 m/s and the wind direction seems to be similar to pattern 1. For pattern 3, the speed and direction are 5–8 m/s and 225–270° (Southwest-West), respectively. The range of wind speeds is similar to pattern 1, however the wind direction is the opposite to that of patterns 1 and 2. Besides, the wind has the highest speed in the afternoon and low wind speed during the night. The speed and direction for pattern 4 are range from 2–4 m/s and 135–180° (Southeast-South), respectively. In this pattern, the wind speed is similar to pattern 2 but a bit higher during the afternoon. However, these wind mostly come from the Southeast from the early day till afternoon. The speed and direction of pattern 5 are 5–8 m/s and 180–225° (South-Southwest), respectively. The wind direction is similar to pattern 4 and most of the wind coming from the south. From pattern 5 we can see that the wind has the highest speed (8 m/s) at the afternoon and midnight, and the lowest speed in the morning.

From the produced wind patterns show that it may provide useful information to the users especially to identify the highest wind speed and from where the wind is coming, and also finding out when such pattern occurred in time and at which station. For instance, pattern 5 is the dominant patterns and the speed range is from 5 to 8 m/s that pass the usual cut in speed (5 m/s) for small scale wind turbine. However, wind speed at 10 m height above the ground is not suitable for a wind turbine to generate energy. Theoretically the wind speed increases with increasing height above the ground, hence stations that belong to this pattern will have better speed when the speed is extrapolated to the actual operating hub height of the wind turbine. Besides, these locations are likely to be high-wind areas for future efficient energy harvesting.

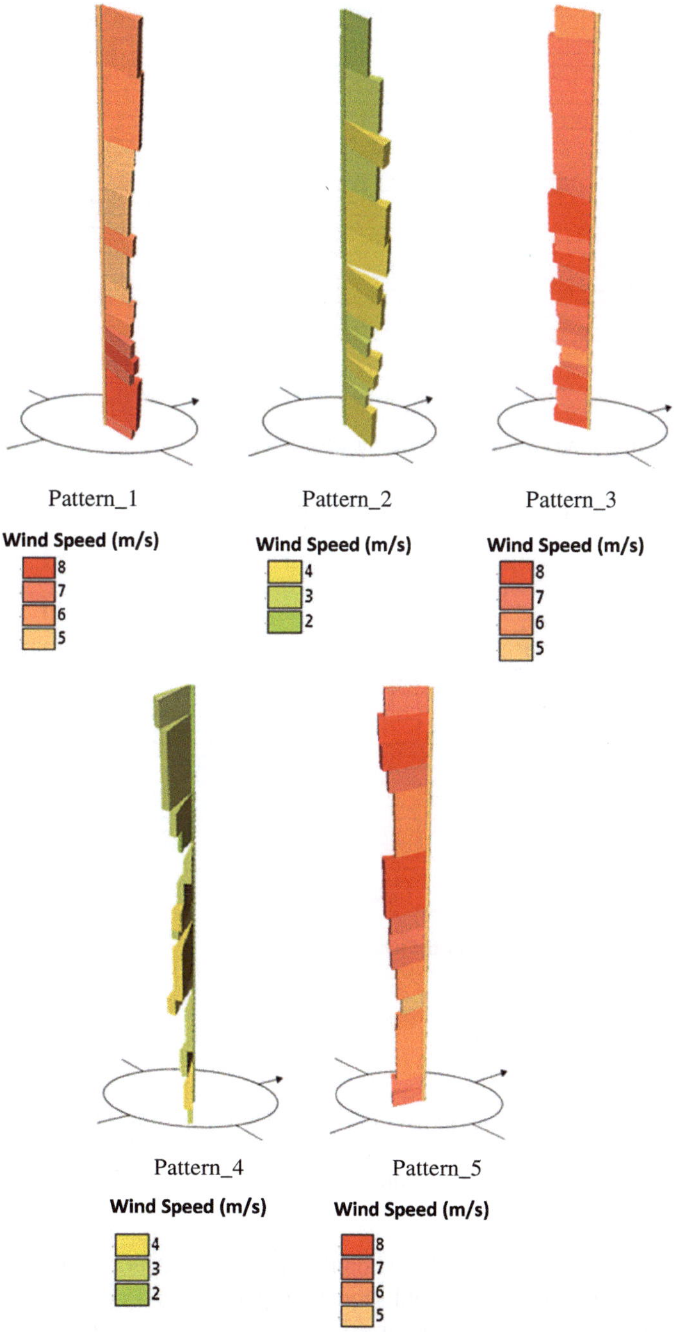

Fig. 9 Sequence of wind speed and direction in hours (24 h)

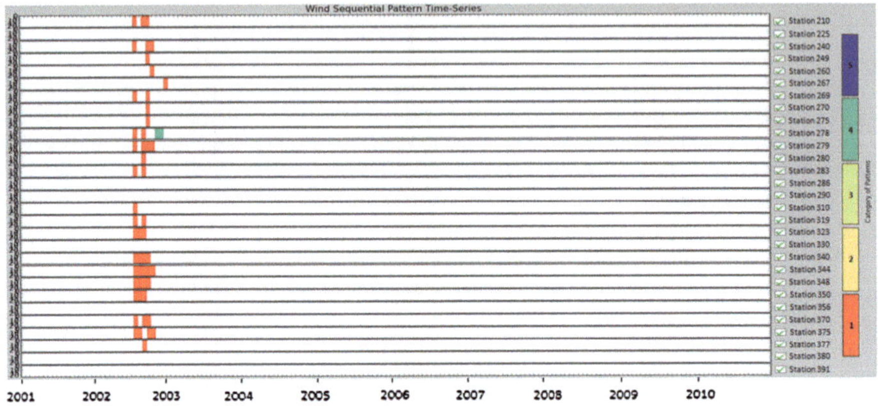

Fig. 10 Occurring wind sequential patterns in selected time

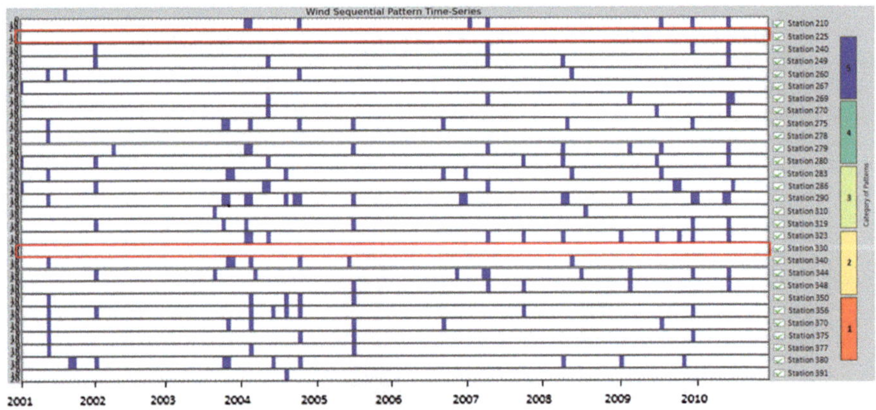

Fig. 11 Distribution of pattern 5 for the whole stations

4.3 Patterns Exploration

Four different type of questions can be formulated to extract some of the detailed information from the wind sequential patterns (i.e. to gain a better understanding of the wind characteristics). The formulated questions (see Sect. 3.2.3) and answers are explained as following:

(a) To find interesting patterns based on the specific time. For example, if we want to visualize the occurring wind sequential patterns that occurred in December 2002. The *TileVis* answers the time based question by visualizing the recurring patterns that fall in this month in every station (Fig. 10). From the *TileVis* there were two patterns occurred in that time, namely 1 and 4. However, pattern 1 is the most frequent pattern.

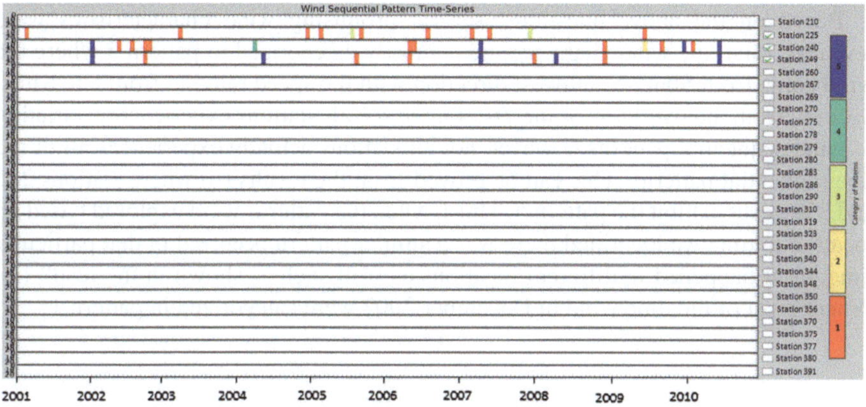

Fig. 12 Wind sequential patterns for stations that located in North Holland

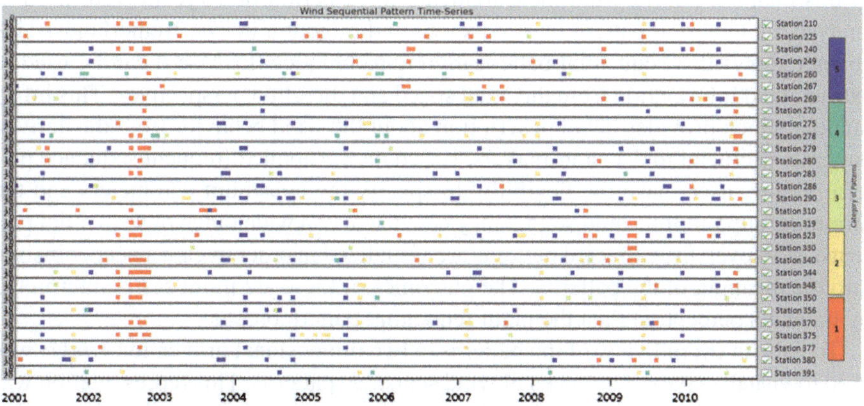

Fig. 13 Recurring wind sequential patterns between 9:00 and 17:00 h

(b) To visualize a specific sequential pattern in all stations. For attribute based question such which station exhibits pattern 5? From Fig. 11 we can identified that most of the stations exhibit this wind pattern except for stations 225 and 330 that are highlighted in the red box.

(c) To identify sequential patterns those belong to a specific area. If one want to know how the wind sequential patterns look like for stations that are located in the North Holland province? There are three stations in this province: 225, 240 and 249 (see Fig. 3). The *TileVis* shown in Fig. 12 illustrates the distribution of wind sequential patterns for these three stations.

(d) To identify how often the patterns occur in a particular time slot. Questions such how often do the wind sequential patterns occur from 9:00 to 17:00h? Fig. 13 illustrates this for every station in the *TileVis*.

Based on the above questions, several promising wind characteristics were revealed. First, the wind characteristic is mostly categorized to pattern 5 where the wind come from South-Southwest direction with the speed ranging from 5 to 8 m/s. Second, from the overview of the *TileVis*, we see all stations experiencing variation of wind sequential patterns however, not all stations actually experiences all of the five sequential patterns. This can be proved from the results (b) where stations 225 and 330 never exhibit the pattern 5. Third, when the patterns were explored based on location in time we identified there was a similar pattern occurred in that particular time but it happened in different locations as well such as shown in the result (a). Where most of the time wind in December 2002 were categorized in pattern 1.

5 Conclusion

In this chapter, hourly wind measurements for a period of 10 years were used to analyse the temporal pattern of wind speed and direction in The Netherlands. This chapter shows that sequential pattern mining techniques (LCMseq) can be used to extract wind information from the wind speed and direction perspectives. This information, which provides detail description about the wind behavior from the derived patterns, is crucial to decision-makers. Based on the applied frequent sequence pattern mining, users have the flexibility to specify the constraint parameters and the minimum support threshold so that they can discover a wide range of sequential patterns. Moreover, the extracted patterns are visualized in order to gain a deeper insight into the meaning of the patterns. The outcomes of these are more easily interpreted by human's perception than the plain sequential pattern output.

To gain an overall view of the patterns, a two dimensional grid called *TileVis* was used to visualize the occurrences of wind sequence patterns. This representation was selected because it is able to present a lot of information in a single display. The *TileVis* is able to visualize the time, location and the sequence patterns for every hour in a day. From the *TileVis*, different task for pattern exploration were proposed to help the users to interpret the patterns. These tasks were performed by formulating question that consist these three elements namely time (at two granularities), location and attributes. By incorporating the *TileVis* with the specified questions, several valuable wind characteristics were revealed. Moreover, the extracted wind sequence patterns were presented with the 3D wind rose in order to gain a better insight on the meaning of the wind sequential patterns. The 3D wind rose shows the changeability of wind speed and direction per hour. By integrating the 3D wind rose with time axis (z-axis), which cannot be done in standard wind rose, can exploit the variability of wind speed and direction for the discovered patterns. Hence, this information could improve the understanding of temporal characteristics of wind resources.

This chapter has presented a novel way to illustrate the discovered wind sequential patterns from the visualization perspective. By incorporating these two different visualization methods show that the generated information can gain better insight by providing more detail of the wind variability through time. In future, several enhancements will be implemented to the existing techniques in order to improve the usefulness of the visualization to the users such as integrating the 3D wind rose to geographical map in multi-dimensional environment. In addition, future work also needed to perform formal user usability evaluation to evaluate the proposed visualization methods against with other visualization techniques.

In addition, the advantages of the proposed geovisual analytics approach are; (1) decrease the time processing over large dataset for discovering hourly wind sequential pattern, (2) discovering the wind patterns with multiple perspectives including wind variables (speed and direction), time and location of the stations, and (3) provide interactive strategy to facilitate the wind sequence patterns discovering process.

Acknowledgments This work was supported by the Malaysia Fellowship (Ministry of Education Malaysia) and Universiti Teknologi Malaysia (UTM). The authors would also like to acknowledge The Royal Netherlands Meteorological Institute (KNMI) for providing the wind data.

References

Agrawal R, Srikant R (1995) Mining sequential patterns In: 11th International Conference on Data Engineering, IEEE Computer Society, Los Alamitos, Taipei, Taiwan, Mar 1995, pp 3–14

Andrienko N, Andrienko G (eds) (2006) Exploratory analysis of spatial and temporal data—a systematic approach. Springer, Berlin

Barszcz T, Bielecka M, Bielecki A, Wójcik M (2012) Wind speed modelling using weierstrass function fitted by a genetic algorithm. J Wind Eng Ind Aerod 109:68–78. doi:10.1016/j.jweia.2012.06.007

Bertin J (1967) Semiologie graphique. Les diagrammes, les reseaux, les cartes. Haye-Paris, Mouton et Gouthier-Villar, 2 ed. 1973

Buddhakulsomsiri J, Zakarian A (2009) Sequential pattern mining algorithm for automotive warranty data. Comput Ind Eng 57(1):137–147. doi:10.1016/j.cie.2008.11.006

Cabello M, Orza JAG (2010) Wind speed analysis in the province of alicante, Spain. Potential for small-scale wind turbines. Renew Sustain Energ Rev 14(9):3185–3191. doi:10.1016/j.rser.2010.07.002

Chang JH (2011) Mining weighted sequential patterns in a sequence database with a time-interval weight. Know Based Syst 24(1):1–9. doi:10.1016/j.knosys.2010.03.003

Chen E, Cao H, Li Q, Qian T (2008) Efficient strategies for tough aggregate constraint-based sequential pattern mining. Inf Sci 178(6):1498–1518. doi:10.1016/j.ins.2007.10.014

Colak I, Sagiroglu S, Yesilbudak M (2012) Data mining and wind power prediction: a literature review. Renew Energ 46:241–247. doi:10.1016/j.renene.2012.02.015

de Prada Gil M, Gomis-Bellmunt O, Sumper A, Bergas-Jané J (2012) Power generation efficiency analysis of offshore wind farms connected to a SLPC (single large power converter) operated with variable frequencies considering wake effects. Energy 37(1):455–468. doi:10.1016/j.energy.2011.11.010

Erdem E, Shi J (2011) ARMA based approaches for forecasting the tuple of wind speed and direction. Appl Energ 88(4):1405–1414. doi:10.1016/j.apenergy.2010.10.031

Floratou A, Tata S, Patel JM (2010) Efficient and accurate discovery of patterns in sequence datasets. In: Data Engineering (ICDE), 2010 IEEE 26th International Conference on, 1–6 Mar 2010, pp 461–472. doi:10.1109/ICDE.2010.5447843

Grosser H, Britos P, García-Martínez R (2005) Detecting fraud in mobile telephony using neural networks. In: Ali M, Esposito F (eds) Innovations in Applied Artificial Intelligence. Lecture Notes in Computer Science, vol 3533. Springer, Berlin, pp 613–615. doi:10.1007/11504894_85

Han J, Kamber M (2001) Data mining: concepts and techniques. In: The Morgan Kaufmann Series in Data Management Systems JG, Series Editor (ed). Morgan Kaufmann Publishers, San Diego: Academic Press, p 550

Herrera JL, Piedracoba S, Varela RA, Rosón G (2005) Spatial analysis of the wind field on the western coast of galicia (NW Spain) from in situ measurements. Cont Shelf Res 25(14):1728–1748. doi:10.1016/j.csr.2005.06.001

Ho Q, Lundblad P, Åström T, Jern M (2011) A Web-enabled visualization toolkit for geovisual analytics. In: Proceedings of SPIE, the International Society for Optical Engineering: SPIE: Electronic Imaging Science and Technology, Visualization and Data Analysis 7868: doi:10.1117/12.872250

Huang H, Ooka R, Liu N, Zhang L, Deng Z, Kato S (2009) Experimental study of fire growth in a reduced-scale compartment under different approaching external wind conditions. Fire Saf J 44(3):311–321. doi:10.1016/j.firesaf.2008.07.005

Jian P, Jiawei H, Mortazavi-Asl B, Jianyong W, Pinto H, Qiming C, Dayal U, Mei-Chun H (2004) Mining sequential patterns by pattern-growth: the PrefixSpan approach. Know Data Eng IEEE Trans 16(11):1424–1440. doi:10.1109/TKDE.2004.77

Jung J, Tam K-S (2013) A frequency domain approach to characterize and analyze wind speed patterns. Appl Energ 103:435–443. doi:10.1016/j.apenergy.2012.10.006

Keim DA (2002) Information visualization and visual data mining. Vis Comput Graph IEEE Trans 8(1):1–8. doi:10.1109/2945.981847

Keim DA, Hao MC, Dayal U, Hsu M (2002) Pixel bar charts: a visualization technique for very large multi-attribute data sets†. Inf Vis 1(1):20–34. doi:10.1057/palgrave.ivs.9500003

Kuo RJ, Chao CM, Liu CY (2009) Integration of K-means algorithm and AprioriSome algorithm for fuzzy sequential pattern mining. Appl Soft Comput 9(1):85–93. doi:10.1016/j.asoc.2008.03.010

Li C, Yang Q, Wang J, Li M (2012) Efficient mining of gap-constrained subsequences and its various applications. ACM Trans Knowl Discov Data 6(1):1–39. doi:10.1145/2133360.2133362

MacEachren AM (ed) (2004) How maps work: representation, visualization, and design. Guilford Press, New York

Mukulo BM, Ngaruiya JM, Kamau JN (2014) Determination of wind energy potential in the mwingi-kitui plateau of kenya. Renew Energ 63:18–22. doi:10.1016/j.renene.2013.08.042

Nakahara T, Uno T, Yada K (2010) Extracting promising sequential patterns from RFID data using the LCM sequence. In: Setchi R, Jordanov I, Howlett R, Jain L (eds). Knowledge-Based and Intelligent Information and Engineering Systems. Lecture Notes In Computer Science, vol 6278. Springer, Berlin, pp 244–253. doi:10.1007/978-3-642-15393-8_28

Raïssi C, Pei J (2011) Towards bounding sequential patterns. Paper presented at the Proceedings of the 17th ACM SIGKDD international conference on Knowledge discovery and data mining, San Diego, California, USA, 21–24, Aug 2001

Şahin AD (2004) Progress and recent trends in wind energy. Prog Energ Combust Sci 30(5):501–543. doi:10.1016/j.pecs.2004.04.001

Sallaberry A, Pecheur N, Bringay S, Roche M, Teisseire M (2011) Sequential patterns mining and gene sequence visualization to discover novelty from microarray data. J Biomed Inf 44(5):760–774. doi:10.1016/j.jbi.2011.04.002

Samuel G, Alexander L, Nils G, Hanspeter P, Marc S (2013) LineUp: visual analysis of multi-attribute Rankings. IEEE Trans Visual Comput Graph (InfoVis '13) 19(12)

Schumann H, Tominski C (2011) Analytical, visual and interactive concepts for geo-visual analytics. J Vis Lang Comput 22(4):257–267. doi:10.1016/j.jvlc.2011.03.002

Soler-Bientz R, Watson S, Infield D (2009) Preliminary study of long-term wind characteristics of the mexican yucatán peninsula. Energ Convers Manage 50(7):1773–1780. doi:10.1016/j.enconman.2009.03.018

Uno T, Kiyomi M, Arimura H (2005) LCM ver.3: collaboration of array, bitmap and prefix tree for frequent itemset mining. Paper presented at the Proceedings of the 1st international workshop on open source data mining: frequent pattern mining implementations, Chicago, Illinois, USA, 21–24, Aug 2005

Zaki M (2001) SPADE: an efficient algorithm for mining frequent sequences. Mach Learn 42(1–2):31–60. doi:10.1023/A:1007652502315

Zaki MJ, Carothers CD, Szymanski BK (2010) Vogue: a variable order hidden markov model with duration based on frequent sequence mining. ACM Trans Knowl Discov Data 4(1):1–31. doi:10.1145/1644873.1644878

STCode: The Text Encoding Algorithm for Latitude/Longitude/Time

Jan Ježek and Ivana Kolingerová

Abstract Encoding the geographic coordinates into the compact string expression is an important problem in geosciences, therefore, several algorithms exist. Such a string is used for several purposes. One of the most frequent is probably its use as a part of a URL or as a hashtag of ad hoc data. An important part of spatially referenced data can be also a timestamp, but current encoding methods do not allow to encode the temporal dimension. In this chapter we propose a new encoding algorithm focused on a point data expressed by latitude, longitude and timestamp coordinates. The algorithm details are described and its use is shown on examples.

1 Introduction

Content of ad hoc messages collected by social media such as Twitter, Google+, Facebook etc... has become a very important source of information. The basic concept of specifying the metadata and keywords of a particular message on these media is based on the so-called hashtags. A hashtag is a word or a phrase prefixed with the symbol # and is basically used to provide a means of grouping particular messages.

The concept of hashtags works well for general keywords, but it is not useful for specifying spatial and temporal information. The location can be specified by coordinates and time by date, but the format of such expression is ambiguous as the coordinates as well as a date can have many formats. And, obviously, we cannot define place and time just by one term as it is usual for other keywords.

In this chapter we propose an algorithm for encoding the latitude, the longitude and the time to a reasonably short and unique string. We call this string STCode that stands for a spatiotemporal code. Such a string offers a possibility to encode position

J. Ježek (✉) · I. Kolingerová
Department of Computer Science and Engineering, University of West Bohemia in Pilsen, Pilsen, Czech Republic
e-mail: jezekjan@kiv.zcu.cz

J. Huerta et al. (eds.), *Connecting a Digital Europe Through Location and Place*, Lecture Notes in Geoinformation and Cartography, DOI: 10.1007/978-3-319-03611-3_10,

and time with an arbitrary precision, where the nearby data (in the time and space) shares the same prefix. Unlike current possibilities, our method groups together the function of spatial encoding with the encoding of the temporal dimension. The STCode string also provides possibilities to be used as a spatiotemporal database index as we will demonstrate on basic spatiotemporal queries.

In Sect. 2 the related work is discussed and the state of art algorithms and relevant techniques are mentioned. Section 3 describes our algorithm in detail. Section 4 provides results, an overall discussion and suggests future work. Finally the conclusion is given in Sect. 5.

2 Related Work

The topic of encoding spatiotemporal data is closely related to the problem of space partitioning and multidimensional data structures. Historically the space partitioning has been a topic of several studies. For example Samet and Webber (1989), Hjaltason and Samet (2002), Gargantini (1982) discuss the representation of pointers for quadtrees with respect to complexity of searching and memory efficiency. Aref and Ilyas (2001) represents implementation of space partitioning and a trie-based structure for spatial databases and emphasize the importance of such a structure for current problems. The use of a general trie for data mining is also discussed in Bodo and Rónyai (2003). Besides the work focused on partitioning the 2d plane space there are also many studies focused on partitioning the sphere (e.g. Szalay et al. (2007), Dutton (1996)).

There are also many techniques based on an auxiliary and one-dimensional data structure that is used as a pointer to particular nodes of a spatial tree. Examples of this approach are applications such as Slippymap project (2013) or even Tile Map Service Foundation (2012) standard. These methods are generally focused on a simple retrieval of the so-called map tiles from the directory structures where such tiles essentially compose a spatial quadtree. It is worth mentioning that these methods are a basics of well-known mapping projects such as OpenStreetMap or Googlemaps API.

Unlike the location of mostly static map tiles, we consider the location of ad hoc data on the Web like pictures, emails, Tweets etc. where the amount of such a data can dynamically grow in time. For that purpose a Geohash[1] service can be used. Part of that service is an encoding and a decoding system for linking a location on the Earth to the sequence of characters. Geohash encoding became very popular in the past and is widely implemented in many geographic information systems for different purposes (e.g. PostGIS). It is also used as a spatial index method of some NoSQL database management systems (e.g. MongoDB). Geohash has been put to public domain and became widely accepted and well-known. Its description has not been published in the form of scientific chapter, but it is provided by the Geohash author

[1] http://geohash.org.

Gustavo Niemeyer through Wikipedia. Besides the Geohash there are also similar (mostly patented) methods available Beatty (2003) but, as far as we are aware, their use is not so wide.

Besides the methods that are focused separately on 2D spatial information there are also many studies focused on the management of a spatiotemporal data, e.g., Wolfson (2002), Chon et al. (2001), Agarwal et al. (2003). For example, Chon et al. (2001) discusses the problem of managing the dynamically changing spatiotemporal data with respect to the insertion operation and range queries. This study proposes the method based on the grid space partitioning and gives a demonstration of its advantages for the mentioned purposes.

Last but not least inspiring work, there are many case studies analysing a spatiotemporal aspect of messages collected by the social media such as Twitter Kamath et al. (2013), Kumar et al. (2011). These studies definitely show a great potential of spatiotemporal analysis of big amount of dynamic and ad hoc data. On the other hand, such studies are usually limited to the analysis of position and time where particular message has been sent and do not concentrate on the space and time that relates to a particular content of such a message as spatiotemporal context of the message is usually hard or impossible to retrieve.

3 Algorithm Description

Let us assume the input data in the form of geographic coordinates and timestamp. Before the encoding process is applied, the data indexing is performed. For the indexing there is a method based on the space partitioning tree described in Sect. 3.1. We define an auxiliary one-dimensional data structure that acts as a pointer to that three-dimensional spatial tree in Sect. 3.2. As a next step there is an encoding process proposed that converts the items from the auxiliary data structure into sequences of characters described in Sect. 3.3. The one-dimensional data structure and the encoding is designed in a specific way that allows some of the usual tree operations to be solved by dealing directly with the string representation of a node. Such operations are described in Sects. 3.4 and 3.5. Finally there are some operations derived that are targeting to most common use cases.

3.1 Spatiotemporal Data Structure

The proposed indexing structure is based on a region quadtree. This term denotes a spatial data structure based on a disjoint regular partitioning of space Hjaltason and Samet (2002). The data structure used in our approach is customization of the region quadtree for three dimensions, so it can be called a region octree. Each node of the tree points to a region with the bounds in 2D space and in time.

The 2D space and the time are defined by these coordinates:

- Latitude $\lambda \in \langle 0, 180\rangle$—a geographic coordinate based on the WGS84 coordinate reference system. The domain of that coordinate is from -90 to $+90$. We added the value of 90 to get just positive values.
- Longitude $\varphi \in \langle 0, 360\rangle$—a geographic coordinate based on the WGS84 coordinate. The domain is -180 to $+180$. Similary to Latitude we convert it to 0–360.
- Time t is a compound of minutes within 1 year $m \in \langle 0, 527040\rangle$, and the year. The values of minutes are chosen to cover the whole year including a leap-year ($527040 = 366 \cdot 24 \cdot 60$). Coordinated Universal Time (UTC) standard is chosen. The encoding algorithm considers just the minutes part of the date and the year is handled separately.

For explanation of the encoding algorithm we can simplify the problem and consider just one dimension, e.g. latitude, the other dimensions are analogous.

The root node has the bounds equal to the latitude dimension (that is 0–180). A parent node has always a left child node and a right one (in one dimension), where each child node has bounds equal to the half of the interval of the parent node. When we want to identify the node of a particular input value in that tree with respect to the required level, we can traverse the tree from the top to the bottom and test if the value belongs to the node bounds of the particular level.

Similarly we can handle all three dimensions. By recursive halving the domains we will get particular bounds of the nodes. For the time domain we will work just with the m part (time within 1 year).

The node of the particular input in the tree is defined by a binary number where 0 stands for the left child node and 1 stands for the right child node in the particular level. The result is that we encode the input value to a binary number (say t_b) with the maximal error equal to the half of the node size.

The example of the bit assignment for one dimension is demonstrated in Fig. 1, where we search for a node for the input value $\varphi = 120°$. When testing the first child of the root node, our item is located in the child with bounds 90–180, so the first digit of t_b will be 1. On the second level our input is located in the left child as it falls within the range of 90–135, so the second digit of t_b will be 0. Similarly the third digit will be 1. So when considering the three-level tree, we get $t_b = 101$.

Other possibility how to convert input i to the binary value t_b is to use the decimal value t_d as an intermediate step. Let

$$t_d = floor\left(\frac{i}{c}\right) \tag{1}$$

where c is the size of the node on the bottom level. This size can be calculated as:

$$c = \frac{N_{rs}}{2^l}, \tag{2}$$

where N_{rs} is the size of the root node and l is the required tree level.

As an example let us have again $\varphi = 120°$ and the three-level-based encoding. The size c of the target node is then:

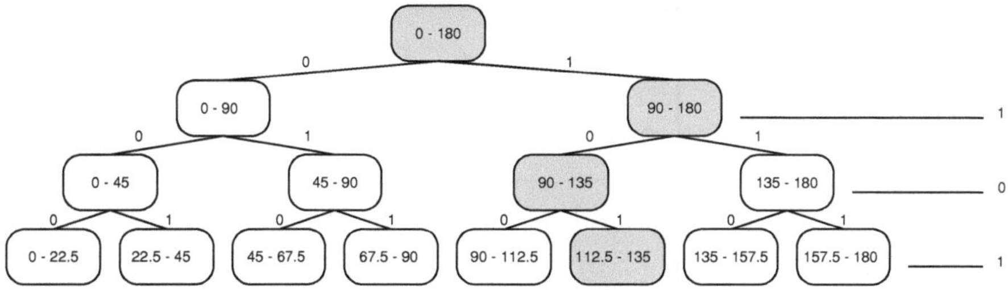

Fig. 1 Tree structure

$$c = \frac{180}{2^3} = 22,5 \tag{3}$$

and t_d is:

$$t_d = floor\left(\frac{120}{22.5}\right) = floor(5.333) = 5, \tag{4}$$

finally by converting the value 5 to binary we get the same value of $t_b = 101$ as described in the previous part. Such an approach is also described in Morton (1966) in more detail.

If we consider all three dimensions and apply the described process on the input specified by latitude, longitude and time, we will get three bit sequences, where the number of bits in each sequence corresponds to the number of levels of the tree.

In that way we can approximate the input value by the center of the particular node, where the worst case error σ is obviously equal to the half of the size of that node (in the particular dimension $\sigma = c/2$). The tree level l can be understood as a variable that corresponds to the precision of such an approximation. That implies that:

$$l = floor\left(\frac{\ln(N_{rs}/\sigma)}{\ln(2)}\right). \tag{5}$$

3.2 Bit Interleaving

For the final key that points to a particular node in all three dimensions we use a bit interleaving technique. As we have explained above, the node of the tree is specified by three bit sequences. The final expression is composed by consecutive bits from the sequence assigned to the longitude, latitude, and time. In this way we get one bit sequence (the key in the binary format) that defines a particular node in all dimensions.

Connecting the nodes according to the numeric values of their key produces the so-called Z-order curve (see Morton (1966) for more). This curve (in 2D) is illustrated as the grey line in Fig. 2. The difference in our case is that such a curve will be produced

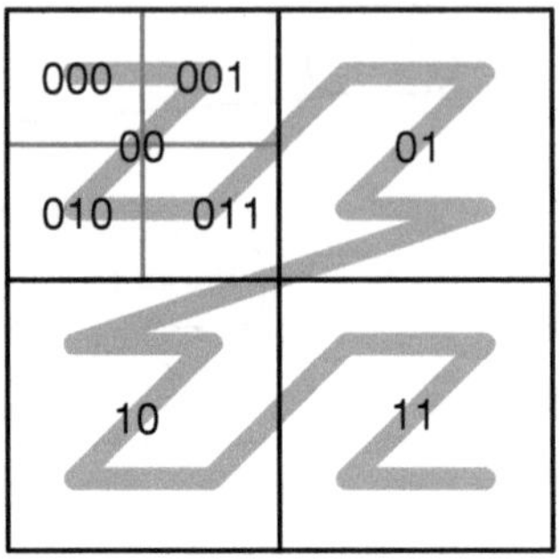

Fig. 2 Z-order curve

in three dimensions. Such a key of the node in the binary form is afterwards encoded into a sequence of characters (String) as described in the following section.

3.3 String Representation

The string representation of the binary format is created by a substitution of every six-bit group by a particular character. The used character set is the binary-to-text encoding schema also referenced as base64. The used characters and corresponding bit assignments are shown in Table 1. From Table 1 it can be seen that the order of original values corresponds to the lexicographical order of the characters, so the lexicographically sorted keys of all nodes still compose the Z-orderd curve.

As has already been mentioned, the order of the bits is set as consecutive bits from latitude, longitude and time which implies that each three bits express the level of the tree in all three dimensions. Then obviously the six bits correspond to two levels. As we are using one character to encode six bits, we can say that each character represents two levels of the tree. That implies that every upcoming character of the string reduces the node size four times in all dimensions. With respect to that we can also customize the calculation of the node c size (2):

$$c = \frac{N_{rs}}{4^m}, \tag{6}$$

where $m = l/2$, and $l \in \langle 2, 4, 6, 8 \ldots \rangle$. The m is called a reduced tree level.

Similarly for m:

$$m = floor\left(\frac{\ln(N_{rs}/\sigma)}{\ln(4)}\right) \tag{7}$$

As a consequence the encoding algorithm deals just with even levels.

Table 1 Base64 encoding

0	–	000000	1	–	000001	2	–	000010	3	–	000011
4	–	000100	5	–	000101	6	–	000110	7	–	000111
8	–	001000	9	–	001001	:	–	001010	A	–	001011
B	–	001100	C	–	001101	D	–	001110	E	–	001111
F	–	010000	G	–	010001	H	–	010010	I	–	010011
J	–	010100	K	–	010101	L	–	010110	M	–	010111
N	–	011000	O	–	011001	P	–	011010	Q	–	011011
R	–	011100	S	–	011101	T	–	011110	U	–	011111
V	–	100000	W	–	100001	X	–	100010	Y	–	100011
Z	–	100100	_	–	100101	a	–	100110	b	–	100111
c	–	101000	d	–	101001	e	–	101010	f	–	101011
g	–	101100	h	–	101101	i	–	101110	j	–	101111
k	–	110000	l	–	110001	m	–	110010	n	–	110011
o	–	110100	p	–	110101	q	–	110110	r	–	110111
s	–	111000	t	–	111001	u	–	111010	v	–	111011
w	–	111100	x	–	111101	y	–	111110	z	–	111111

3.4 Parent and Child Nodes

One of common operations when dealing with a tree data structure is the searching of a parent or child node. In the case of STCode we can derive the key of a parent or child node directly from its string representation. According to Table 1 every character in the string corresponds to six bits which represent two levels of the tree. With respect to that, the code of the parent node (considering just even levels) is derived simply by omiting the last character of a particular code.

Similarly all the child codes can be derived be extending a particular code by all the characters from base64 set.

Concrete example and spatial and temporal visualisation of parent and child nodes is demonstrated in Sect. 4.2.

3.5 Codes of Adjacent Nodes

To search the adjacent code of a particular node, the code needs to be converted back to the binary value and the bit sequences of each particular coordinate have to be decomposed (a particular example of the bit interleaving is given in Sect. 4.2). Once this is done, we can simply add the corresponding shift (e.g. 1 for the first upcoming node) using the binary addition or subtraction and then compose the key in the string format again. For example, let us assume the derivation of the node that is the first upcoming node in the direction of latitude of a node represented by the code u8XUhjL9X. The binary representation of the latitude of such a code is then

110001110001110001. As we are searching the first adjacent node, we add the value 1, so the binary value representing the latitude coordinate of the adjacent node is 110001110001110010. By interleaving the bits with the rest of the coordinates and encoding them to the string, we get the code of the adjacent node as u8XUhjL9k.

4 Results and Experiments

The described algorithm has been implemented as a Java library named STCode.[2] The library is available under the opensource licence. The upcoming parts demonstrate practical impacts and the use of the algorithm. The performance will be briefly discussed and an example of encoding particular coordinates will be given. Some suggestions about the usage in databases will be also briefly mentioned.

4.1 Performance

According to Sect. 3.1 the STCode string can be composed by using the mathematical division operation and the power operation, where the power exponent is equal to the half of the tree level. Such a level with relation to the encoding precision is displayed in Table 2. According to this relation the complexity of encoding is $O(\sigma)$, where σ is the required precision of the encoding.

4.2 Encoding Example

Let us assume the input values: $\varphi = 50.0000$, $\lambda = 14.0000$, $t = 22.8.2013$ 18:10:00 UTC, and the required time error of the encoding $\sigma_{time} <= 1.5$ min. Before we start the encoding, we will customize the values according to the definition in Sect. 3. So $\lambda_r = 140$, $\varphi_r = 194$, $m = 336610$.

By using Eq. (7) we can get the reduced tree level m:

$$m = \frac{\ln(366 \cdot 24 \cdot 60/1.5)}{\ln(4)} = \mathit{floor}(9.209) = 9 \tag{8}$$

From Eq. (2) we get the node size in each dimension as:

$$\begin{aligned} c_\varphi &= 0.000686645 \\ c_\lambda &= 0.00137329 \\ c_{time} &= 2.00500488 \end{aligned} \tag{9}$$

[2] http://code.google.com/p/stcode/.

Table 2 Nodes sizes

Level	Size in longitude [^{o}]	Size in latitude [^{o}]	Size in time [minutes]	Size in east west direction [m]	Size in south north directions [m]
0	360.00000	180.00000	527040.00	40074155.89	20037077.94
1	180.00000	90.00000	263520.00	20037077.94	10018538.97
2	90.00000	45.00000	131760.00	10018538.97	5009269.49
3	45.00000	22.50000	65880.00	5009269.49	2504634.74
4	22.50000	11.25000	32940.00	2504634.74	1252317.37
5	11.25000	5.62500	16470.00	1252317.37	626158.69
6	5.62500	2.81250	8235.00	626158.69	313079.34
7	2.81250	1.40625	4117.50	313079.34	156539.67
8	1.40625	0.70313	2058.75	156539.67	78269.84
9	0.70313	0.35156	1029.38	78269.84	39134.92
10	0.35156	0.17578	514.69	39134.92	19567.46
11	0.17578	0.08789	257.34	19567.46	9783.73
12	0.08789	0.04395	128.67	9783.73	4891.86
13	0.04395	0.02197	64.34	4891.86	2445.93
14	0.02197	0.01099	32.17	2445.93	1222.97
15	0.01099	0.00549	16.08	1222.97	611.48
16	0.00549	0.00275	8.04	611.48	305.74
17	0.00275	0.00137	4.02	305.74	152.87
18	0.00137	0.00069	2.01	152.87	76.44
19	0.00069	0.00034	1.01	76.44	38.22
20	0.00034	0.00017	0.50	38.22	19.11
21	0.00017	0.00009	0.25	19.11	9.55
22	0.00009	0.00004	0.13	9.55	4.78
23	0.00004	0.00002	0.06	4.78	2.39

Then we calculate the decimal and the binary value of each coordinate:

$$\begin{aligned} t_\lambda &= floor((50+90)/s_\lambda) = 141266 = 100010011111010010 \\ t_\varphi &= floor((14+180)/s_\varphi) = 203889 = 110001110001110001 \\ t_{time} &= floor(527040/s_{time}) = 167884 = 101000111111001100 \end{aligned} \tag{10}$$

By interleaving the bits we get the final key in the binary format. Afterwards we will use the base64 encoding:

$$\underbrace{111010}_{u}\ \underbrace{001000}_{8}\ \underbrace{100010}_{X}\ \underbrace{011111}_{U}\ \underbrace{101101}_{h}\ \underbrace{101111}_{j}\ \underbrace{010110}_{L}\ \underbrace{001001}_{9}\ \underbrace{100010}_{X}$$

So the final STCode value is **u8XUhjL9X**. The approximation based on the center of the node and the precision of such a key is

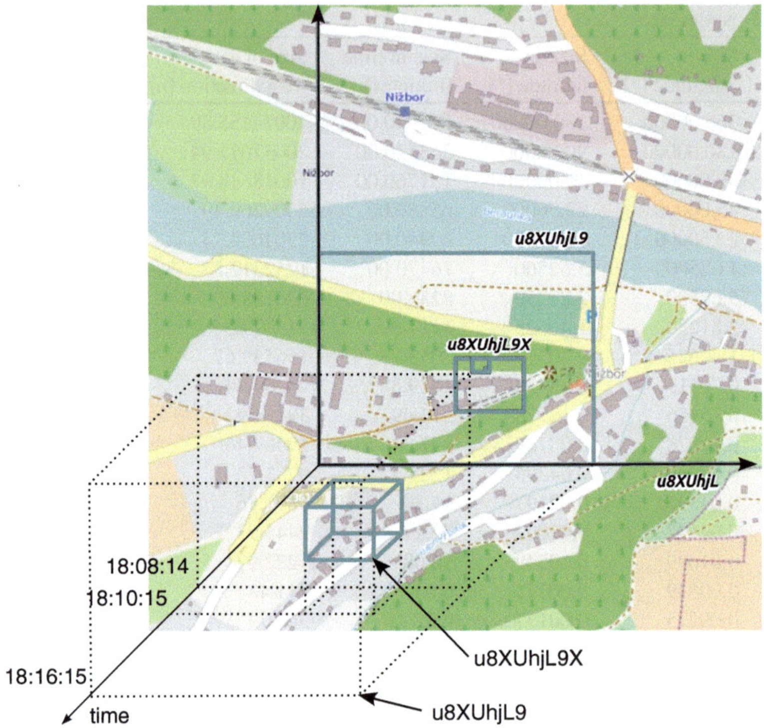

Fig. 3 Code bounds in big scale

$$time = 22.8.18 : 09 : 14 \pm 1.002 \text{ min} \tag{11}$$

$$\varphi = 49.99981 \pm 0.00034^o (\approx \pm 38m) \tag{12}$$

$$\lambda = 14.00001 \pm 0.00069^o (\approx \pm 49m) \tag{13}$$

The visualization of the node including the parent code is shown in Fig. 3. Farther parent codes are shown in Fig. 4.

4.3 Use for a Data Storage

Let us assume the common scenario, where we have a collection of records expressed by time and position. We encode every record of that collection by six character long STCode string for each item. Let us assume 10 random records from a certain area and time interval shown in Table 3. A visualization of records on the map is available in Fig. 5.

Now we are going to demonstrate the usage of STCode in common databases.

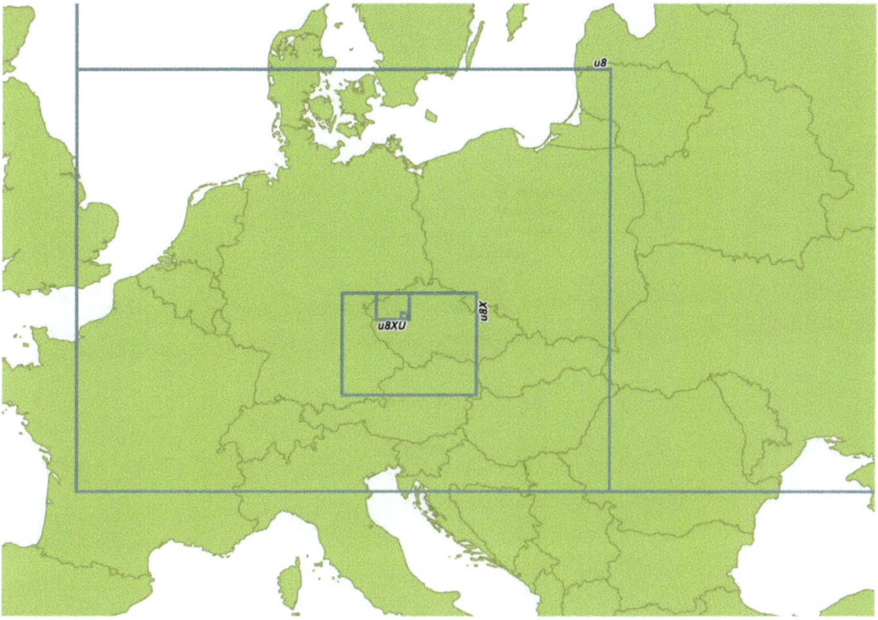

Fig. 4 Code bounds in small scale

Table 3 Random records

Latitude	Longitude	Time	STCode
50.00000	14.00000	2013-08-22 18:10:00	u8XUhjL9
50.00722	13.99805	2013-08-22 18:13:00	u8XUhjLQ
50.01500	13.99186	2013-08-22 18:17:00	u8XUhx0j
50.01737	13.98699	2013-08-22 18:22:00	u8XUhx1k
50.02262	13.98202	2013-08-22 18:30:00	u8XUhx35
50.03154	13.97854	2013-08-22 18:36:00	u8XUhx3P
50.03723	13.97210	2013-08-22 18:45:00	u8XUhtpx
50.03875	13.97088	2013-08-22 18:53:00	u8XUhtwo
50.04161	13.96464	2013-08-22 19:04:00	u8XUhtwn
50.04647	13.96312	2013-08-22 19:12:00	u8XUhtyD

4.3.1 Trie Construction

As it has been shortly mentioned in Sect. 1, one of useful data structures in databases is a trie. According to Sect. 3.4, all the keys of child nodes share the same prefix. This property enables a direct construction of a trie for the collection of STCodes. An example of a trie designed for the ten random records from the Table 3 is visualized in Fig. 6.

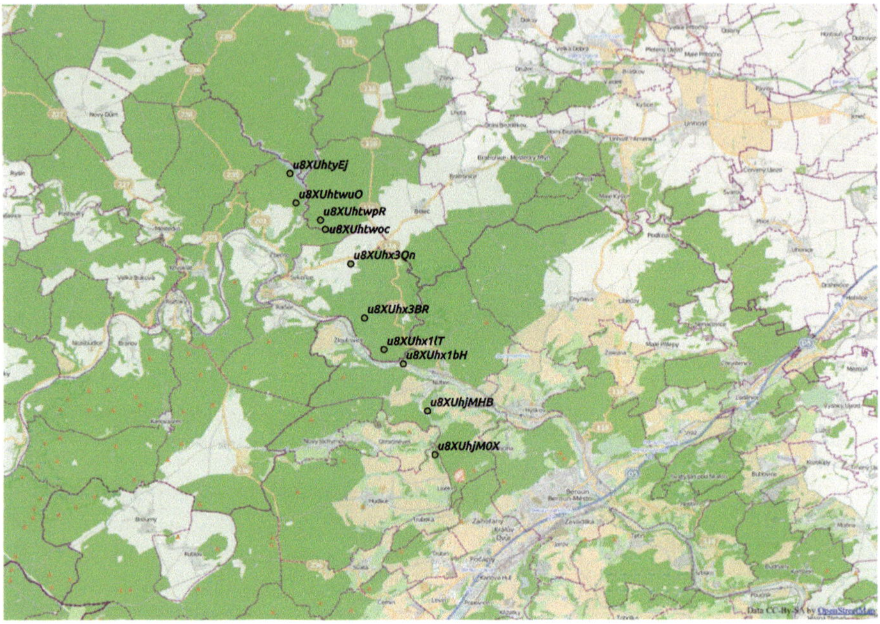

Fig. 5 Map of random records

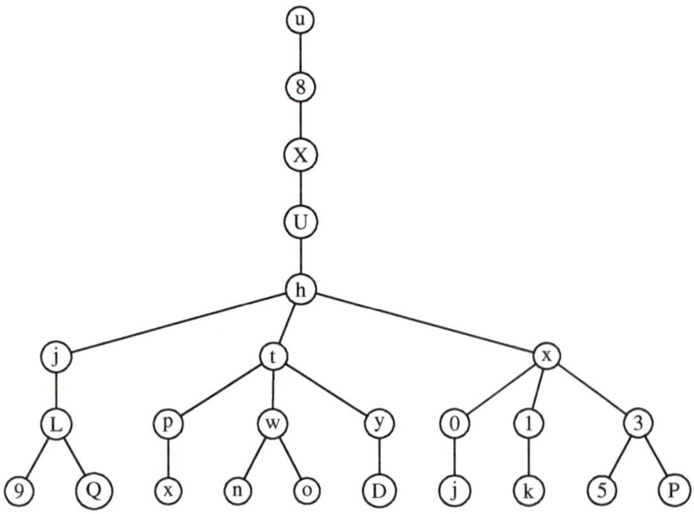

Fig. 6 Trie

Table 4 Aggregation result

Count	STCode
4	u8XUht
2	u8XUhj
4	u8XUhx

The Trie data structure can be used as a displacement of a hash table or for a data compression. The fact that the structures based on the region quadtree are a type of trie is generally known. Unlike other encoding algorithms (e.g Beatty (2003)), STCode encoding exploits this fact and assigns the strings to the node in the way that preserves the possibility of the straightforward trie construction.

4.3.2 SQL Examples

When we consider databases with the SQL support, we can use the STCode value as a helper column. The STCode value can be used for a searching of the data located in a particular node. When searching all the data located in a particular node (say the node with u8XUht), we can define the query as:

```
select * from records where substring(stcode,0, 7)= 'u8Xuht'
```

The column with the STCode can also be used to aggregate the data according to their spatiotemporal proximity. The query that aggregates all the records according to the key of a certain level (say 6) will look like:

```
select count(*), substring(stcode,0,7) from records group by
substring(stcode,0,7)
```

The result is shown in Table 4, where the count column expresses the number of points that lie in a particular node.

Such an example demonstrates mainly convenience of work with the STCode strings. With respect to the lexicographical ordering of STCode, the traditional indexing (e.g. btree) can be applied on the STCode column to reach a better performance of such queries.

It is worth to mention that such an aggregation is often used in databases by using the Geohash algorithm as it presents a very simple, but handy way of a data clustering based on spatial proximity. In contrast to Geohash, STCode considers also a temporal proximity so it might be more convenient for the cases where a clustering according to the temporal dimension should be taken into account.

4.4 Discussion and Limitations

Similar to the Geohash algorithm, there is one limitation, when STCode string is used to find points in mutual spatiotemporal proximity by using a common STCode prefix. For example, two close points, but on the opposite side of the Equator, will result in STCode, with no common prefix. Such a drawback is caused by known limitations when there is the multidimensional space represented by the space filling curve (Z-order curve in our case). The solution for such an issue is the calculation of a surrounding STCodes (see Sect. 3.5) of a particular STCode in corresponding dimension for such purpose.

Another limitation is caused by the use of geographical coordinates that are not isometric. For example the metric distance that corresponds to one degree in latitude on Equator is different in one degree for the parallel near the pole. On other hand, one degree difference in meridian expresses still the same distance. This causes that the real size of the node in the longitude direction is decreasing in the direction to the poles. Such a limitation can be eliminated by using isometric geographic coordinates, using an appropriate cartographic projection or using a sphere-based partitioning algorithm, but, on the other hand, such a calculation increases the complexity of the encoding process.

As has been mentioned in Sect. 3.1, the encoding deals just with the time in the scope of one year and the year itself is not considered to be encoded. One of the reasons for the choice of one year as the initial interval was the size of the low level node (see e.g. level 23 in Table 2) as such a node size can be precise enough to represent the data collected by custom GPS receivers.

The year information can be handled for example as a prefix of STCode (e.g `2013!u8XUhjL9X`). However, the existence of a leap year will causes that the the same STCode can be interpreted as one day different date when it is referenced to the leap year and to the not-leap year. The STCode library deals with that and provides suitable methods that can encode and decode the data with respect to a particular year. Thanks to the reviewers comments one of the consideration for the future work is to make the algorithm more generic and to enable customization of the initial time interval to an arbitrary size.

5 Conclusion

This chapter proposes an algorithm for encoding longitude/latitude/timestamp coordinates to a compact sequence of characters. The example of use of the algorithm are: A unique spatiotemporal identifier, representation spatiotemporal point data in a database and spatiotemporal tagging. Comparing to other related encoding algorithms, the STCode elaborates also the temporal dimension. Described working examples have shown the convenience of the STCode for encoding particular its important advatages. Such advatages are a possibility of arbitrary precision, a simple trie construction and a simple derivation of nearby codes. In the future work, we aim to apply the STCode encoding in new applications.

Acknowledgments The work was supported by the UWB grant SGS-2013-029 Advanced Computer and Information Systems and by Ministry of Education, Youth, and Sport of Czech republic—University spec. research—1311.

This action is realized by the project EXLIZ – CZ.1.07/2.3.00/30.0013, which is co-financed by the European Social Fund and the state budget of the Czech Republic.

References

Agarwal PK, Arge L, Erickson J (2003) Indexing moving points. J Compt Syst Sci 66(1):207–243 (Special Issue on PODS 2000)

Aref WG, Ilyas IF (2001) An extensible index for spatial databases. In: Proceedings thirteenth IEEE international conference on scientific and statistical database management SSDBM 2001, pp 49–58

Bryan B Compact text encoding of latitude/longitude coordinates. United States, Patent Application Publication, 2003. Original Assignee: Microsoft Corporation

Bodo F, Rónyai L (2003) Trie: an alternative data structure for data mining algorithms. Math Comput Model 38(7–9):739–751

Chon HD, Agrawal D, Abbadi AE (2001) Using space-time grid for efficient management of moving objects. In: Proceedings of the 2nd ACM international workshop on data engineering for wireless and mobile access, MobiDe '01, New York, NY, USA, pp 59–65

Dutton G (1996) Encoding and handling geospatial data with hierarchical triangular meshes. In: Proceeding of 7th International symposium on spatial data handling, vol 43. Talor & Francis, The Netherlands

The Open Source Geospatial Foundation (2012) Tile map service specification. Technical report, The Open Source Geospatial Foundation

Gargantini I (1982) An effective way to represent quadtrees. Commun ACM 25(12):905–910

Hjaltason GR, Samet H (2002) Speeding up construction of pmr quadtree-based spatial indexes. VLDB J 11(2):109–137

Kamath KY, Caverlee J, Lee K, Cheng Z (2013) Spatio-temporal dynamics of online memes: a study of geo-tagged tweets. In: Proceedings of the 22nd international conference on World Wide Web, WWW '13, International World Wide Web Conferences Steering Committee. Republic and Canton of Geneva, Switzerland, pp 667–678

Kumar S, Barbier G, Abbasi MA, Liu H (2011) Tweettracker: an analysis tool for humanitarian and disaster relief. In: In: Fifth international AAAI conference on weblogs and social media ICWSM

Morton GM (1966) A computer oriented geodetic data base and a new technique in file sequencing. International Business Machines Company, Ottawa

OpenStreetMap project. Slippy map tilenames. Technical report, OpenStreetMap project, 2013.

Webber RE (1989) A comparison of the space requirements of multi-dimensional quadtree-based file structures. Visual Comput 5(6):349–359

Szalay AS, Gray J, Fekete G, Peter ZK, Kukol P, Thakar A (2007) Indexing the sphere with the hierarchical triangular mesh. arXiv, preprint cs/0701164

Ouri W (2002) Moving objects information management: the database challenge. In: Fifth workshop on next generation information technologies and systems NGITS 2002. Springer, pp 75–89

Fast SNN-Based Clustering Approach for Large Geospatial Data Sets

Arménio Antunes, Maribel Yasmina Santos and Adriano Moreira

Abstract Current positioning and sensing technologies enable the collection of very large spatio-temporal data sets. When analysing movement data, researchers often resort to clustering techniques to extract useful patterns from these data. Density-based clustering algorithms, although being very adequate to the analysis of this type of data, can be very inefficient when analysing huge amounts of data. The Shared Nearest Neighbour (SNN) algorithm presents low efficiency when dealing with high quantities of data due to its complexity evaluated in the worst case by $O(n^2)$. This chapter presents a clustering method, based on the SNN algorithm that significantly reduces the processing time by segmenting the spatial dimension of the data into a set of cells, and by minimizing the number of cells that have to be visited while searching for the *k*-nearest neighbours of each vector. The obtained results show an expressive reduction of the time needed to find the *k*-nearest neighbours and to compute the clusters, while producing results that are equal to those produced by the original SNN algorithm. Experimental results obtained with three different data sets (2D and 3D), one synthetic and two real, show that the proposed method enables the analysis of much larger data sets within reasonable amount of time.

1 Introduction

Huge amounts of spatio-temporal data are collected nowadays with the support of current positioning and sensing technologies. For the analysis of this vast amount of data, researchers often resort to clustering techniques as an important tool to extract

A. Antunes · M. Y. Santos (✉) · A. Moreira
ALGORITMI Research Centre, University of Minho, Guimarães, Portugal
e-mail: maribel.santos@algoritmi.uminho.pt

A. Antunes
e-mail: armenio.antunes@algoritmi.uminho.pt

A. Moreira
e-mail: adriano.moreira@algoritmi.uminho.pt

J. Huerta et al. (eds.), *Connecting a Digital Europe Through Location and Place*, Lecture Notes in Geoinformation and Cartography, DOI: 10.1007/978-3-319-03611-3_11,

patterns from these data. The Shared Nearest Neighbour (SNN) algorithm, previously proposed by Jarvis and Patrick (1973), and later improved by Ertoz et al. (2002), is a density-based clustering algorithm which is able to identify clusters with convex and non-convex shapes, with different sizes and densities, and capable of dealing with noise. With SNN, the number of clusters emerges directly from the data under analysis, thus being independent from the knowledge or experience of the analyst. Control over the clustering results is obtained through three input parameters.

Due to its characteristics, the SNN clustering algorithm is often used in the analysis of spatial data (Moreira et al. 2010). However, the ever increasing size of those data sets makes it prohibitive due to the time needed to compute the clusters. With respect to performance, SNN has very low efficiency when dealing with high quantities of data due to its complexity evaluated in the worst case by $O(n^2)$ (Ertoz et al. 2002). Besides its complexity, one single data analysis process often requires several runs of the algorithm, with different input parameters, to tune the clustering process and the obtained results. The algorithm complexity, added to this need of several runs in order to obtain satisfactory results, leads to a clustering process that can last for hours or days, depending on the size of the data set and the complexity of the analysis task. A more efficient solution is then needed to enable the exploratory analysis of large data sets within reasonable time.

A deeper analysis of the SNN algorithm, combined with experimental results obtained by processing a few real data sets, revealed that the major bottleneck of the algorithm is due to the need to compute the list of k-nearest neighbours of each point. Fortunately, when dealing with spatial data, some optimizations are possible. This chapter presents a clustering approach that divides the spatial domain of the data to be clustered into a set of cells and then limits the searching procedure to a subset of the cells, aiming to minimize the time required to compute the k-nearest neighbours of each point. The proposed approach proved to reduce the average time to perform the clustering of a dataset by a factor of 14–24 (for 8000 points, and improving for larger data sets), while preserving all the advantages of the original SNN algorithm.

This chapter is organized as follows: Sect. 2 presents the SNN algorithm and its characteristics, as well as alternative solutions being proposed by other authors to improve the efficiency. Section 3 describes the proposed approach for the computation of the k-nearest neighbours of each point. In Sect. 4, the proposed approach is validated and evaluated using both synthetic and real data sets. Section 5 presents the analysis of a real data set, showing how the proposed algorithm can be used in an iterative analytical context. Section 6 concludes with some remarks and guidelines for future work.

2 Related Work

The SNN algorithm is a density-based clustering algorithm that was proposed by Ertoz et al. (2002) and that is able to find clusters of varying sizes, shapes and densities, even in the presence of large amounts of noise. It has also the capability to handle multidimensional data, and automatically determines the number of clusters.

This algorithm requires three input parameters: *k*, *Eps* and *MinPts*. *k* is used to set the number of neighbours that must be identified for each point based on a given distance function. In its simplest form, the Euclidean distance between two points is used. Using the *k*-nearest neighbours, the number of shared neighbours, or reflexive links, is computed for each point. Two points are considered reflexive if they are in each other's *k*-nearest neighbour list and share *Eps* or more nearest neighbours. The number of reflexive relations of a point is used to calculate its density. If a point's density is at least *MinPts*, it is classified as a core point. Clusters are formed around these core points, by joining two core points in the same cluster if they are reflexive. Any non-core point, which is reflexive to a core point, is considered to be a border point of the same cluster as this core point. All the remaining points are classified as noise points. In summary, while *k* is used to set the size of the nearest neighbours' list, *Eps* is the reflexivity threshold and *MinPts* is the density threshold that enables a point to be classified as core.

The SNN algorithm, while presenting several characteristics that makes it attractive for the analysis of spatial data, is very inefficient when dealing with large datasets. The complexity of this algorithm is $O(n^2)$ due to the need to calculate the similarity matrix between all points (Bhavsar and Jivani 2009).

To optimize the algorithm and decrease the processing time required for a given set of points, Bhavsar and Jivani (2009) conceptualized the "Shared Nearest Neighbour Algorithm with Enclosures" (SNNAE). The strategy of this algorithm is to reduce the number of computations needed to identify the *k*-nearest neighbours of a point by partitioning the data into overlapping subsets called 'enclosures', and then by computing the distance only among the points within the same enclosure. This way, the algorithm can be more efficient as a point is not compared against all the other points in the data set when looking for its neighbours.

In SNNAE, the radius *r* of these overlapped enclosures is obtained from a ratio between the area of a circle containing all points in the dataset, and the area of a rectangle containing the same objects. The first point of the dataset is considered the centre of the first enclosure. All points at a distance *r* or less to the centre of the enclosure are considered to be the *nearest adjacent*. If a point's distance to the centre of the enclosure is less than or equal to $1.5 \times r$ then it is considered to be a *nearest far adjacent* point. All the points whose distance is greater than $1.5 \times r$ and less than $2 \times r$ to the enclosure's centre are considered to be the centre of the next enclosure, ensuring the overlapping of enclosures.

After the creation of enclosures, the parameter *Eps* is defined dynamically for each enclosure, enabling, at this stage, the calculation of the nearest neighbours' lists within each enclosure and the execution of the remaining steps of the clustering process as in the original SNN algorithm.

In their chapter, Bhavsar and Jivani (2009) discuss the performance of their proposal. However, only a very small dataset is used (209 points). The described results are not conclusive about the real performance gains of this solution, and the gain compared to the original SNN is less than 50 % in most tests.

Another proposal for improving the efficiency is the Fast Clustering Algorithm (FCA) that is able to find clusters with different sizes, shapes and densities, characteristics inherited from SNN, and aiming to achieve an almost linear time complexity (Li et al. 2009). The FCA solution uses the concept of 'nearest neighbour', but not in a direct way as in the SNN algorithm. First, a single-pass algorithm is applied, by partitioning the dataset into *k* clusters. All these clusters have approximately the same radius. The second phase of the algorithm treats each cluster obtained previously as an object, and merges them with a version of the SNN algorithm capable of dealing with categorical data. This version is designed to be of greater efficacy with data in discrete spaces.

However, this algorithm needs more input parameters. Besides *k*, *Eps* and *MinPts* as in SNN, it is necessary to define the radius *r*, a parameter required for the single-pass algorithm, used to restrict the radius of each cluster. Although an automated way of finding this value for each dataset is provided, it introduces additional work for the analyst, meaning that *r* may have different values for two analyses tasks of the same dataset.

When SNN is applied, on the following step, the initial clusters can only be joined to form larger clusters when, actually, splitting some of these initial clusters could provide a better final result. This approach may then deteriorate the clustering results. Regarding the efficiency of this algorithm, its improvements come from the reduced number of elements that have to be dealt with by SNN. However, there is still a need to evaluate if the gain is substantial with larger datasets, given the cost of the first step of this approach.

The Fast Density-based Clustering Algorithm (FDCA) (Tripathy et al. 2011) is an algorithm specifically designed for spatial databases, such as the algorithm in which it is based on, the Density-Based Spatial Clustering of Applications with Noise (DBSCAN) (Ester et al. 1996). DBSCAN initially requires two input parameters, *Eps* and *MinPts*. In turn, FDCA only requires the *MinPts* input parameter. The *Eps* parameter is automatically adjusted during the process. It is initially defined based on *MinPts* for each cluster centre, then being reduced as distant points in the same cluster are processed. This approach, while reducing the involvement of the analyst, allows also the identification of local clusters that are very close to each other.

Firstly, the FDCA defines the new concept of *Reachability Density* (RD). This represents the density around a point. A high RD means high density around a point. It creates a new cluster for a point that has high RD, and its *Eps* neighbourhood does not intersect the points already existing in another cluster previously created. There is also the notion of point *accessible* or *connected* to another point, similarly to DBSCAN. For a given value of *MinPts*, the points within a cluster have a RD higher than the points located near the border of the cluster.

In this algorithm, the construction of clusters begins from the interior to the exterior. Over the expansion of a cluster, the RD value of its points decreases, meaning the value of *Eps* used to process the same points also reduces, thereby preventing it to include points which belong to other nearby clusters.

To improve the performance of this algorithm, there is no need to process all the points within a circle of *Eps* radius centred at a point being processed at a given time, but only those that are near the boundary of this circle. Thus, a smaller amount of points are processed, increasing the efficiency of this algorithm.

The several methods described above, while improving the processing time for clustering the data, provide different clustering results than those produced by SNN, meaning that the quality of the obtained clusters must be evaluated. In the method described in the following section, we aim at producing clusters as with the original SNN algorithm, thus with the same quality. Regarding processing time, although some improvements were achieved by the alternative approaches described above, the reported tests on the literature review were carried out mainly with small data sets and, therefore, the gains on the processing time are not conclusive. The approach proposed in this chapter goes beyond those achievements being able to deal with very large data sets.

Regarding the inefficiency of finding the nearest neighbours of every point in the data set, as needed in the SNN approach, metric data structures might be used to index data objects, optimizing the search for the k-nearest neighbours. A metric space integrates a set of valid objects and a distance function that measures the distance between these objects (Chávez et al. 2001). Metric data structures are used to index data in a metric space, supporting similarity queries and giving the possibility of minimizing the number of distance evaluations. Indexing a data set with a metric data structure takes a significant runtime. However, once built, the structures can be reused avoiding the time required to use or update the structure.

Faustino (2012) proposed the use of metric data structures in primary or secondary storage, to index spatial data and support the SNN in querying for the k-nearest neighbours. The experimental results were undertaken using the kd-Tree (Bentley 1975), for primary storage, and the DF-Tree (Traina et al. 2002), for secondary storage. The results showed good performance gains when the metric data structure is in primary storage. In secondary storage, and using data sets with more than 128.000 objects, the performance is comparable with the performance of the original implementation of the SNN (Ertoz et al. 2002).

3 The F-SNN Approach

Despite the several advantages of the SNN algorithm, it presents major drawbacks in the analysis of large data sets mainly related to performance issues. This algorithm can be divided into three main steps: (i) calculation of distances between points to build the similarity matrix; (ii) checking for reflexive points and classifying them as core or border points; and, (iii) building clusters around these core points. In Fig. 1, the computing time used by each one of these steps is compared for two different sets of values of the input parameters. These results have been obtained with our own implementation, in Java, of the original SNN algorithm. In these tests, several data sets of randomly generated points, and increasing size, were used (see Sect. 4 for further details).

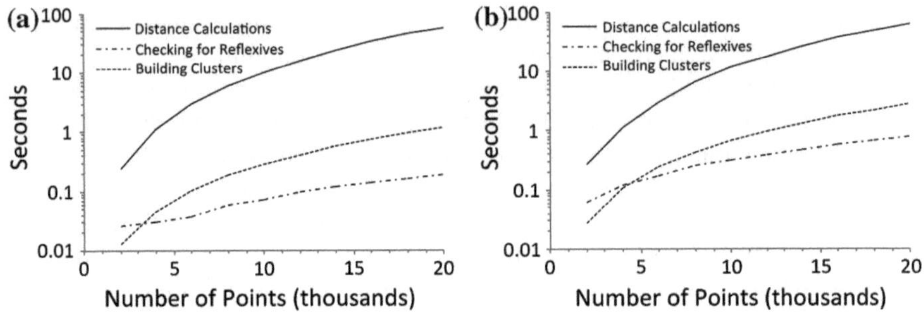

Fig. 1 Processing times of each step of SNN, for different quantities of points and two sets of parameters: **a** k = 10; Eps = 3; MinPts = 7, **b** k = 20; Eps = 7; MinPts = 14

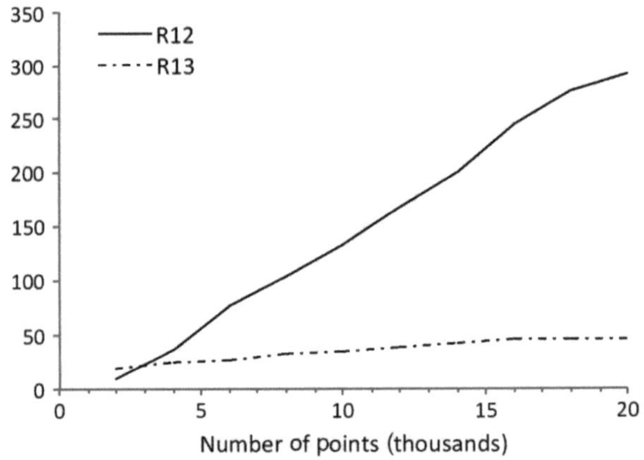

Fig. 2 Ratios between the time used in Step 1 to the time used in Step 2 (R12) and Step 3 (R13)

As the size of the data set increases, the time needed to perform each one of these steps also increases. Clearly, the total time is dominated by the first step: creating the similarity matrix, which involves computing the distances between each point and all other points in the data set (the Euclidean distance has been used). Therefore, $N \times (N - 1)/2$ distance calculations are needed, where N is the number of points, making the time complexity of the SNN algorithm quadratic with the size of the input data set, $O(n^2)$. Actually, not only the weight of the first step is clearly dominant (note the logarithmic scale in Fig. 1), but it also increases with the number of points, as shown in Fig. 2.

Being notorious that the bottleneck of the SNN algorithm is the identification of the k-nearest neighbours of a point, and that it has an increasing weight as the size of the data set increases, this chapter proposes an approach based on an iterative expansion of the nearest neighbours search procedure. This approach produces the same results as the original SNN algorithm, while significantly reducing the processing time needed to compute the clusters.

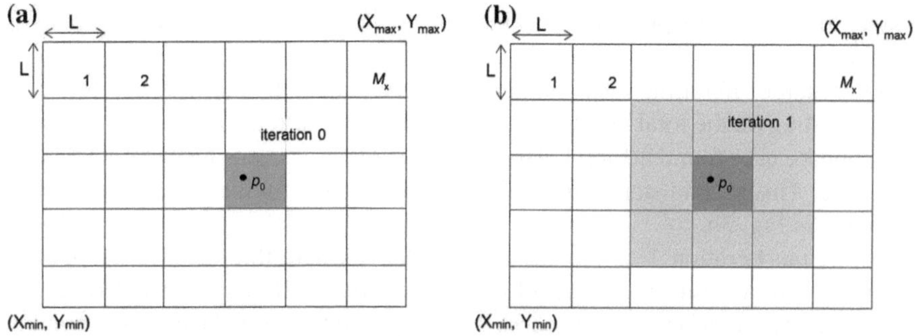

Fig. 3 Dividing the geographic domain into a set of square-shaped cells: searching for neighbours within the current cell (**a**) and within the surrounding cells (**b**)

3.1 Bi-Dimensional Objects

Movement data analysis is strongly influenced by the spatial dimension, meaning that the position of the objects, described by a pair of coordinates, usually have great relevance in the analysis. Therefore, the clustering process could be optimized by limiting the geographic area where to search for the *k*-nearest neighbours of a point. The approach here proposed is based on dividing the geographic, or geometric, domain in a set of square-shaped cells and on processing one cell at a time. We start by describing the process for a bi-dimensional dataset where the points are represented in the geographic domain only and then we extend it to more dimensions on the next section.

Let (X_{min}, Y_{min}) and (X_{max}, Y_{max}) be the coordinates of the two points defining the rectangle containing all the points in the data set, and M_x the number of cells in the X dimension, as shown in Fig. 3.

The length of each cell side is then:

$$L = \frac{X_{max} - X_{min}}{M_x} \tag{1}$$

The goal is to compute the *k*-nearest neighbours of each point as the list of *k* points that are at the shortest distances from them. For each point, the process starts by searching for the *k*-nearest points within its own cell, and then by expanding the search to the neighbouring cells only if necessary.

Let p_0 be the point for which we want to find the *k*-nearest points, and let $D(p_a, p_b)$ be the Euclidean distance between points p_a and p_b (other functions f can be used as soon as they satisfy the following: $f(p_a, p_a) = 0; f(p_a, p_b) \geq 0; f(p_a, p_b) = f(p_b, p_a)$).

The proposed procedure includes several steps. In Step 1, the list of the *k*-nearest points is obtained by computing the distance between p_0 and all the other points inside the same cell, and by selecting the *k* shortest distances. We name this Iteration

0 since the searching procedure occurs only within the cell to which p_0 belongs (Fig. 3a).

The result of Iteration 0 is a list of $n \leq k$-nearest neighbours, with $n \in \{0, 1, \ldots, k\}$, depending on the total number of points inside the current cell. Since less than k neighbours can be found in Step 1, a second step is required to ensure that k neighbours are found. This second step is simply based on searching in the neighbouring cells, as shown in Fig. 3b.

We define Iteration 1 as the searching process where the distance between p_0 and the points within the cells enclosing the current cell are evaluated (Fig. 3b). Consequently, Iteration i is the process where the points within the cells surrounding the cells of Iteration $i - 1$ are evaluated.

The number of Iterations required for Step 2 depends on the spatial distribution of the points within the cells enclosing the current cell. Actually, no additional Iterations are required if the number of points within the current cell is larger than k. Step 2 stops when at least k neighbours of p_0 are found after evaluating all the points within each enclosure. Obviously, Step 2 does not assure that the obtained list of k-nearest points is the absolute list of k points closest to p_0, as illustrated in Fig. 4a for $k = 3$, since point B is closer to p_0 than point A (in this case after Iteration 1).

Step 3 of the process is, therefore, responsible for finding the absolute list of k-nearest points of p_0, or for confirming that Steps 1 and 2 have already resulted in that list.

Let us define $P_i = (p_1, p_2, \ldots, p_k)$ as the vector of the k-nearest points found at the end of Step 2 after Iteration i, and $d_m = D(p_0, p_m)$, $m \in \{1, 2, \ldots, k\}$, so that $d_{m+1} \geq d_m$ (in other words, p_k is the point, among the k-nearest, that is farther way from p_0).

At the end of Step 2, it is possible to confirm if P_i is the absolute list of the k-nearest neighbours of p_0, but not if it is not. P_i is the absolute list of k-nearest neighbours if:

$$\forall m \in \{1, 2, \ldots, k\}, d_m < d_{wall} + i \times L \tag{2}$$

where d_{wall} is the distance from p_0 to the nearest boundary of its own cell, as shown in Fig. 4b. Otherwise, as is the case in Fig. 4b, it is not possible to confirm if P_i is the absolute list of k-nearest neighbours of p_0 since there might exist one or more points outside the current enclosure that are closer to p_0, as is the case of point B. If this is the case, the cells in enclosure $i + 1$ must be visited to verify if any of its points is closer to p_0 than any of the points in P_i. After being visited, P_i is updated and the verification procedure in Eq. (2) is repeated. This iterative process is repeated until it is possible to ensure that P_i is the absolute list of the k-nearest neighbours of p_0.

At the end of Step 2, it is possible to calculate an upper bound on the number of iterations required in Step 3 to obtain the absolute list of k-nearest neighbours. As shown in Fig. 4b, it is not possible to find points closer to p_0 than those already in P_i outside of the dashed circle area. Therefore, it is useless to visit cells outside that region, including the cells marked with an X. From this point on, iterations are made under the shape of a cross instead of a square, extending the visited area up, down, left and right, as shown in Fig 4c.

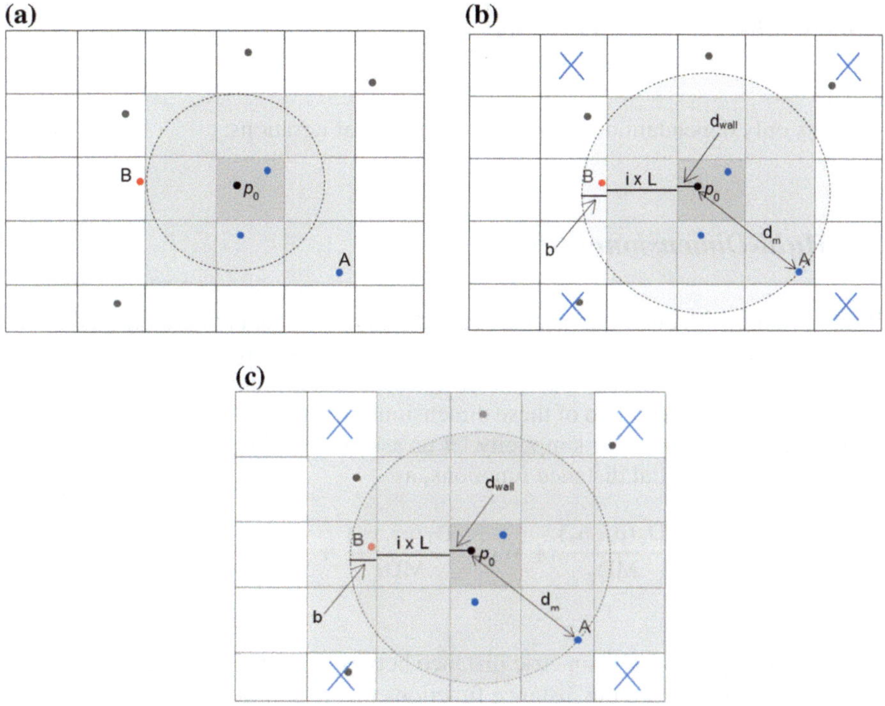

Fig. 4 After finding the *k*-nearest neighbours, additional iterations might be required: **a** point B is closer to p_0 than point A; **b** calculating the maximum number of additional iterations; **c** iterations executed under a cross shape

Let us define $j \in \{0, 1, \ldots, j_{max}\}$ as the number of additional iterations required in Step 3. Form Fig. 4b one can derive that, for an arbitrary point p_0, the maximum possible distance d_{max} from p_0 to the current *k*-nearest neighbours is given by:

$$d_{max} = \sqrt{2} \times L \times (i + 1) \quad (3a)$$

and, therefore, it is not possible to find nearest points at longer distances from p_0. Therefore, the maximum number of iterations required to find points potentially closer to p_0 is given by:

$$j_{max} = \text{Ceiling}\left(\frac{b}{L}\right) \quad (3b)$$

where *Ceiling* (*x*) is the smallest integer not smaller than *x*, and with:

$$b = d_{max} - i \times L \quad (3c)$$

and, therefore:

$$j_{max} = \text{Ceiling}\left(\left(\sqrt{2} - 1\right) \times i + \sqrt{2}\right) \tag{3d}$$

which is only dependent on the previous number of iterations.

3.2 Multi-Dimensional Objects

When dealing with other dimensions, in addition to the spatial coordinates, different criteria are needed to guarantee that the absolute list of *k*-nearest neighbours is found.

Let us define let D'(p_a, p_b) the distance between points p_a and p_b in a M-dimensional space, with two of these dimensions being the coordinates of each point in the geometric space. For simplicity let us assume that D'(p_a, p_b) is defined as a weighted sum of partial distance functions, as:

$$D'(p_a, p_b) = W_g \times \frac{D_g(p_a, p_b)}{MD_g} + W_1 \times \frac{D_1(p_a, p_b)}{MD_1} + \cdots + W_{M-2} \times \frac{D_{M-2}(p_a, p_b)}{MD_{M-2}} \tag{4a}$$

where $D_g(p_a, p_b)$ is the geometric distance between two points (e.g. the Euclidean distance), $D_i(p_a, p_b)$ are the distance functions that take into account the additional dimensions, MD_i are the normalization factors to ensure that each partial distance takes values between 0 and 1, w_g is the weight of the geometric distance to the overall distance, and w_i are the weighting parameters for the other dimensions. All weighting parameters take values between 0 and 1, and their sum is equal to one. This distance function can be simplified by grouping all non-geometric components into a single distance function as:

$$D'(p_a, p_b) = W_g \times \frac{D_g(p_a, p_b)}{MD_g} + W_{ng} \times \frac{D_{ng}(p_a, p_b)}{MD_{ng}} \tag{4b}$$

For the multi-dimensional case, the same three steps are required. Steps 1 and 2 are equal to those described in the previous section. However, the test described in Eq. (2), and used to verify if P_i contains the absolute list of *k*-nearest neighbours, must be updated to take into account the non-geometric dimensions. Using the same rationale, at the end of Step 2, and at the end of each additional iteration in Step 3, P_i contains the absolute list of *k*-nearest neighbours if:

$$\forall m \in \{1, 2, \ldots, k\}, d'_m < W_g \times \frac{d_{wall} + i \times L}{MD_g} \tag{5}$$

where we assumed that the worst case occurs when the non-geometric component of d_m is zero.

As for the bi-dimensional case, it is also possible to calculate an upper bound on the additional number of iterations needed in Step 3. From Fig. 4b, one can derive that, for an arbitrary point p_0, the maximum possible distance d'_{max} from p_0 to the current k-nearest neighbours is given by:

$$d'_{max} = w_g \times \frac{\sqrt{2} \times L \times (i+1)}{MD_g} + W_{ng} \times 1 \tag{6a}$$

where the non-geometric component of the distance is the maximum possible. Therefore, the maximum number of iterations required to find points potentially closer to p_0 is given by:

$$j'_{max} = \text{Ceiling}\left(\frac{b}{L}\right) \tag{6b}$$

with:

$$b = d'_{max} - w_g \times \frac{(i \times L)}{MD_g} \tag{6c}$$

and, therefore:

$$j'_{max} = \text{Ceiling}\left(w_g \times \frac{(\sqrt{2}-1) \times i + \sqrt{2}}{MD_g} + \frac{w_{ng}}{L}\right) \tag{6d}$$

In this case, the upper bound dependents not only on the previous number of iterations but also on the relative weights of the geometric and non-geometric components of the distance. Note that, if $w_g = 1$ and $w_{ng} = 0$, the upper bound is the same as the one in Eq. (3d). On the other hand, if $w_{ng} >> w_g$, then it might not make sense to adopt the approach here described since the additional number of iterations might grow up to the point where all cells have to be visited.

4 Validation and Performance Analysis

The initial SNN implementation has been modified to integrate the method described in the previous section for computing the list of k-nearest neighbours of each object.

For the validation of the correctness of the new approach and for the performance analysis, three datasets were used. The first data set, named randomDataset, is an artificial synthetic data set of 200,000 points, created by assigning uniformly distributed random values to the X and Y coordinates, within the interval [0, 100[. Uniformly distributed random values were also assigned to a third dimension, within the interval [0, 360[, representing the bearing (heading) of geo-referenced movement vectors. Table 1 shows an extract of this data set.

Table 1 First rows of the randomDataset

Id	X coordinate	Y coordinate	Bearing
0	55.879	38.755	324.633
1	0.844	28.266	285.327
2	53.108	55.240	258.103

Table 2 First rows of kidsDataset

Id	X coordinate	Y coordinate	Bearing	Timestamp	Kid_Id
2965	628912.06	5804220.93	229.39	14:00:07	893
2966	629373.92	5804219.69	146.92	14:00:04	751
2967	628914.63	5804226.56	4.27	14:00:28	893

Table 3 First rows of parkDataset

Id	Timestamp	X coordinate	Y coordinate	Bearing	Speed
1	06-08-2006 08:34	733085.78	5857745.59	0	0
2	06-08-2006 08:35	733091.45	5857748.59	62.11	0.37
3	06-08-2006 08:38	733094.15	5857781.96	23.6	3.19

The second and third datasets, named kidsDataset and parkDataset respectively, contain real movement data obtained through the use of several GPS receivers. These receivers were carried out by a group of pedestrians for some amount of time. For building the kidsDataset, recordings were taken at an almost constant rate of one point every 10 s, while for the parkDataset the most frequent recording period was between 3 and 4 s. The kidsDataset contains 13,507 points and its structure is depicted in Table 2. The parkDataset contains 140,446 points, with the structure shown in Table 3. The spatial distribution of both datasets is presented in Fig. 5. For calculating the distance between points, a function similar to the one in Eq. (4b) was used, using the Euclidean distance for the geometric component, and a BearingDistance function for the non-geometric component defined as:

$$D_{ng}(p_a, p_b) = \begin{cases} |b_1 - b_2|, |b_1 - b_2| \le 180 \\ 360 - |b_1 - b_2|, |b_1 - b_2| > 180 \end{cases} \tag{7}$$

Each one of these data sets were processed with the original SNN algorithm and then with F-SNN, using the same values for the input parameters, in this case k=10, $Eps = 3$, $MinPts = 7$. Two sets of tests were conducted. One using only the two spatial dimensions, and another using the spatial dimensions and the additional (bearing) dimension, with $w_g = 95\,\%$ and $w_{ng} = 5\,\%$ (as suggested in (Moreira et al. 2010)).

Figure 6 shows the processing time required to cluster an increasing number of records for the randomDataset. The value of M_x was set to 800 due to general good

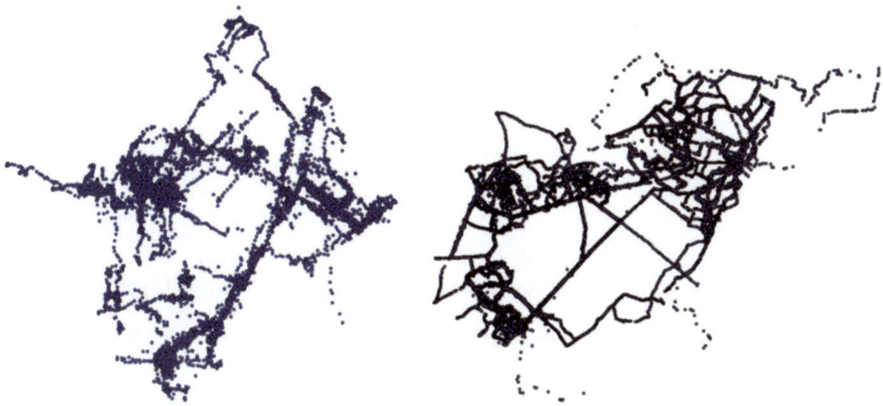

Fig. 5 Spatial distribution of the points in the kidsDataset (*left*) and parkDataset (*right*) data sets

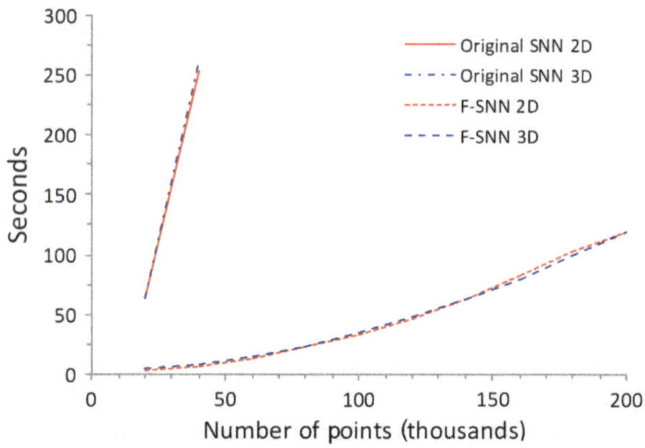

Fig. 6 Processing times for the randomDataset using the original SNN and the F-SNN algorithms (2D and 3D cases)

performance, as perceived on previous tests. Similar results are shown in Fig. 7 for the other two datasets.

The gains, in the processing time, are very clear and increase with the number of points. For example, in the case of the kidsDataset, the total time for clustering 8000 points reduces from 8.6 to 0.4 s, corresponding to a gain of around 24 for the 2D case, and a gain of 14 for the 3D case. Obviously, the obtained gain is dependent on the total number of points and on the number of cells defined by the M_x parameter. A higher value for M_x means a smaller size for each cell of the matrix that divides the data set. This causes fewer distances to be calculated in each expansion, but potentially increases the number of expansions to be made and introduces a new overhead due to the need to compute more expansions. Choosing the right M_x means choosing a compromise between these factors. The optimum value for M_x may vary for different

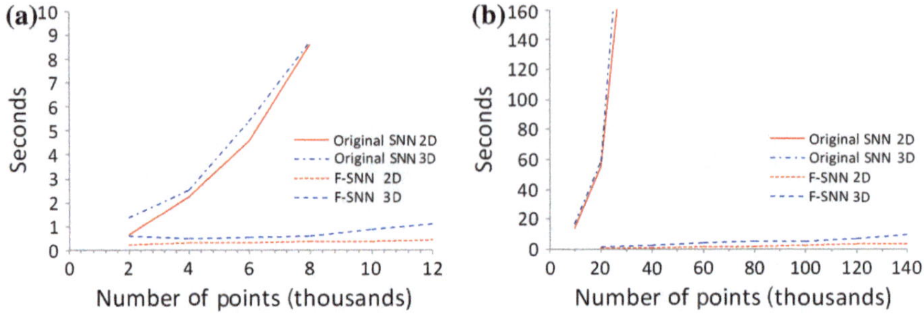

Fig. 7 Processing times for the two real data sets using the original SNN and the F-SNN algorithms (2D and 3D cases): **a** kidsDataset; **b** parkDataset

data sets. For the data sets described above, results obtained with values ranging from 300 to 1200 do not lead to significant changes in the processing time, meaning that the algorithm is not too sensitive to the value of this parameter. We also observed that the more uniformly the points are distributed in space, the less sensitive is the algorithm to the value of M_x. Note, though, that the value chosen for M_x only impacts the processing time, and not the clustering results.

5 Iterative Movement Data Analytics

Previous section presented three data sets, with different sizes and points' densities, which were used to assess the performance of the F-SNN algorithm. This section presents the analysis of the parkDataset illustrating how this algorithm can be used to analyse a geospatial data set and extract meaningful information from it. This demonstration case makes use of the parkDataset. As the size of the data set increases, the difficulty of identification of the appropriate input parameters, and tuning them to the analytical context, also increases.

For this data set, and having in mind that it integrates position readings for a group of pedestrians, this demonstration case will show how it can be clustered with the aim of identifying movement flows. The characteristics of the flows will vary attending to the input parameters that are used as well as the weighting factors that are set for each dimension of analysis.

The identification of the appropriate input parameters to start the clustering task is usually a trial and error process as no guidelines have been proposed to guide the analyst. Previous works on this matter identified some relationships between the input parameters (Ester et al. 1996, Ertoz et al. 2003), but those were tested in small data sets. When applied to large data sets those guidelines showed to be inappropriate. Recently, (Moreira et al. 2013) proposed specific guidelines that guide the identification of the input parameters of the SNN algorithm for large data sets. Using those guidelines, for a data set with 140,446 points, the first trial of the clustering process can start with $k = 210$, $MinPts = 197$ and $Eps = 37$. For the weighting factors, two

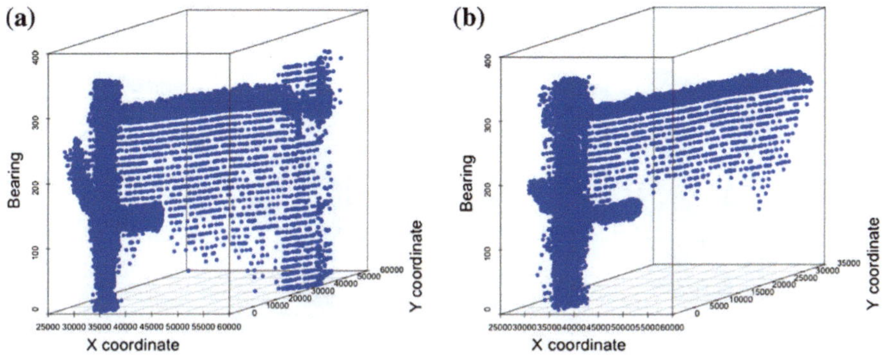

Fig. 8 Obtained clusters for the parkDataset, $k = 210$, $MinPts = 197$ and $Eps = 37$: **a** $w_{g1} = 95\,\%$, $w_{ng1} = 5\,\%$ (19,682 points), **b** $w_{g2} = 90\,\%$, $w_{ng2} = 10\,\%$ (15,620 points)

sets will be used to show the impact of the weighting factor in the clustering process, ($w_{g1} = 95\,\%$, $w_{ng1} = 5\,\%$) and ($w_{g2} = 90\,\%$, $w_{ng2} = 10\,\%$).

Clustering the parkDataset with these values allow the identification of 111 clusters, in both cases, with different sizes, shapes and densities. Plotting all the clusters, although possible, adds no analytical value due to the difficulty in having, in a 3D plot, different colours for each cluster. Instead, the clusters with more points in each case are plotted (Fig. 8).

These clusters are plotted in a 3D perspective that emphasises points including several bearing values. The identified flows include intersection zones where pedestrians can follow different paths. In the analysed area, one of these paths is followed until another intersection point is reached. As can be seen, given more weight to the bearing component diminishes the variety of bearing values in zones of low points density.

The main role of the k input parameter is to control the granularity of the clusters. As k increases, the size of the clusters would also increase. At this point, the analyst can change the input parameters and adjust the clustering results. This process can be done in a trial and error approach, or the guidelines suggested by Moreira et al. (2013) can again be used. In this work, (Moreira et al. 2013) shown that the input parameters present a repetitive pattern that maintains the characteristics of the clusters. If we want to maintain the characteristics presented above, only reducing the granularity of the clusters, the analyst needs to divide the initial adopted parameters by a factor of 2. In this case, with $k = 105$, $MinPts = 98$ and $Eps = 18$, 193 clusters are obtained for (w_{g1}=95%, $w_{ng1} = 5\,\%$) and 170 clusters for ($w_{g2} = 90\,\%$, $w_{ng2} = 10\,\%$). As the number of points in each cluster decreases, the number of different clusters must increase. After the analysis of the several clusters, Fig. 9 presents the two obtained larger clusters.

The several results obtained with this demonstration case show the importance of the input parameters in the clustering process, as well as the impact of the weighting factors in the clustering results. Moreover, the processing time needed to run several trials, with different parameters' values, also depends on the input parameters. As k

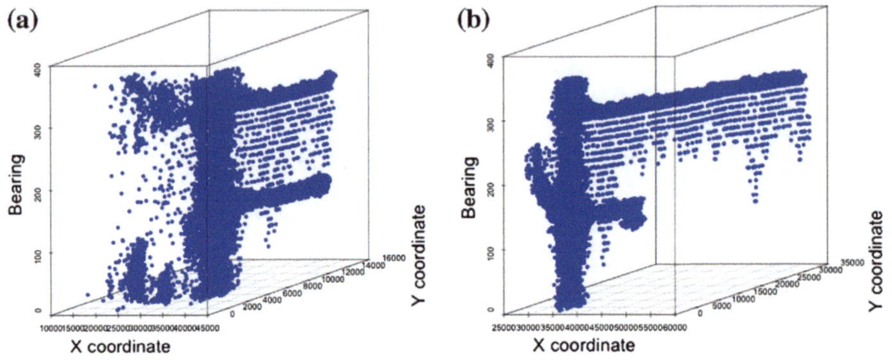

Fig. 9 Obtained clusters for the parkDataset, $k = 105$, $MinPts = 98$ and $Eps = 18$: **a** $w_{g1} = 95\,\%$, $w_{ng1} = 5\,\%$ (11,974 points), **b** $w_{g2} = 90\,\%$, $w_{ng2} = 10\,\%$ (14,990 points)

Table 4 Processing times for different sets of the parameters' values

	$w_g = 95\,\%$, $w_{ng} = 05\,\%$	$w_g = 90\,\%$, $w_{ng} = 10\,\%$
k = 210, MinPts = 197, Eps = 37	37 m 36 s	40 m 19 s
k = 158, MinPts = 147, Eps = 27	16 m 39 s	17 m 00 s
k = 105, MinPts = 98, Eps = 53	04 m 58 s	05 m 08 s
k = 79, MinPts = 74, Eps = 14	02 m 19 s	02 m 27 s
k = 53, MinPts = 49, Eps = 9	00 m 53 s	01 m 01 s

increases, the time needed to compute the clusters also increases. Table 4 shows the processing time required for the several runs.

These results show the importance of having a fast algorithm for the clustering, as it provides the flexibility to test several sets of parameters' values during the analysis task. Although the processing time depends on the used parameters, they are short enough to allow this kind of iterative analysis procedure.

6 Conclusions and Future Work

Density-based algorithms, like SNN, have good applicability in spatial data analysis and proved to find valuable results. However, spatial data sets often reach huge sizes, which allied to the quadratic complexity of the SNN algorithm, make the clustering task very time consuming. The approach proposed in this chapter allows the identification of the same clustering results while significantly reducing the time needed to compute the clusters. Moreover, this work shows how more than two dimensions can be included in the clustering process without compromising the efficiency of the algorithm. The major contribution is on the possibility of performing the analysis of very large data sets, by experimenting with several sets of parameters' values, within a reasonable amount of time.

Future work includes the automatic identification of the optimum value for the M_x parameter, which is related to the data set size and the spatial distribution of points.

The code used to implement the algorithm described in this chapter, and used to obtain the results reported in Sects. 4 and 5, is freely available for other researchers at http://ubicomp.algoritmi.uminho.pt/projects/f-snn. Although it was not used for computing the results reported here, this implementation is capable of exploiting the several cores available in more recent microprocessors, further improving the processing time through the parallel processing of cells.

Acknowledgments This work was supported by FCT– *Fundação para a Ciência e Tecnologia*, within the Project Scope: PEst-OE/EEI/UI0319/2014.

References

Bentley JL (1975) Multidimensional binary search trees used for associative searching. Commun ACM 18(9):509–517

Bhavsar HB, Jivani A G (2009) The shared nearest neighbor algorithm with enclosures (SNNAE). In: 2009 WRI world congress on computer science and information engineering. doi:10.1109/CSIE.2009.997

Chávez E, Navarro G, Baeza-Yates R, Marroquín JL (2001) Searching in metric spaces. ACM Comput Surv 33(3):273–321

Ertoz L, Steinbach M, Kumar V (2002) A new shared nearest neighbor clustering algorithm and its applications. In: Proceedings of the workshop on clustering high dimensional data and its applications at 2nd SIAM international conference on data mining, p 105–115

Ertoz L, Steinbach M, Kumar V (2003) Finding clusters of different sizes, shapes, and densities in noisy, high dimensional data. In: Proceedings of the 3rd SIAM international conference on data mining

Ester M, Kriegel HP, Sander J, Xu X (1996) A density-based algorithm for discovering clusters in large spatial databases with noise. In: Proceedings of 2nd international conference on knowledge discovery and data mining, KDD 1996, p 226–231

Faustino B (2012) Implementation for spatial data of the shared nearest neighbour with metric data structures. M.Sc. thesis, New University of Lisbon

Jarvis RA, Patrick EA (1973), Clustering using a similarity measure based on shared near neighbors. IEEE Trans Comput C 22(11):1025–1034. doi:10.1109/T-C.1973.223640

Li X, Jiang SY, Su XK (2009) A novel fast clustering algorithm. In: Proceedings of the international conference on artificial intelligence and computational intelligence, pp 284–288

Moreira A, Santos MY, Wachowicz M, Orellana D (2010) The impact of data quality in the context of pedestrian movement analysis. In: Painho M, Santos MY, Pundt H (eds) Geospatial thinking. Springer, Berlin, pp 61–78

Moreira G, Santos MY, Moura-Pires J (2013) SNN Input Parameters: how are they related? In: Proceedings of the 19th IEEE international conference on parallel and distributed systems (ICPADS'2013). Seoul, Korea, pp 15–18 (IEEE computer society, December)

Traina C, Traina A, Faloutsos C, Seeger B (2002) Fast indexing and visualization of metric data sets using slim-trees. IEEE Trans Knowl Data Eng 14(2):244–260

Tripathy A, Maji SK, Patra PK (2011) FDCA: A fast density based clustering algorithm for spatial database system. In: Proceedings of the 2nd international conference on computer and communication technology, pp 21–26

RSS and Sensor Fusion Algorithms for Indoor Location Systems on Smartphones

Laia Descamps-Vila, A. Perez-Navarro and Jordi Conesa

Abstract Location-based applications require knowing the user position constantly in order to find out and provide information about user's context. They use GPS signals to locate users, but unfortunately GPS location systems do not work in indoor environments. Therefore, there is a need of new methods that calculate the location of users in indoor environments using smartphone sensors. There are studies that propose indoor positioning systems but, as far as we know, they neither run on Android devices, nor can work in real environments. The goal of this chapter is to address that problem by presenting two methods that estimate the user position through a smartphone. The first method is based on euclidean distance and use Received Signal Strength (RSS) from WLAN Acces Points present in buildings. The second method uses sensor fusion to combine raw data of accelerometer and magnetometer inertial sensors. An Android prototype that implements both methods has been created and used to test both methods. The conclusions of the test are that RSS technique works efficiently in smartphones and achieves to estimate the position of users well enough to be used in real applications. On the contrary, the test results show that sensor fusion technique can be discarded due to bias errors and low frequency readings from accelerometers sensor.

L. Descamps-Vila · A. Perez-Navarro (✉) · J. Conesa
IT, Multimedia and Telecommunications Department, Universitat Oberta de Catalunya, Rambla del Poblenou 156, 08018 Barcelona, Spain
e-mail: aperezn@uoc.edu

L. Descamps-Vila
e-mail: ldescamps@uoc.edu

J. Conesa
e-mail: aperezn@uoc.edu

J. Huerta et al. (eds.), *Connecting a Digital Europe Through Location and Place*, Lecture Notes in Geoinformation and Cartography, DOI: 10.1007/978-3-319-03611-3_12,

1 Introduction

Location-based applications for smartphones require knowing the location of the user in order to provide information of user's context. That information can be easily gathered in outdoor environments by using Global Positioning System (GPS). However, GPS does not work in indoor environments. Inside a building, the GPS signal is attenuated and scattered by the walls. Therefore, there is a need of alternative location-sensing systems that are able to run on smartphones within indoor environments. Singh et al. 2013, Al Nuaimi and Hesham 2011, Ingram et al. 2004 or Ubisense (Woniak et al. 2013) present different techniques to develop indoor positioning systems.

Among the indoor-location systems, many are wireless driven, i.e. they use of Wireless Local Area Networks (WLAN) present in buildings to position the user: the measurement of Received Signal Strength (RSS) from WLAN Access Points (APs) available in the surrounding space allows to estimate the position of the user.

Fingerprinting is a technique that uses wireless technology to locate users in indoor environments. As Melkonyan 2011 explains, fingerprinting measures, in a preliminary stage called calibration, the RSS from APs of the area at known locations. Such readings are stored as a radio-map database. After that, users at any location can measure same signal features and try to find the statistical match with the radio-map entries and then to find their location on the map.

An algorithm used to estimate position using RSS fingerprinting technique is the nearest reference node that uses the euclidean distance metric (Teuber and Eissfeller 2006). This method compares the RSS values from different APs measured at well-know positions in calibration stage with RSS values measured in an unknown position. The difference between recorded values and current data can be computed as an euclidean distance. It is possible to determine which calibration node is closest to the current position and then convert this distance to a coordinate system.

Gansemer and Pueschel (2010) adapted the basic Euclidean distance algorithm to an environment with changing sets of base stations. However, under specific circumstances still individual heavily incorrect location estimations occur. In addition, the system is not tested on a smartphone.

Ferris et al. (2007) presented Wifi-SLAM project based on a Gaussian Process Latent Variable Model to determine locations using signal strength data. Their library was also used to develop a GIS indoor positioning application (Descamps-Vila et al. 2013). Even though Wifi-SLAM could be integrated in an Android application, calibration data and most of system functionalities were hosted in a web server owned by a company that no longer offers the service. Thus, today the system is no longer available.

As Jung et al. (2012) explain, WiFi-based indoor positioning involve some risks, since in a real situation the infrastructure of WLAN is not controlled. This may lead to different drawbacks: (1) the number of APs emitting is uncontrolled, some APs may be active during the calibration phase but may be inactive when the user estimates the position (see Lee et al. 2013); (2) RSS of each AP is variable and unstable due to reflections, diffractions, multi-path effects, the amount of people or simply the placement of the furniture (Gansemer and Pueschel 2010); (3) user may hold on the smartphone in different position/direction during the calibration and during the measurement stage, which derives in different RSS values (Shala and Rodriguez 2011).

Although there are several methods to estimate indoor location through RSS signals (Lee et al. 2013), (Teramoto et al. 2011), (Hui et al. 2007), as far as we know, there is not any indoor positioning system ready-to-use in smartphones as end-user application and not only in a test and controlled space. To face this problem, the main objective of this work is to design indoor positioning algorithms that can be used as location-sensing systems in smartphones within any indoor space.

The chapter proposes two different algorithms to estimate the user's position in indoor environments. The first one follows a fingerprinting technique, using RSS from WLANs present in the building. This means that the system must overcome problems of WiFi-based indoor positioning and problems of operation performance on smartphones when positioning through WLAN (see Melkonyan 2011).

It is important, also, to preserve accuracy when estimating position. Lee et al. (2013) show that the accuracy of Euclidean distance algorithm can be improved if it is possible to reduce the instability of AP signal strength. Hence, our algorithm has as starting point the algorithm of the Euclidean distance with additional improvements to avoid unstable signals. The second method proposed to estimate user location is the sensor fusion technique, which merge data obtained from different inertial sensors (Woodman 2007), such as accelerometers, gyroscopes and magnetometers, to estimate the movement of the user.

The chapter is structured as follows: Sect. 2 explains the proposed indoor positioning algorithms and an Android prototype that implement them. Section 3 presents the tests done to each system to determine whether they can be used in real situations. Finally, Sect. 4 presents the conclusions and further work.

2 Indoor Algorithm

The indoor location algorithm uses two different techniques: fingerprinting and sensor fusion. The first subsection details how the RSS algorithm estimates user position and the second subsection explains in detail how raw data obtained from smartphone sensors can be combined to estimate user location.

2.1 RSS Algorithm

As explained in Sect. 1, fingerprinting technique is based on the measurement of RSS from different APs. RSS from APs can fluctuate for several reasons such as: a change on the number of people in a room, collisions of radio waves, orientation of mobile receiver, etc. These fluctuations can cause errors on estimating the position. To avoid them, we propose to store only stable data in front of those fluctuations.

As fingerprinting technique is based on two phases, calibration and location, following subsections detail how the fluctuations on both stages can be avoided.

2.1.1 Calibration

On calibration stage there are measurements of RSS at different locations of the building, where there are multiple APs that emit radio signals with different intensities, namely, different levels of RSS.

We call *Node*, N, every location where there is a measurement of RSS during calibration stage. Each node is related to a location in 3D (x, y, z coordinates): $N \in \Re^3$. Every node has associated the measurements of several AP signals: $N_i = [lc_1, ..., lc_k]$ where i is the index of the node N, k is the number of APs visible from the node N and lc_j (level calibration) is the RSS level for each AP (index j) measured in dBm.

In order to store reliable values as calibration data, the algorithm uses two thresholds to discard unstable values. The first one is based on the standard deviation in multiple measurements and the second one limits the number of stored values.

Standard deviation upper threshold

Since values of RSS are very unstable, the measurement of RSS values is repeated n times in each node N_i. Thus, there are n measurements of each AP from each node:

$$N_i(n) = [[lc_{i,1}{}^1, ..., lc_{i,1}{}^n], ..., [lc_{i,k}{}^1, ..., lc_{i,k}{}^n]] \tag{1}$$

where index i represents each Node and index j each AP.

Considering n measurements of RSS values, the standard deviation of RSS values of each AP (index j) by each node (index i) is:

$$\sigma_{i,j} = \sqrt{\frac{1}{n-1}\sum_{r=1}^{n}(lc_{i,j}{}^r - \bar{lc}_{i,j})^2} \tag{2}$$

where the mean is defined by:

$$\bar{lc}_{i,j} = \frac{\sum_{r=1}^{n} lc_{i,j}{}^r}{n} \tag{3}$$

The standard deviation is a measure of statistical dispersion. Hence, the algorithm discards APs with high values of $\sigma_{i,j}$ because it means that those APs have many fluctuations. Only the results of APs that are under a threshold are used for the calibration. This threshold has been calculated through the following test.

Table 1 APs standard deviation

	AP1	AP2	AP3	AP4	AP5	AP6
σ (m)	2.134	2.438	2.174	1.569	2.007	2.038

We did 20 measurements of RSS values at a static position. We stored data of six controlled APs. Then, we calculated the standard deviation of each AP. Results are displayed in Table 1.

Results show that all standard deviations are below 3.0 m. Then, the upper threshold to discard unstable measures is 3.0:

$$\textit{if } \sigma_{i,j} \geqslant 3.0 \Rightarrow \bar{lc}_{i,j} \textit{ values discarded} \quad (4)$$

$$\textit{if } \sigma_{i,j} < 3.0 \Rightarrow \bar{lc}_{i,j} \textit{ used for calibration} \quad (5)$$

Note that the value of AP signal stored for the calibration map is the mean obtained in Eq. 3.

Limit the number of APs by Node

In some indoor spaces there are many AP signals detected from a single Node, N, sometimes there are over 40 AP signals. Although the algorithm discards the APs measurements that have a standard deviation value over a limit, the calibration map does not need so much data for each Node. On the other hand, as the system runs in a smartphone, the calibration matrix should be of a dimension as low as possible in order to fit the memory requirements of smartphones. Thus, we need to reduce the number of data of calibration and therefore we choose the more stable candidates to calibration data.

To address this issue, standard deviations $\sigma_{i,j}$ of measured APs in a single node N_i are ordered in descendant order. Note that each AP is defined by AP_j and the number of visible APs is defined by k. Thus, the algorithm stores a maximum of k_{max} of AP_j values.

As a result, the $\bar{lc}_{ij}$ with standard deviation $\sigma_{i,j}$ that are in a lower position than k_{max} are discarded because are more unstable than the others. This filtering avoids memory problems and increases performance efficiency, because it limits the size of the calibration matrix.

Calibration matrix

The measurements of RSS that passed both thresholds, define the calibration map, which is modelled as a matrix with the means of RSS levels:

$$Cal\_Matrix = \bar{lc}_{ij} \quad i \in [0, s],\ j \in [0, k] \quad (6)$$

where index i refers to the Node, index j refers to the index of AP, s is the total number of nodes and k the number of APs. Note that the number of nodes in the calibration stage can be defined by the user, but the number of detected APs depends on the WLAN infrastructure of the building and their surroundings.

As explained previously, each Node i has a maximum of k_{max} associated APs. However, each Node could detect different APs. Then, the matrix dimension will not be k_{max}, it will have a variable dimension. In addition, whether a Node, N_i, does not detect an AP_j that is already detected by another Node, the RSS value lc_{ij} will be null, because the Node N_i does not have RSS from the AP.

2.1.2 Location

Once the calibration map is created it is possible to estimate user location. When the user is at an unknown location, the system measures RSS values of APs from the unknown position.

We define the unknown position of the measurement as P, which is associated to a location in x, y, z coordinates: $P \in \Re^3$. Every position P has associated the measures of different AP signals and it is defined as: $P = [lp_1, ..., lp_m]$ where m is the number of APs visible from position P and lp_m (level position) is the RSS value for each AP measured in dBm.

As it is done in calibration stage, the algorithm applies the standard deviation threshold. The system takes n measurements of RSS from a single position P:

$$P(n) = [[lp_1^1, ..., lp_1^n], ..., [lc_m^1, ..., lc_m^n]] \tag{7}$$

$$= \sum_{j=1}^{m} (lp_j{}^1, ..., lc_j{}^n) \tag{8}$$

where index j represent each AP.

To discard unstable measurements, the algorithm apply Eqs. 2–5standard deviation threshold, but only by the index $i = 1$, because the system does measurements from one position P when estimating the location. In this case, instead of a matrix, the results are a list with the means of RSS levels:

$$Measure\_List_j = \bar{lp}_j \quad j \in [0, m] \tag{9}$$

where index j refers to the index of AP and m is the number of APs that passed the standard deviation threshold.

2.1.3 Positioning Through RSS Values

As Teuber and Eissfeller (2006) explains, the RSS signal ratio differences can be expressed in a signal ratio difference vector in which the number of elements represents the number of AP. The ratio difference between nodes N_i and position P can be obtained by euclidean metric:

$$\mathrm{l(P, N_i) = l_i} = \sqrt{\sum_{j=1}^{r} (\bar{lc}_{ij} - \bar{lp}_j)^2} \tag{10}$$

where $\bar{lc}_{ij}$ is the data from the calibration map of Eq. 6, $\bar{lp}_j$ is the location data obtained from Eq. 9 and i is the index of each node used in the calibration stage.

As this algorithm is designed for any indoor environment, it may happen that the detected APs from a node/location may not be always the same. For instance, whether the system makes five measurements in each Node, Node 1 (i=1) may have 5 values for AP_1 (j = 1): $N_1(5) = [lc_{1,1}{}^1, lc_{1,1}{}^2, ..., lc_{1,1}{}^5]$ and four values for AP_2 (j=2): $N_1(4) = [lc_{1,2}{}^1, lc_{1,2}{}^2, ..., lc_{1,2}{}^4]$ because one measurement has not detected the signal of AP2.

Thereby, the algorithm only introduces values in Eq. 10 when there are RSS values from the same APs. This restriction is important to avoid adding non-existent distances. Hence, in Eq. 10, index r defines the coincident APs with signals measured both in calibration and location stages: $AP_r = AP_k \cap AP_m$.

The drawback of taking only the r coincident APs is that we can obtain a smaller distance l_i when there are few APs coincidents than when there are many. Thus, each distance l_i is divided by the number of r values in order to normalize l_i value according to the number of coincident APs. If there is not any coincidence, namely, AP_r = 0, the distance cannot be calculated and it is necessary to perform new RSS measures.

Finally, to estimate the position P of a particular location with respect to the coordinate system, Teuber and Eissfeller (2006) suggest using the weighted mean of the coordinates of the closest q calibration Nodes. The selected q Nodes are the ones that have a lower value of l_i. Equations 11 to 13 show how to calculate the coordinates of the position:

$$x = \frac{1}{\sum_{i=1}^{q}(\frac{1}{l_i})} \cdot \sum_{i=1}^{q}(\frac{x_i}{l_i}) \tag{11}$$

$$y = \frac{1}{\sum_{i=1}^{q}(\frac{1}{l_i})} \cdot \sum_{i=1}^{q}(\frac{y_i}{l_i}) \tag{12}$$

$$z = \frac{1}{\sum_{i=1}^{q}(\frac{1}{l_i})} \cdot \sum_{i=1}^{q}(\frac{z_i}{l_i}) \tag{13}$$

where l_i is obtained from Eq. 10 and x_i, y_i, z_i are the coordinates associated to each Node N_i when doing the calibration.

Thus, the position obtained using RSS algorithm is:

$$\mathbf{r}_{RSS} = (x, y, z) \tag{14}$$

At this point we have a first guest of the position. To improve it, in the next step we introduce data from inertial sensors to develop a sensor fusion system.

2.2 Sensor Fusion

Inertial sensors such as accelerometers, gyroscopes and magnetometers can be used to determine the user's movement and location. Thus, these inertial motion sensors may provide data to improve the accuracy of the position obtained with Eq. 14.

Following subsections explain how to estimate the distance traveled and the direction of movement of users using sensor fusion of accelerometer and magnetometer data.

2.2.1 Position Estimation

The accelerometer of a smartphone offers the values of the acceleration (m/s^2) of the device through measurement of the forces applied on the sensor. It provides the values of acceleration in three dimensions of space (x, y, z). Google (R) explains on the documentation that internally the accelerometer sensor used this equation for every single axis:

$$a_i = g_i - \sum_{j=1}^{q} \frac{F_i^j}{m} \tag{15}$$

where i represents each of the coordinate axis, g is the value of the force of gravity, F the force acting on the device, m is the mass of the device and q the number of forces that act on the device.

Equation 15 shows that accelerometer row data includes the gravity force. To measure the real acceleration of the device, the contribution of the force of gravity must be removed from the accelerometer data. The simplest way is to use a high-pass filter to isolate the force of gravity and obtain what is called the linear acceleration, then:

$$a_i' = a_i - g_i \tag{16}$$

where the index i represent each coordinate axis.

Once there is a vector with the acceleration on the three axes, we can estimate the position after some time. It is necessary to apply the equations of cinematics of an object that has a linear movement accelerated. It is shown in Eq. 17:

$$\mathbf{r}_a = \sum_{i=1}^{n} \mathbf{r}_{i-1} + \frac{1}{2}\mathbf{a'}_i \Delta t_{0,i}^2 \tag{17}$$

where $\mathbf{r}_a = (x, y, z)$ is the position after a time Δt, $\mathbf{r_0}$ is the initial position, and a is the acceleration obtained from Eq. 16.

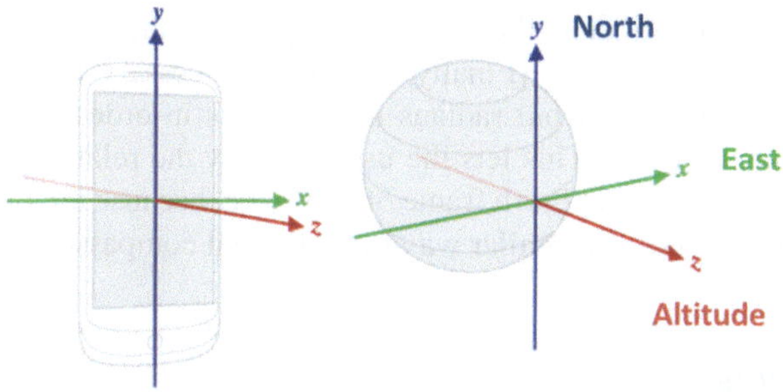

Fig. 1 Smartphone coordinate system (*left*). World coordinate-system defined by Google (*right*)

To obtain the initial position, $\mathbf{r_0}$, we consider that we are still and just in a point where we can obtain the position via GPS or just in a calibration node. Thus, from the starting point and with the accelerometer, we can obtain a new guess of the position: $\mathbf{r_a}$.

2.2.2 Movement Direction

Besides the position estimation, the accelerometer sensor could be useful to determine orientation of movement. The magnetometer measures the strength of the ambient magnetic field in $\mu Tesla$, in x, y, z axes. The combination of both sensors data provides information from the orientation of the device.

Magnetometer and accelerometer measure data in the device coordinates system, but to know the direction the user is moving, it is necessary to know values respect real-world coordinate system. Thereby, raw data obtained from the smartphone must be transformed into another system, which can be used to position the user. This system is defined by Google as *worlds coordinate system*.

The left side of Fig. 1 represents the smartphone's coordinate system. On the right side there is the representation of the coordinate system that Google defined as world's coordinate system. In this coordinate system, x is tangential to the ground at the device's current location and points approximately East; y is tangential to the ground at the device's current location and points toward the geomagnetic North Pole and z points toward the sky and is perpendicular to the ground plane and represents altitude. Both drawings are extracted form Google documentation.

Then, using a rotation matrix defined as an Android internal method, it is possible to transform any vector from the smartphone's coordinate system to the world's coordinate system or vice-versa. The processing of data with the rotation matrix derives into the orientation of the smartphone respect world's coordinate system.

However, the positioning system requires to know the azimuth angle. This is the angle between magnetic north and the device's y axis (Fig. 1). When the device's y axis is aligned with magnetic north this value is 0, if device's y axis is pointing south this value is 180°, when pointing east the value is 90° and when it is pointing west this value is 270°.

To obtain the azimuth angle the system can apply another internal method from Android, called getOrientation. Finally, it is only necessary to transform values obtained from this method from radians into degrees in order to know the angle respect the magnetic north. This lets the user to know the relative direction that is oriented the device respect to Earth frame of reference, because the Earth's magnetic north is known. It works in a similar way to traditional compasses.

2.3 Prototyping

We implemented the algorithms and methods described in this section on an Android prototype. The objective is to test whether all of them can be used in an Android smartphone and whether it is possible to estimate the position in an indoor environment.

The main functionalities of the prototype are: to display the available floor plan of buildings; to calibrate the floor plan with RSS values (see Sect. 2.1); to measure the position of the Android device using the accelerometer (see Sect. 2.2.1); to show the direction of movement of the user respect north magnetic pole like a compass (see Sect. 2.2.2); to show the estimated user position over the floor map.

3 Testing

The smartphone used to perform the tests is a Galaxy Nexus with Android 4.3. The first subsection presents tests performed on the prototype with the RSS algorithm and the second subsection shows the tests performed using sensor fusion.

3.1 RSS Tests

This subsection shows the tests performed with the prototype of the RSS positioning algorithm described in Sect. 2.1 within the Android device. We did positioning tests in two different buildings in order to analyze the behavior of the system in two different environments. One place is a flat (Building A) and the other is an office building (Building B). In these tests we used only one floor for each building, what means that the coordinate z is always equal to 0. In addition we dissociated the Wi-Fi connection of the smartphone.

The metrics for the tests are the same in each building. For the calibration we measured 40 nodes: $s = 40$. In each node we repeated the RSS measurement 5 times: $n = 5$. The limit of AP signal values stored by the matrix is 15 for node, $k_{max} = 15$. The standard deviation threshold is 3.0, as defined in Sect. 2.1.1.

We measured 10 different positions in each building. In order to have statistical values, we estimated the position of each one 10 times (therefore, we had 100

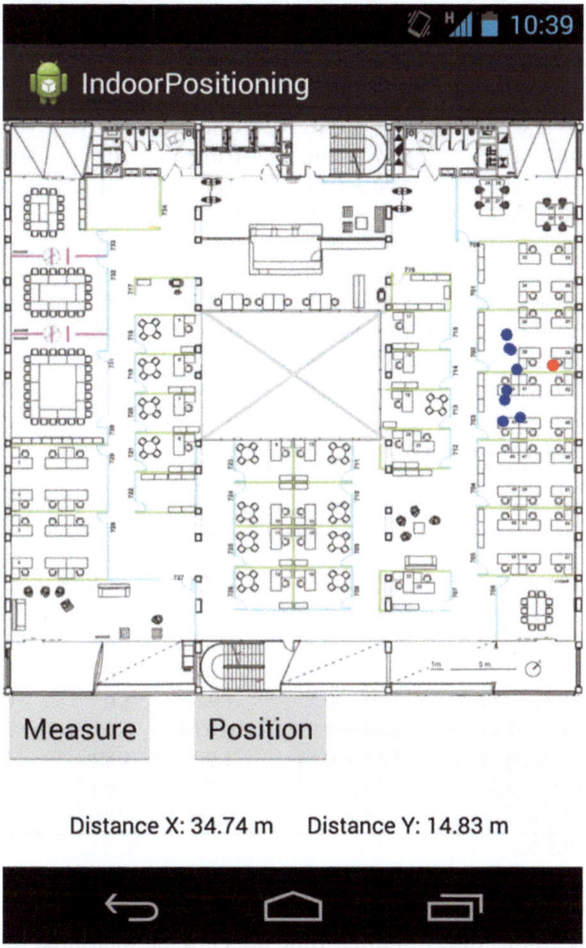

Fig. 2 Building B—Position 4

measurements for building). For each estimation of the position, the algorithm took 5 measurements, $n = 5$ and it used the fourth nearest nodes to calculate the x, y coordinates, $q = 4$.

3.1.1 Results

Table 2 presents the results of building A and Table 3 presents the results of Building B. Tables show the mean of the 10 measurements for each position, the error, the standard deviation and the accuracy respect to the theoretical x, y position.

Table 2 Positioning tests in building A

Position	Mean ± error (m)		σ (m)		Accuracy (m)	
	x_p	y_p	σ_x	σ_y	x	y
1	6.94 ± 0.09	9.31 ± 0.16	0.30	0.52	−0.29	0.38
2	3.40 ± 0.01	9.97 ± 0.01	0.01	0.01	−0.53	−0.96
3	4.79 ± 0.02	7.70 ± 0.01	0.06	0.01	−0.23	−0.25
4	5.53 ± 0.22	3.99 ± 0.10	0.68	0.33	1.25	0.42
5	8.65 ± 0.35	3.97 ± 0.19	1.11	0.61	−0.78	−0.65
6	6.42 ± 0.14	8.42 ± 0.09	0.44	0.29	−1.27	1.06
7	5.33 ± 0.18	5.11 ± 0.17	0.57	0.55	−0.24	1.18
8	6.92 ± 0.04	11.07 ± 0.01	0.13	0.02	0.42	−0.26
9	8.03 ± 0.18	1.83 ± 0.09	0.59	0.29	−1.23	0.66
10	6.89 ± 0.08	3.25 ± 0.08	0.26	0.24	0.32	1.30

Table 3 Positioning test building B

Position	Mean ± error (m)		σ (m)		Accuracy (m)	
	x_p	y_p	σ_x	σ_y	x	y
1	35.04 ± 0.31	7.01 ± 0.08	0.99	0.26	−3.02	−0.43
2	32.81 ± 0.32	10.25 ± 0.58	0.92	1.64	0.64	−1.08
3	30.59 ± 0.32	12.48 ± 0.42	1.00	1.31	3.59	2.76
4	34.92 ± 0.12	17.34 ± 0.76	0.37	2.41	−3.08	5.01
5	34.50 ± 0.32	26.76 ± 0.90	1.01	2.84	−3.33	−1.18
6	36.05 ± 0.57	32.90 ± 0.52	1.82	1.63	−0.84	−2.77
7	7.15 ± 1.91	22.63 ± 1.01	6.05	3.19	3.43	−2.26
8	5.85 ± 0.61	15.52 ± 0.59	1.92	1.89	1.79	1.80
9	28.65 ± 0.26	26.36 ± 0.82	0.82	2.60	−0.18	−4.20
10	23.75 ± 0.08	23.58 ± 0.32	0.25	1.01	3.42	−1.36

Figures 2 and 3 show two screen-shots of the Android prototype with the tests on two different positions of a Building. Blue dots are the measured positions (results of Tables 2 and 3) and red dot is the theoretical position.

3.1.2 Analysis

The algorithm designed to estimate the position through RSS values provide reasonably good results. In building A results show an accuracy below 1.5 m, however, in building B accuracy is around 2–3 m, except in two positions where the accuracy is around 4–5 m.

The standard deviation of all results, except position 7 of building B, are within 1–2 m of deviation. Thus, the scattering of results is low, as intended in the design of the algorithm. Standard deviation is also better in building A than in building B.

Results show the position measurement is more precise and more accurate in building A. This may be due to two reasons: (1) the number of nodes calibrated is the same in the two buildings, $s = 40$, however building A is smaller than building

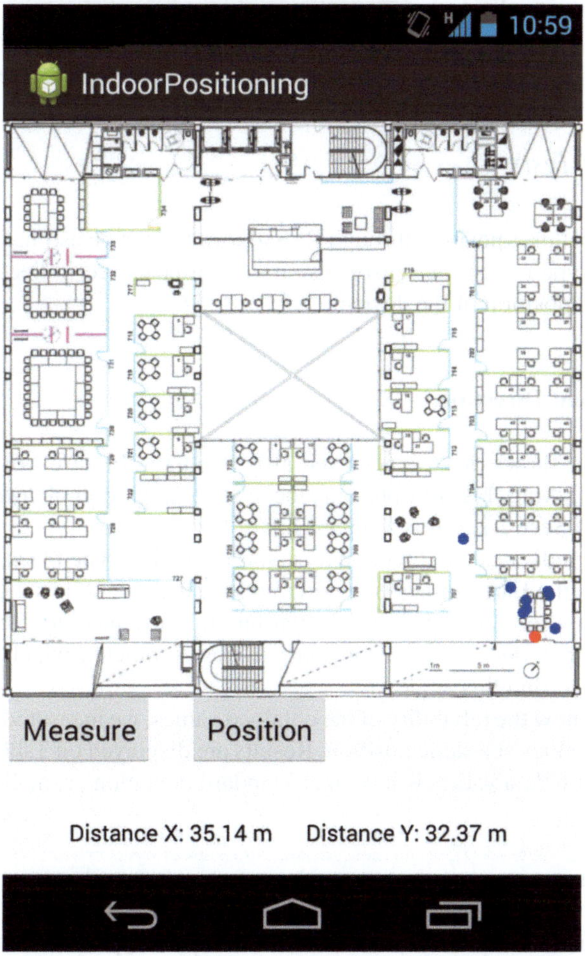

Fig. 3 Building B—Position 6

B and therefore there are more nodes calibrated by square meter which improves the calibration; (2) we have seen that in building A there are less number of APs detected, the ones detected are stronger and are always the same. Hence, as stated on Introduction section, the more stable and controlled are the APs, the better is the position estimation.

In addition, note that the theoretical point is not within the margins of error of the measured point. Therefore there are sources of error that are not controlled.

Table 4 Orientation measurements at rest

Mean of angle respect north (degrees)	σ
65.4795	0.8391

3.2 Sensors Tests

Once shown that user position through RSS algorithm can be estimated, this section uses the prototype to test the sensor fusion system proposed in Sect. 2.2 to get the orientation and position of the device.

3.2.1 Movement Direction

This subsection shows that we achieved also to obtain the direction of the device while user is walking. Figure 4 presents the system that we implemented following explanations of Sect. 2.2.2. It is a screenshot of the prototype.

The system indicates the direction a person is moving. The black line inside the circle indicates the Earth's magnetic north and the number shows the orientation to the north. In this case, the head of the smartphone is oriented at 65.5°. In addition, Fig. 4 displays raw data of accelerometer and magnetometer, which are the source data necessary to find this angle.

In order to know the reliability of the obtained values, we measured 2,000 samples of orientation values at a static position. Results are displayed on Table 4. They show the dispersion on data values is low since standard deviation is small.

3.2.2 Position

This subsection shows the tests performed with the prototype to estimate the distance traveled by the user, using the equations described in Sect. 2.2.1.

The prototype measured the acceleration while a user is holding the device horizontally and walking in only one direction. Hence, Eq. 17 is used only in one axis, in this case y axis of the smartphone coordinate system (refer to Fig. 1).

Figure 5 presents a graphic with the acceleration raw data measured in the three axes. Note that the raw acceleration from the sensor is the one defined in Eq. 15. The z axis shows the acceleration produced by the force of gravity over the device. y axis shows the acceleration values on the direction of movement because we have shown this direction as the moving axis, and x axis is perpendicular to the direction of movement. The variations shown on the graphic are due to step movement of the user. Hence, it is clear that this data is influenced by the force of gravity.

After isolating the force of gravity and applying Eq. 17 we measured 5 times the same distance.

The distance traveled by the user that is holding the device is 6 m. Results of the tests gave a distance equal to $d_a = 10.801 \pm 0.119\ m$. Results are far away from the

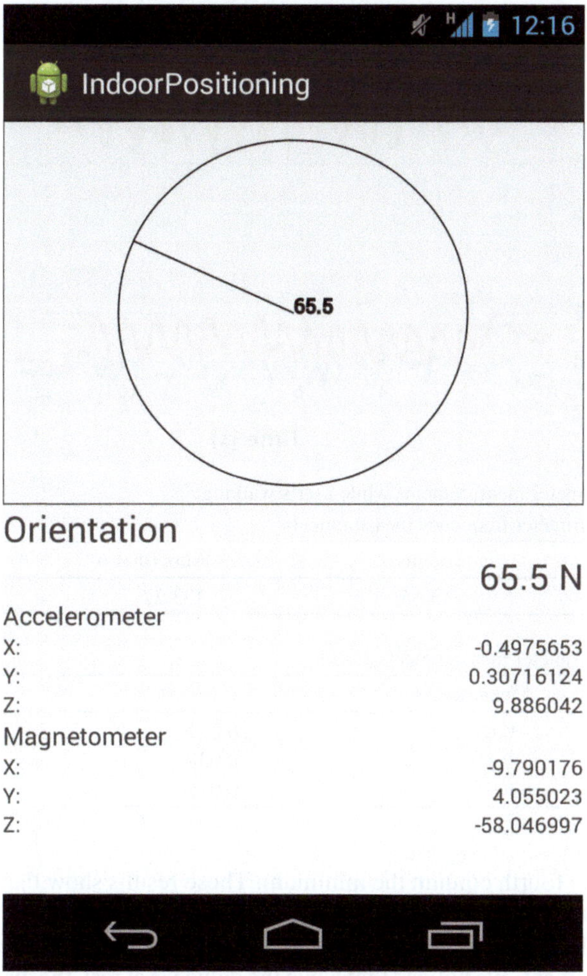

Fig. 4 Android prototype that shows device movement direction

real distance. These bad results can be caused by a low rate of update readings and bias errors.

If the frequency of data collection is low, the system loses information. For instance, imagine the last measured acceleration is low and then there is a significant increase. In case the sensor takes time to read these data, the information is lost and increases the error.

To test this hypothesis, we took measurements of 2,000 samples of accelerometer values in a static position. Table 5 shows the rate of readings of accelerometer values. The first column displays the mean of the frequency, which is about 60 ms, the second column shows the standard deviation, the third column the maximum value of update

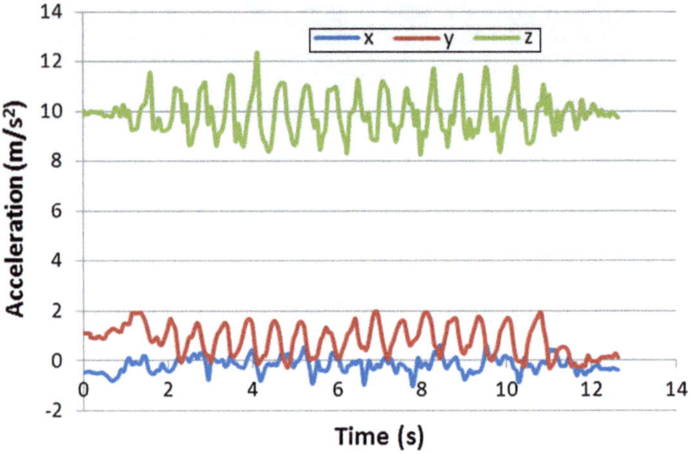

Fig. 5 Accelerometer measurements while user's walking

Table 5 Accelerometer frequency measurements

Mean (ms)	σ (ms)	Max (ms)	Min (ms)
63.73	13.65	133.27	3.81

Table 6 Accelerometer measurements at rest

Axis	Mean (m/s^2)	σ (m/s^2)	Accuracy (m/s^2)
x	−0.497	0.037	−0.497
y	0.119	0.038	0.118
z	9.919	0.049	0.109

reading and the fourth column the minimum. These results show that accelerometer sensors have a low rate of update readings and it is very unstable.

Additionally, the measured distance varies greatly depending on the time it takes to make the movement. We noticed that when the device is at rest, the distance increases constantly, instead of remaining constant. This may be due to the accelerometer have a initial bias error that corresponds to the offset of its output signal from the true value (see report from (Woodman 2007)). A constant bias error of accelerometer causes an error in position which grows quadratically with time.

We measured the bias of the accelerometer from the 2,000 samples measurements defined previously. Table 6 shows the results.

The average of acceleration values is different from the theoretical values (0, 0, 9.8). This is what makes the distance to increase constantly increases even though the device is at rest, because there are a constant bias of the accelerometer values that are integrated over time, the error is cumulative.

In order to avoid this issue, it is therefore necessary to subtract the bias error (mean values of Table 6) during the measurements. However, in a real application, it is not possible to calibrate each smartphone constantly to reduce the bias, because we have seen that the bias is not always the same for the same accelerometer.

After doing several sensor tests, we have seen it is possible to know the direction of movement, however, the accelerometer sensors have a low update reading results and bias error which derive into very bad distance measurement results. Hence, at the moment we avoid using accelerometer to estimate user position.

4 Conclusions

In this chapter two indoor positioning algorithms that can run entirely in a smartphone device have been presented. One of them uses RSS values of WLANs present in a building to estimate the user position. The algorithm is based on the euclidean distance method and it is designed to be used in a smartphone, taking into account the performance limitations of these devices. The other algorithm uses the sensor fusion technique, which combines inertial sensors raw data to detect the direction of movement and distance travelled by the user.

Both algorithms have been implemented in an Android prototype in order to demonstrate that they can work efficiently in such devices and to test their effectiveness. With the prototype, the RSS algorithm has been tested in two different buildings, giving good results. In fact, results show that most of the position measurements have accuracies around 1–3 m and standard deviations of 1–2 m. Comparing results of two buildings, we have seen that, the more nodes are measured during calibration and the stronger and stable the APs of the building, the better the results.

Sensor fusion technique allowed to find out what direction the user is moving, but tests show very bad results when calculating the user position with the accelerometer sensors. The problem seems to be that the accelerometer sensor has a big bias error and a low and unstable frequency of reading results, which derives on big errors when calculating the distance traveled by a user.

Therefore, the main contributions of this chapter are two fold. On the one hand, it presents a couple of algorithms that use smartphone sensors to locate users in indoor environments and demonstrate that they can be implemented and work efficiently in smartphones. On the other hand, it test both algorithms and identify some problems that worsen the quality of location estimation when using data from inertial sensors.

As future work, we plan to do tests on different floors of the two tested buildings, taking into account the z axis in order to analyze the accuracy and precision of the system over different floors. We also would like to establish a calibration standard that works as a guide for the users and to study different methods to address the errors in sensors such as the accelerometer.

Acknowledgments This article has been developed with the support of the Internet Interdisciplinary Institute of the Universitat Oberta de Catalunya.

References

Al Nuaimi K, Hesham K (2011) A survey of indoor positioning systems and algorithms. In: 2011 international conference on innovations in information technology, pp 185–190. doi:10.1109/INNOVATIONS.2011.5893813

Descamps-Vila L, Perez-Navarro A, Conesa J (2013) Integración de un sistema de posicionamiento indoor en aplicaciones SIG para dispositivo móvil. In: VII Jornadas SIG Libre de Girona, 1. http://www.sigte.udg.edu/jornadassiglibre/uploads/articulos_13/a29.pdf

Ferris B, Fox D, Lawrence N (2007) WiFi-SLAM using gaussian process latent variable models. In: International joint conference on artificial intelligence IJCAI, pp 2480–2485. http://www.aaai.org/Papers/IJCAI/2007/IJCAI07-399.pdf

Gansemer S, Pueschel S (2010) Improved RSSI-based euclidean distance positioning algorithm for large and dynamic WLAN environments. Int J Comput 9(1):37–44. http://archive.nbuv.gov.ua/portal/natural/computing/2010_1/PDF/10SGADWE.pdf

Hui L, Darabi H, Banerjee P, Jing L (2007) Survey of wireless indoor positioning techniques and systems. IEEE Trans Syst Man Cybern Part C 37(6):1067–1080

Ingram SJ, Harmer D, Quinlan M (2004) Ultrawideband indoor positioning systems and their use in emergencies. In: Position location and navigation symposium, PLANS 2004, pp 706–715. doi:10.1109/PLANS.2004.1309063

Jung WR, Bell S, Petrenko A, Sizo A (2012) Potential risks of WiFi-based indoor positioning and progress on improving localization functionality. In: International workshop on indoor spatial awareness. http://dl.acm.org/citation.cfm?id=2442621

Lee JY, Yoon CH, Hyunjae P, So J (2013) Analysis of location estimation algorithms for WiFi fingerprint-based indoor localization. In: The 2nd international conference on software technology, vol 19, pp 89–92. http://onlinepresent.org/proceedings/vol19_2013/23.pdf

Melkonyan A (2011) Integrity monitoring and thresholding-based WLAN indoor positioning algorithm for mobile devices. In: The 9th international system of systems engineering conference, pp 191–196. doi:10.1109/SYSOSE.2011.5966596. http://ieeexplore.ieee.org/lpdocs/epic03/wrapper.htm?arnumber=5966596. http://ieeexplore.ieee.org/xpls/abs_all.jsp?arnumber=5966596

Shala U, Rodriguez A (2011) Indoor positioning using sensor-fusion in android devices. Ph.D. thesis, school of health and society department computer science embedded systems. http://hkr.diva-portal.org/smash/get/diva2:475619/FULLTEXT02.pdf

Singh R, Sharma S, Mohan R (2013) A study and analysis of wireless based localization and motion processing systems for healthcare applications. Int J Emerg Technol Adv Eng 3(7):406–414. http://www.ijetae.com/files/Volume3Issue7/IJETAE_0713_68.pdf

Teramoto Y, Sato A, Asahara A, Tomita H (2011) Indoor positioning based on radio signal strength distribution modeling using mirror image method. In: Workshop on indoor spatial awareness, Nov, pp 15–22. http://dl.acm.org/citation.cfm?id=2077361

Teuber A, Eissfeller B (2006) WLAN indoor positioning based on Euclidean distances and fuzzy logic. In: Proceedings of the 3rd workshop on positioning, navigation and communication, pp 159–168. http://www.wpnc.net/fileadmin/WPNC06/Proceedings/31_WLAN_Indoor_Positioning_Based_on_Euclidean_Distances_and_Fuzzy_Logic.pdf

Woniak M, Odziemczyk W, Nagrski K (2013) Investigation of practical and theoretical accuracy of wireless indoor positioning system ubisense. Rep Geodesy Geoinformatics 95(0). http://www.reports.gik.pw.edu.pl/index.php/reports/article/view/233

Woodman O (2007) An introduction to inertial navigation. Technical Report UCAM-CL-TR-696. University of Cambridge, Computer Laboratory. http://www.cl.cam.ac.uk/techreports/UCAM-CL-TR-696.pdf

An Image Segmentation Process Enhancement for Land Cover Mapping from Very High Resolution Remote Sensing Data Application in a Rural Area

M. Vitter, P. Pluvinet, L. Vaudor, C. Jacqueminet, R. Martin and B. Etlicher

Abstract In this chapter, we describe a procedure for enhancing the automatic image segmentation for land cover mapping from Very High Resolution images. The increased need for large scale mapping (1:10000) for local territorial monitoring led to think about mapping production. Nowadays mapping production for land cover and land use (LUC) is mainly performed with human photo-interpretation. This approach can be extremely time consuming, expensive and tedious for data producers. This is confirmed from the evidence of rural areas where the use of the GIS database for LUC is less numerous than the GIS urban database. In the last decade, Geographic Object-Based Image Analysis (GEOBIA) has been developed by the image processing community. This new paradigm builds on theory, methods and tools for replicating the human photo-interpretation process from remote sensing data (Hay and Castilla 2008). However, the GEOBIA community is still fragile and suffers from a lack of protocols and standards for operational LUC mapping applications. Currently, human photo-interpretation seems a safer option. The objective of this research is to find an alternative to this time consuming and expensive use of human expertise. We explored the limits of GEOBIA to propose an automatic image segmentation enhancement for an operational mapping application. Questions behind this study were: What is a good segmentation? How can we obtain it?

M. Vitter (✉) · C. Jacqueminet · B. Etlicher
University of Saint-Étienne, ISTHME UMR 5600 EVS, 6 rue Basse des Rives, 42023 Saint-Étienne Cedex 02, France
e-mail: maxime.vitter@univ-st-etienne.fr

M. Vitter · P. Pluvinet · R. Martin
ASCONIT Consultants, 6-8 espace Henry Vallée, parc scientifique Tony Garnier, 69366 Lyon Cedex 07, France

L. Vaudor
ENS of Lyon, UMR 5600 EVS, 15 parvis René Descartes BP 7000, 69342 Lyon Cedex 07, France

J. Huerta et al. (eds.), *Connecting a Digital Europe Through Location and Place*, Lecture Notes in Geoinformation and Cartography, DOI: 10.1007/978-3-319-03611-3_13,

1 Introduction

From the onset of GIS in 1960, the needs for land cover and land use mapping (LUC) have been ever increasing. This is particularly evidenced by the creation of national and international mapping programs such as Copernicus, and from the updating of global repositories of land cover/use such as Corine Land Cover (CLC), Urban Atlas or, in France, the RGE *(Référentiel à Grande Échelle)* from the IGN (*Institut Géographique National*).

Currently, CLC is the European land cover reference for many studies on a scale of 1:100000. However, for the most part its scale and/or its typology are not convenient for local territorial management. With Very High Resolution (VHR) sensor development since 2000 and newly acquired drone technology, the scale needs for studies are between 1:10000 and 1:25000. Nomenclature should be more specific and data quicker to refresh. The mapping production of these recent requirements in environmental studies or mapping surveys are performed mainly by manual digitizing of aerial and/or satellite imagery. This method can be extremely expensive in terms of both time and money. This is confirmed by evidence from rural areas where the use of the GIS in connection with land cover or land use is less numerous than the GIS urban database. Therefore, questions about the accuracy and updating of spatio-temporal data over a large area have become essential research topics.

In the last decade, a new automatic image processing method called the Geographic Object-Based Image Analysis (GEOBIA) has been developed to adapt to VHR data and to improve the land mapping (Blaschke 2010). According to Hay and Castilla (2008), GEOBIA is

> "a subdiscipline of Geographic Information Science (GIScience) devoted to developing automated methods to partition remote sensing imagery into meaningful image-objects, and assessing their characteristics through spatial, spectral and temporal scales, so as to generate new geographic information in GIS-ready format."

While traditional pixel-based classification approaches have been widely used to map general land cover and detect changes for urban, forest, water or agricultural monitoring from High and Low Resolution satellites sensors (Landsat, SPOT), object-oriented approaches are not only dependent on the pixel information (Blaschke and Strobl 2001) but also on spatial image information to extract and identify land use features and man-made structures such as agricultural parcel size, house shape or forest texture (Blaschke 2010).These approaches can be considered finer mapping typologies and achieve satisfactory results when applied using VHR images (Schiewe et al. 2001; Baatz et al. 2004). The object-based approach aims to replicate the approach of classical manual digitizing (Blaschke and Strobl 2001; Schiewe et al. 2001; Benz et al. 2004; Kim et al. 2009). GEOBIA consists of two steps:

- Segmentation: extraction of different homogeneous areas from the image (agricultural land, forest, urban area).
- Classification: characterization of previously identified areas from spectral and spatial information.

Although, GEOBIA has been extremely popular in image processing for many reasons, there is still an important gap between the use of this new paradigm for LUC mapping from VHR images in research and its application in operational studies. The GEOBIA community is still fragile because there is a lack of protocols, formats or standards for a robust segmentation (Hay and Castilla 2008; Kim et al. 2009; Arvor et al. 2013). It is difficult for a project manager to find an easy and efficient solution to meet mapping expectations that are rarely the same (data, localization, time, aims etc.). All these aspects still lead consulting firms to choose manual approaches rather than semi-automatic solutions. The reasons for this are mainly the lack of skills and fear of using an expensive method without good results. Currently human photo-interpretation seems the safer option.

In their chapter, Hay and Castilla (2008) proposed an analysis to provide insight into the current state of GEOBIA. They described the weaknesses and threats involved in a GEOBIA Project. Of the weaknesses, we can underline several important points. The image segmentation is an ill-posed problem because it is a result of the choice of segmentation scale parameter. As with handmade digitizing, the object delimitation between the different image-segmentations will not be exactly the same (Arvor et al. 2013). Moreover, the relationship between image-objects and landscape-object involves an empirical acceptance. Based on the evidence, we can ask the following questions: What is good image segmentation? And how can we obtain it for an operational mapping application?

In this chapter, we propose to analyse the different steps of image segmentation. We have identified three limiting points in the image-segmentation process for use in an operational land cover and LUC application and derived the following questions:

- What is a suitable segmentation scale parameter?
- How can the geometry of image-objects be simplified?
- How can oversegmentation be reduced?

We consider the key points of image-segmentation key points and these limitations. In Sect. 3, the materials and method we propose are presented. Then, results obtained on each of the points by the method are presented and discussed in Sect. 4. Finally, the discussion and conclusions are in Sect. 5.

2 Image Segmentation

2.1 Image-Objects

Object-based image analysis starts by determining the steps of image segmentation. The image segmentation steps allow us to create “image-objects” (or segments). Castilla and Hay (2008) defined “image-object” as

> "a discrete region of a digital image that is internally coherent and different from its surroundings."

In theory, an image-object has no thematic meaning. It's just a "discrete" or "unique" entity with internal coherence and external contrast with neighbouring objects based on color, tone, texture, shape or size patterns. In practice, a GEOBIA expert has his own idea (implicit or not) of image-object he wants. Indeed, image-object is a result of choice of parameters based on expert knowledge and experience. Thus, image-object is a subjective concept. Unique cutting solution doesn't exist and image segmentation methods are rarely transferable (Arvor et al. 2013). Although, image-segmentation is a crucial step and it influences classification accuracy and quality (Dorren et al. 2003; Meinel and Neubert 2004; Kim et al. 2009).

2.2 Image Segmentation Process

There are two main approaches to the image segmentation process: the top-down method directed by knowledge and the bottom-up method directed by data (Baatz et al. 2004). The first method assumes that the object of study on the image is known; the model tries to find the best way to extract it. It is used to identify one or a few landscape elements. On the other hand, the second bottom-up method assumes that the study objects are not well-known. Image-objects are generated randomly from the whole image. This method can be considered as a clustering method, which means that image-objects have no thematic meaning. At this stage, it is best to call these "primitive image-objects" (Thomas 2005).Then, the identification of "primitive image-objects" is performed by the user. Generally, the bottom-up method is used to map large classes of LUC. It can be used to make a summary of pixels according to the criteria of homogeneity and heterogeneity (Baatz et al. 2004). Our research was based on a bottom-up approach because it seems to be appropriate for meeting most mapping expectations.

However, there are some limitations to the usage of image-segmentation, as has been pointed out. The three main limitations identified that affect the project manager in performing a land cover mapping are considered in turn as follows:

2.2.1 What Is the Suitable Segmentation Scale Parameter?

The image segmentation process requires different configuration settings depending on the software used. We used eCognition® Developer software. In this image analysis software, the segmentation scale parameter is most important, and it corresponds to the level of pixel aggregation. It is expressed as the allowable limit on heterogeneity. The higher the scale factor is, the larger the size of image-objects.

Nowadays, the optimal parameter for image segmentation remains problematic. There is no efficient method to determine a suitable scale for image segmentation according to aims of study (Kim et al. 2008; Drăgut et al. 2010). Usually, the evaluation of the scale parameter is performed by feedback on the study area and image. Around ten segmentations are necessary before confirming one (Meinel and Neubert 2004; Thomas 2005; Kim et al. 2008). Validation is often empirical (visual quality of objects, consistent with the aims of study). Although, several semi-automatic evaluation methods of image segmentation exist. Supervised Evaluation methods compare image segmentation with references usually produced by human interpretation (Neubert and Herold 2008; Marpu et al. 2009). Recently, much effort has been engaged into unsupervised evaluation methods to auto-adjust segmentation parameters (Zhang et al. 2012). Most of them are based on variance measure for each image-object. These values are averaged into global measure to provide an indication about image segmentation suitability. (Woodcock and Strahler 1987; Kim et al. 2008; Drăgut et al. 2010; Zhang et al. 2012; Drăgut et al. 2014)

Perfect image segmentation does not exist, as selecting a good segmentation is often a compromise between oversegmentation and undersegmentation (Castilla and Hay 2008). However, both these terms are subjective notions because they are defined by the practitioner's interpretation. An oversegmentation refers to a low spatial and spectral difference between several contiguous image-objects that should be merged. This phenomenon is accentuated when the segmentation scale parameter is low. By contrast, undersegmentation refers to a high spatial and spectral heterogeneity in an image-object, in which the object contains several landscape elements. However, we consider that oversegmentation is less problematic than undersegmentation, because in practice post-processing image-object aggregation is easier than image-object splitting. Thus, a little oversegmentation seems to be a better segmentation (Castilla and Hay 2008).

2.2.2 How Can the Geometry of Image-Objects Be Simplified?

VHR images are a valuable resource for LUC mapping. Thanks to submetric accuracy, photo-interpreters can detect and identify visually some small landscape entities (isolated habitats, hedge networks, isolated trees). However, during the digitizing process, the photo-interpreter does not create image-objects according to their pixel resolution but according to the Minimum Unit Collection (UMC). This is the minimum size of objects often expressed in square metres and it is usually imposed within the study's aims. Moreover, to set a digitizing scale and it makes impossible a distance between vertexes too short. Usually, manual digitizing from submetric images is metric (between 5 and 10 m). In an automatic image segmentation approach, the delimitation of image-objects follows from the pixel resolution and generates the "tread of a stair" effect. This effect is not well-liked by project managers. However, geometric simplification is necessary for many reasons: first, it helps to keep homogeneity, by cutting out potential manual digitizing adjustments; second, the data is lighter (fewer vertexes); and last, image-objects are more consistent and easier to handle for users of the map data.

2.2.3 How Can Oversegmentation Be Reduced?

According to Castilla and Hay (2008), a little oversegmentation is a good segmentation. Generally, oversegmentation operates mainly on large homogeneous areas such as forests, water or herbaceous areas. The reason is that the segmentation scale parameter induces heterogeneity that limits objects. Therefore, oversegmentation determines the limit size of objects. On a single image segmentation it is unlikely that a forest of several hectares can be represented by a single image-object or an agricultural plot or a house object of a few hundred square meters. Even if oversegmentation is easier to correct than undersegmentation, it can lead to a long and tedious post-processing aggregation.

3 Materials and Methods

3.1 Study Site

The study area is located in the South-East of France. The area, which measures approximately 30 km, is included in the regional natural park at Pilat, 50 km south of Lyon city (Fig. 1). It transects a rural area between the lower slopes of the Pilat mountains and the banks of the Rhône river. This area describes complex plots composed of a discontinuous urbanization, vineyards and a fragmented forest configuration. The choice of this study site in a rural area was voluntary. Generally, rural areas and urban areas are distinguished in mapping production. Indeed, the mapping needs of urban areas are mainly based on land use for the monitoring of urban planning (building permits, transport, activities or industrial areas etc.). In France, there are numerous GIS urban data (updated by IGN or administrative authorities). Nevertheless, the majority of the surface of the territory is rural and the availability of GIS data is limited. Unlike urban landscapes, landscape changes in rural areas are not always submitted to administrative authorities but may be natural phenomenon (landscape enclosure) or concern agriculture monitoring (undeclared croplands). Thus, VHR remote sensing data is really a major opportunity for rural area monitoring.

3.2 Data

In scientific literature, the GEOBIA processes use specific and expensive VHR remote sensing data (Worldview, Quickbird, SPOT). At present, the availability of satellite images is little known to project owners as the financial budget does not allow the purchase of satellite images. In France, project managers mainly use free data like VHR aerial photography (BDORTHO®IGN) or free GIS database (BDTOPO®IGN).

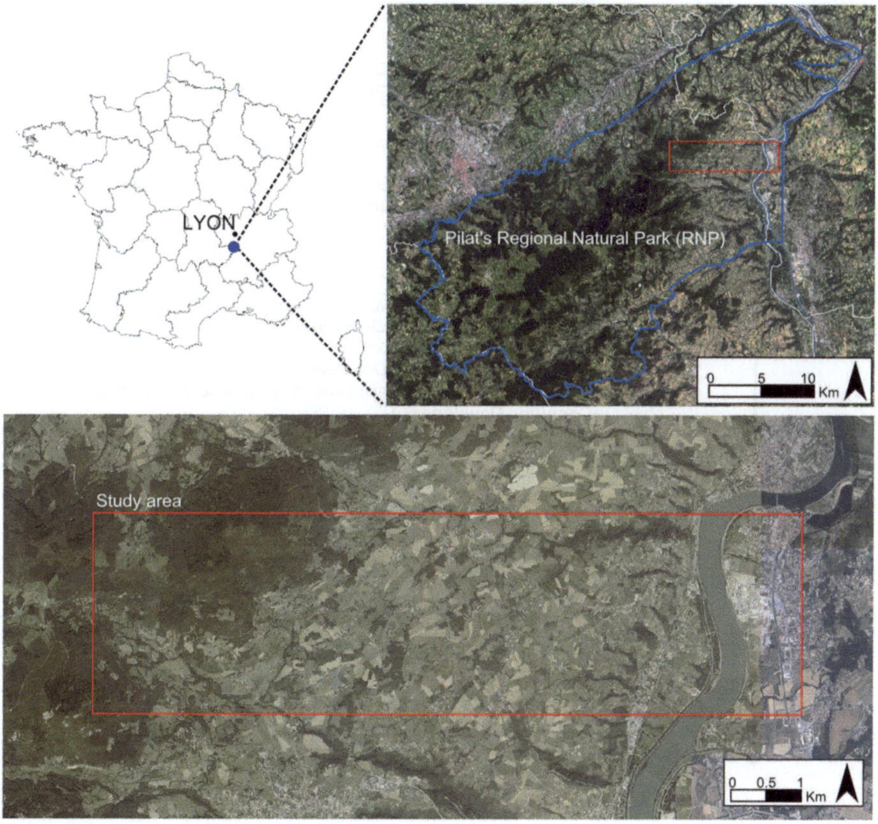

Fig. 1 Location map of study area

Therefore, our experimental data is RGB ortho-photography (BDORTHO® IGN, 2010). BDORTHO® is extremely dense aerial photography data due to its spatial resolution (50 cm) and its spatial extent (5 × 5 km or 100 million pixels). BDORTHO® is ideal data for a human photo-interpretation to detect and extract small landscape entities. However, it is more difficult to handle it with automatic image processing, thus tiling or multiprocessing solutions must be considered to reduce the processing time (Hay and Castilla 2008). In addition, BDORTHO® spectral resolution is too low and unusable for a LUC mapping (Jappiot et al. 2003).

3.3 Proposed Method

The objective of this study is to propose an enhancement to image segmentation. Our purpose is to provide at all times a polygonal "base" close to that a photo interpreter could realize taking a lot more time. Thus, we explored several working processes to

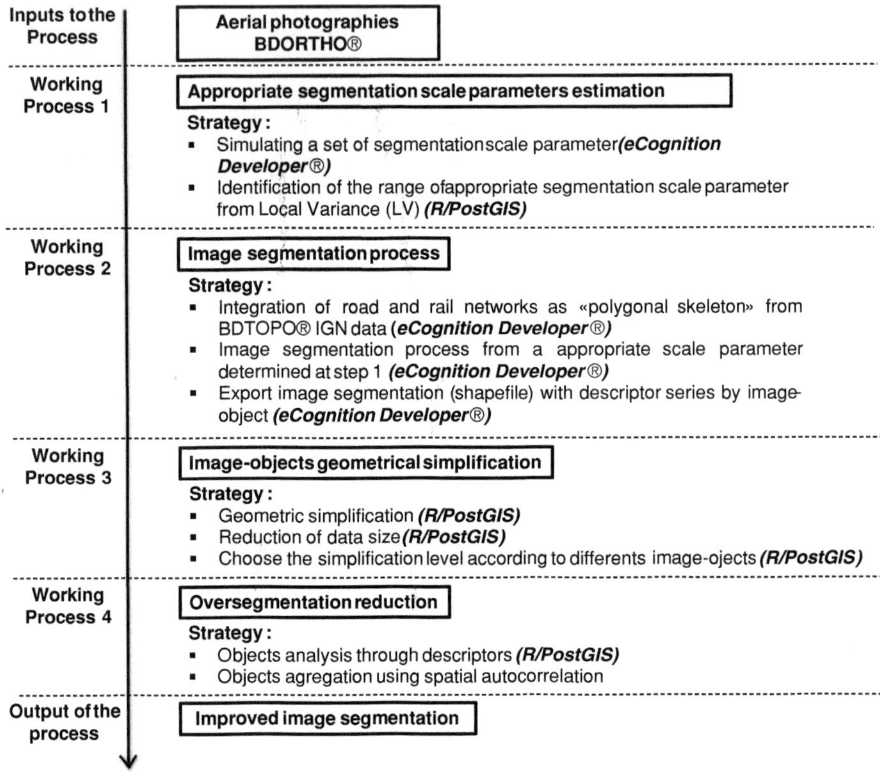

Fig. 2 General methodology for an image segmentation process enhancement

address the image-segmentation limits described in Sect. 3.2 for an operational land cover mapping application.

In this section, we expose our image segmentation method for the LUC to BDORTHO® thanks to eCognition Developer® software and R and PostgreSQL/PostGIS languages (Fig. 2). Our procedure includes the following steps:

1. Appropriate segmentation scale parameter estimation
2. Image segmentation process
3. Image-objects geometric simplification
4. Oversegmentation reduction.

These steps are described in the following.

(1) Appropriate Segmentation Scale Parameter Estimation

In this working process, we explore unsupervised methods to evaluate the image segmentation parameters. We experienced the Estimation Scale Parameter (ESP) tool variant. The ESP tool has been developed by Lucian Dragut (Drăgut et al. 2010). It can be integrated into a package in eCognition Developer® software.

This tool simulates a series of several scale values and provides an indication of segmentation level suitability. The tool is based on the work of Kim et al. (2008). It explores the relationship between image-object Local Variance (LV) and spatial autocorrelation at difference scale parameter. According to Drăgut et al. (2010), the more the scale parameter increases, more the size of image-object grow and more the Standard Deviation (SD) of image-objects increases until several little saturation that it matches the object in the real world (forest stand, houses, agricultural parcel…). To detect theses breaks, the tool combine LV information with another indicator: Rate of Change (ROC). The LV-ROC combination measures the LV change between two scale values. The peaks of the LV-ROC curve indicate the scale value where the image can be segmented in the most appropriate manner (Drăgut et al. 2010).

The simulation of a series of segmentation scale values seems to be a good way to study the segmentation performance. In addition, this provides information about the spatial structures of the landscape. However, it is difficult to justify the choice of a unique scale value for several geographic entities (houses, agricultural parcels, forest) with the LV-ROC curve in a complex landscape configuration. Thus, we developed the ESP tool variant and propose a range of appropriate segmentation scale values.

The LV profile has a logarithmic trend. The more the scale parameter increases, more the LV increases until global saturation is reached. Assuming that a good segmentation is a little oversegmentation (Castilla and Hay 2008), we can define LV curve saturation as being relative stability of image-object creation. Thus we tried to detect scale values just before LV curve saturation.

We used Hubert segmentation (Hubert 2000). The method finds "segments' in the LV curve. The location segments detected in the LV curve were processed with R software. It is an R code adaptation. A post-treatment was performed to identify the LV saturation segment. When the segment was identified, we processed another Hubert segmentation on this segment to isolate a specific rupture just before LV saturation. The results are presented in Sect. 4.

(2) Image Segmentation Process

The image segmentation process was performed with eCognition Developer® software (Fig. 3). Three commands are necessary and described in the following.

(a) **Roads and Railways segmentation.** The compatibility of LUC mapping with other geographic reference data, used in many studies, is very important. Road and rail networks from BDTOPO® IGN are often used as a "polygonal skeleton" where the image segmentation is grafted. There are multiple reasons for its use. First, BDTOPO® is complete and accurate GIS data that references much information including all transport networks with polylines and associated spaces with polygons. Each polyline is identified and prioritized according to an importance level (levels 1–5). Second, this data is extremely structured in the landscape and it is difficult to digitize it (tree cover or road narrowing). This approach was applied to our area of study.

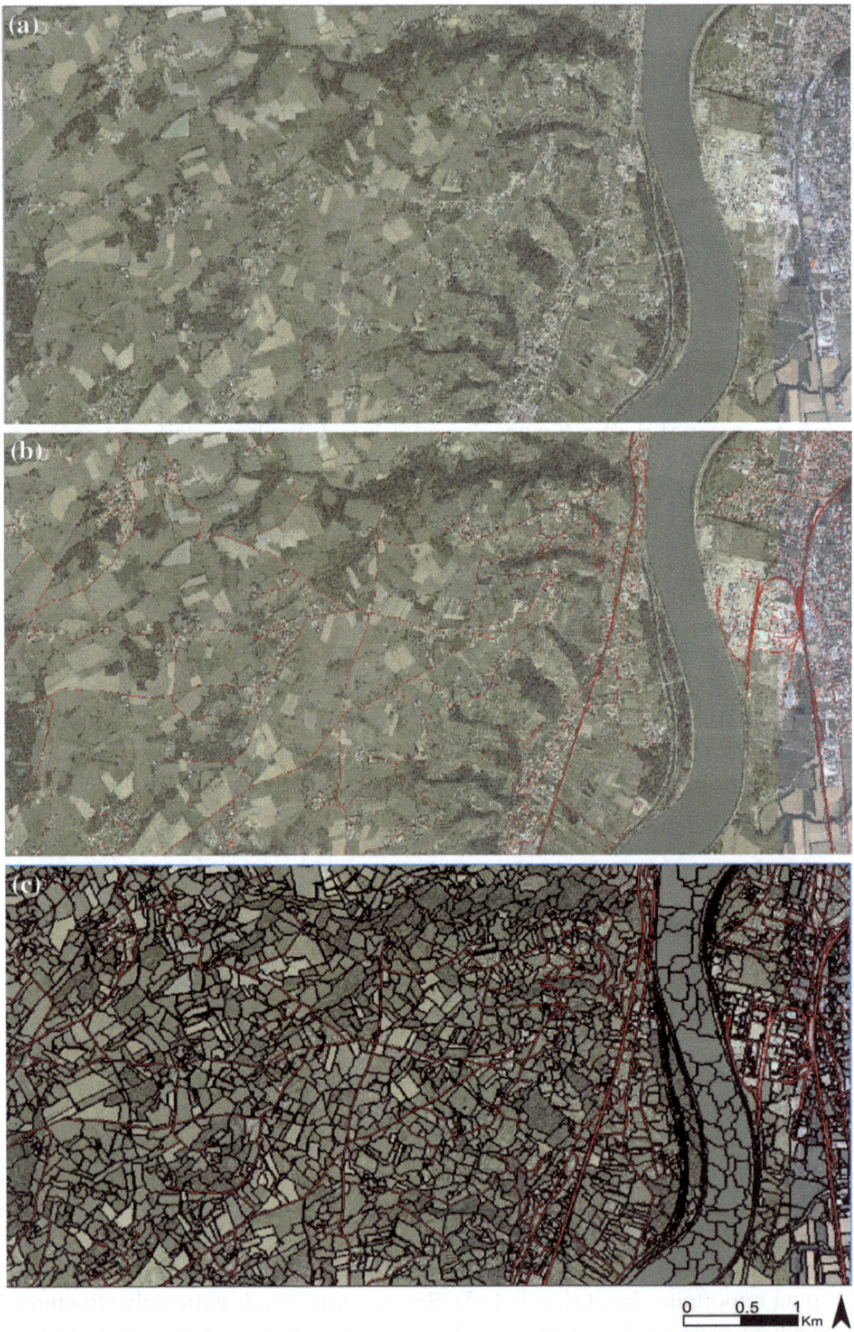

Fig. 3 Image segmentation process with eCognition Developer® software. **a** raw image; **b** roads and railways segmentation (*red*); **c** export image segmentation

(b) **Image segmentation.** This was performed with the eCognition® multiresolution segmentation algorithm. We used the segmentation scale parameter determined in the previous working process.
(c) **Export image segmentation.** eCognition® converts image segmentation into a vector file (shapefile). We can calculate and export several descriptors by image-objects (spectral, spatial or textural descriptors) into the attribute table of each vector file. This step is important for the fourth step of the process concerning oversegmentation reduction.

(3) Image-object geometrical simplification

The objective of this working process is to simplify the image-objects' geometry, erase the "tread of a stair" effect without generating topological errors, and reduce size data.

Several commercial software applications like eCognition Developer® or ArcGIS® propose image-object geometric simplification solutions with more or less efficiency and according to software licence level. We decided to develop a script based on R and POSTGIS.

Our scripts used the SQL simplification script of the Sandro Santilli code (Santilli 2013; http://strk.keybit.net/blog/). Thanks our scripts, we were able to apply a geometric simplification on different image-objects. For example, we decided to simplify road and rail objects at the level of 1 metre and other image-objects at 2 m. Moreover, our SQL script had been optimized to reduce processing time. However, while geometrical simplification is a relatively long process, it is a necessary one. The results are presented in Sect. 4.

(4) Oversegmentation reduction

Generally, at this stage of the object-based approach, the image segmentation is followed by the classification of image-objects' land cover or land use classes. To reduce oversegmentation, contiguous image-objects of the same class are aggregated to form the final image segmentation. The classification is either a computer-aided photo-interpretation process or a semi-automatic process. The former process is very long and tedious; the latter is generally inefficient for several reasons. First, the definition of decision rules from spatial, spectral or textural descriptors for many classes attribution is very difficult. The results are usually dependent on sampling or the threshold of human determination, which generates much class confusion. Second, the spectral resolution of RGB aerial photographies (BDORTHO®) is not adapted to a semi-automatic classification process. Last, decision rules are usually applied to the entire image without taking into account the local variability of landscape.

Spatial autocorrelation is a property often observed during spatial data observation. Two close spatial entities are more similar than two distant spatial entities. This notion is widely regarded in geography (Griffith 1987). By applying these considerations to the over-segmentation problem, it is possible to reduce oversegmentation. The hypothesis is that two neighbouring objects with close descriptors are likely to belong to the same thematic class and should be aggregated as a single object. The descriptors are indicators or properties used to describe each image-object. eCognition Developer® can generate a large number of descriptors. Only two types of indicators are selected. First, colour descriptors refer to the relative brightness or colour in the image-object. These descriptors are important for extracting image-objects. Moreover, tone variations allow some shapes or textures to be identified (Provencher and Dubois 2007). Second, texture descriptors refer to the frequency variation and the arrangement of colour tone in the image. For example, we can distinguish oriented texture like croplands or vineyards entities and homogeneous texture like forest entities (Bloch et al. 2004; Caloz and Collet 2001). The use of texture descriptors is particularly relevant with VHR images (Lefebvre et al. 2008). We used texture descriptors from co-occurrence matrices developed by Haralick et al. (1973).

This working process performed with the R/POSTGIS script was structured as follows. (1) The script identifies neighbours of each image-object. (2) A pre-processing is performed on "urban image-objects"; usually, these consisted of large pixel heterogeneity (houses, parking, gardens, trees). The descriptors have a chaotic distribution. To isolate these objects, building information from BDTOPO® IGN (usually provided) is intersected with image-objects according to a building density threshold. Thus, "urban object" cannot be aggregated with a nearby image-object except with another "urban object". (3) To perform the aggregation of neighbouring image-objects that are considered similar, we explored a statistical multi-criteria analysis such as a Principal Component Analysis (PCA) that allowed us to study distances between image-objects. Instead of analysing all couples of distance image-objects we processed to a clustering on PCA results and in every cluster we determined a maximum distance for considering two image-objects as similar. The results are presented in the following section.

4 Results

In this section we present the results of every working process for the study area described in Sect. 3.

In Fig. 4 we can observe the results of ESP tool variant used to determine a range of appropriate segmentation scale values. The first graph (Fig. 4a) presents the LV curve evolution (in blue) depending on the segmentation scale values. The scale value simulation ranges from 50 to 300 with spacing of 5. The red curve represents the detection of segments with Hubert segmentation. The segment locations are highlight by vertical markers. The green marker localizes the start of LV saturation. Then, we extracted this first saturation segment and we processed a second Hubert segmentation (Fig. 4b). The first segment presents the phase just before saturation. It means there is a little oversegmentation. Thus, we propose this segment or range of segmentation scale values between 90 and 105 as the most appropriate to process segmentation.

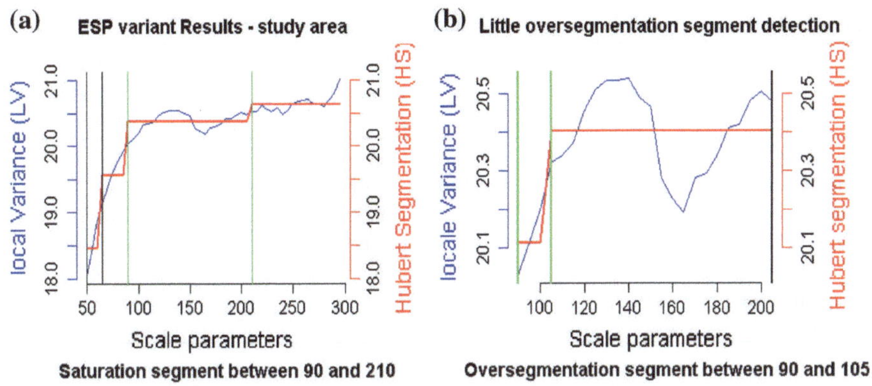

Fig. 4 Determination of appropriate segmentation scale values range

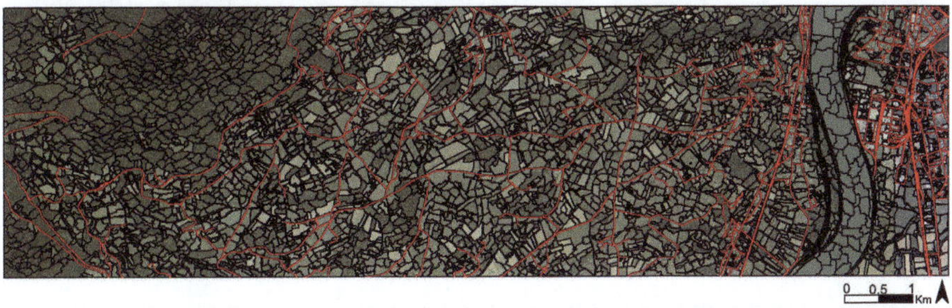

Fig. 5 Image segmentation output of eCognition Developer® with communication network

We did not want to propose a single segmentation scale value that we could not justify. We wanted to just provide an indication about a range of appropriate scale values to obtain a little oversegmentation according to the study area and remote sensing image.

Then, we processed the image segmentation with eCognition Developer®. The result is presented in Fig. 5. We used 90 as scale value. Roadways and railway networks from BDTOPO® were incorporated in the image segmentation process (red image-objects). Next, we performed image segmentation export with descriptors for oversegmentation reduction in the fourth step. Each image-object was specified for each variable. Table 1 presents the statistics concerning vector file features. We observed that the vector file is large (28 338 Ko) because the vertex number is very high. Therefore, the vector file is difficult to handle. Moreover, we observed that the processing time was not very long (10 min) but the study area was small.

To use this vector, we processed the image-objects to obtain geometric simplification. The results are presented in Fig. 6. At this stage, the vector file has a resolution of 50 cm. We decided to process two levels of geometric simplification. The first level is operated on network image-objects. These image-objects come from an accurate database (BDTOPO®). We processed a slight geometric simplification (1 m) to erase the "tread of stair" effect without losing the precision of these image-objects. The

Table 1 Statistics summary of image segmentation enhancement process

	Original image segmentation	After the geometric simplification	After the oversegmentation reduction
Number of image-objects	5,689 (484 "urban objects")	5,689 (484 "urban objects")	3,156
Number of vertexes	1,145,138	739,063	424,647
Size of vector file (Ko)	28,338	8,966	6,812
Processing time (min)	10	35	2

Fig. 6 Geometric simplification of image-objects

Fig. 7 Meeting between image segmentation and buildings information (*in red*) from BDTOPO®IGN

second level is operated on the rest of image-object. We considered that 2 m is a good level of geometric simplification. It can erase the "tread of stair" and reduce considerably the size of vector file. In Table 1 we compare image-segmentation before and after geometric simplification. We can observe that the number of image-objects is the same between the original image segmentation and the geometrically simplified image segmentation.

Last, we processed oversegmentation reduction. First, we identified "urban image-objects" with BDTOPO (Figs. 7 and 8). Second, we performed PCA and then the clustering on the PCA result. We analysed the distances between image-objects in

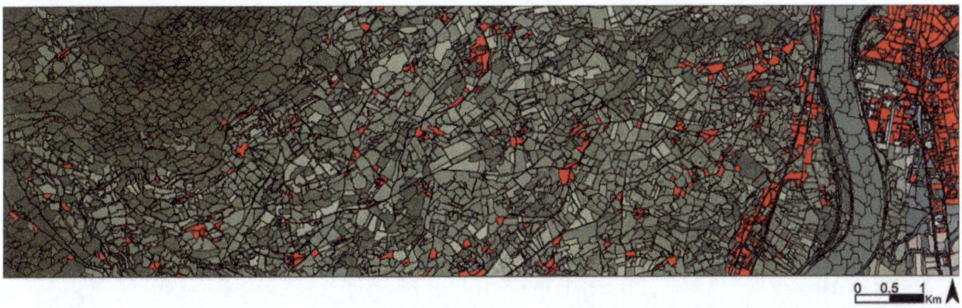

Fig. 8 Identification of "urban image-objects"

Fig. 9 Result of oversegmentation reduction

every cluster. Thus, we could determine more precisely a maximum distance for considering two image-objects as similar. The similarity threshold was set to 0.9. In Figs. 9 and 10 we show the result of the oversegmentation reduction. The statistics for the output vector file are presented in Table 1. The output image segmentation loses 2,533 image-objects and the size of the vector file was reduced by 19,372 Ko. The reduction of image-objects operates especially in the forest area where image-objects are very similar. In open country, the result is more contrasted and oversegmentation is less efficient than for the forest area (Fig. 10). However, the similarity threshold was voluntarily minimized to avoid undersegmentation.

5 Discussion and Conclusions

This chapter presents an automatic image segmentation enhancement for land cover mapping from VHR images. It is based on three points of limitation of the GEOBIA approach to image segmentation used in operational LUC mapping applications.

First, the limitation concerning the determination of suitable segmentation scale parameters remains problematic. It is an ill-posed problem. No objective protocol exists for setting a segmentation scale value. Currently, the scale selection is based on trial-and-error methods. To improve these methods, and assuming that a good segmentation seems to involve a little oversegmentation, we proposed guiding the

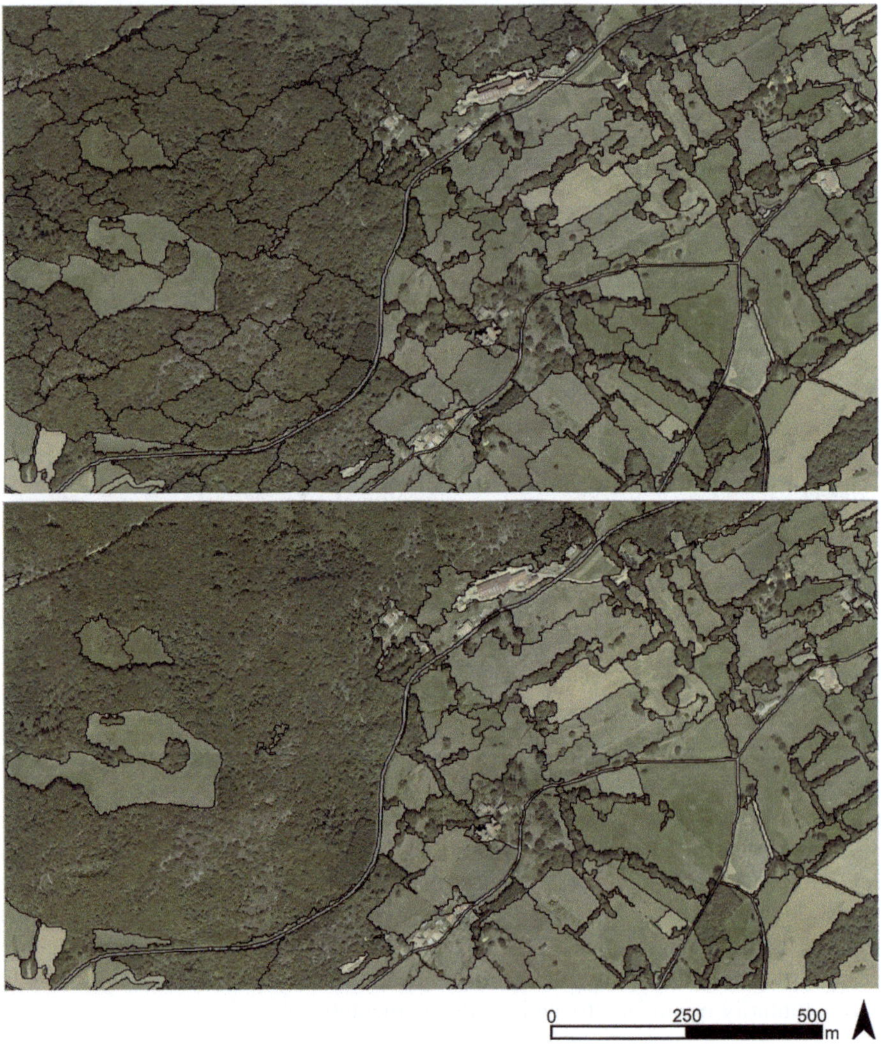

Fig. 10 Zoom of oversegmentation reduction result

user to an appropriate scale range and not to a single scale value. The method is based on an analysis of LV information in a series of scale value simulation. The results are shown in Fig. 4 and present a short scale value range just before LV saturation. This range suggests scale values that produce a little oversegmentation of the image. This approach makes the choice of an appropriate segmentation scale value easier, but does not affirm that it is the best segmentation scale value.

After the image segmentation process, we identified that the image-object's geometry was a main limit for its use in operational LUC application. The "tread of a stair" effect in image-objects delimitation is problematic. We performed efficiency R and

PostGIS scripts to solve this problem (Fig. 6). It is an independent script and it can perform several simplification levels on different image-objects. However, the process time remains long but necessary.

Last, we explored a method to reduce oversegmentation and make it easier to use image segmentation for the classification step. Assuming oversegmentation is less problematic than undersegmentation, post-processing the aggregation of image-objects can be long and tedious. We propose a method based on spatial autocorrelation to automatically aggregate the nearest neighbouring image-objects (Fig. 9). The main advantages are that this method considers the local variability of the landscape and avoids a global LUC classification from ill-adapted remote sensing data like BDORTHO®. The PostGIS script was adapted to minimize the aggregation process and to avoid an undersegmentation result. The method gives satisfactory results on homogeneous areas such as forests or grasslands.

In conclusion, several limitations of the GEOBIA process can lead project managers to consider manual digitizing as safer than automatic image segmentation. GEOBIA processes lack operational application or robust methodologies. Moreover, commercially-oriented software is often overly complicated for a non-specialist user; software provides many black box options and promotes many confusions. Nevertheless, the GEOBIA approach for capturing features automatically from VHR remote sensing data is a major opportunity for LUC mapping in the future for a number of reasons: first, this approach could reduce financial and human production costs; second, it could process large datasets in less time; third, thanks to the increasing number of practitioners in the GEOBIA community, many opportunities will emerge to adapt it to specific mapping needs; and last, it can update geographic data faster.

Our development provides an operational answer to object-based image-segmentation problems for LUC mapping production. It proposes a way to optimize the image segmentation process to have at all times a "polygonal base" close to human production. Future research will be dedicated to testing other study areas with different landscape configurations. Another major challenge will be testing very large datasets and experimental tiling processing solutions on large study areas. Finally, the proposed scheme could provide an interesting framework for classification step of the LUC.

References

Arvor D, Durieux L, Andrés S, Laporte M-A (2013) Advances in geographic object-based image analysis with ontologies: a review of main contributions and limitations from a remote sensing perspective. ISPRS J Photogram Remote Sens 82:125–137

Baatz M, Benz U, Dehghani S, Heynen M, Höltje A, Hofmann P, Lingenfelder I, Mimler M, Sohlbach M, Weber M (2004) eCognition professional user guide 4. Definiens Imaging, Munich

Benz UC, Hofmann P, Willhauck G, Lingenfelder I, Heynen M (2004) Multi-resolution, object-oriented fuzzy analysis of remote sensing data for gis-ready information. ISPRS J Photogram Remote Sens 58:239–258

Blaschke T (2010) Object based image analysis for remote sensing. ISPRS J Photogram Remote Sens 65:2–16
Blaschke T, Strobl J (2001) What's wrong with pixels? some recent developments interfacing remote sensing and gis. Interfacing Remote Sens GIS 6:12–17
Bloch I, Gousseau Y, Maître H, Matignon D, Pesquet-Popescu B, Schmitt F, Sigelle M, Tupin F (2004) Le traitement des images. Polycopié du cours ANIM, Département TSI-Télécom-París, p 370
Caloz R, Collet C (2001) Traitements numériques d'images de télédétection. Précis de télédétection. Presses de l'Université du Québec, Agence universitaire de la Francophonie, Québec, Montréal, p 398
Castilla G, Hay GJ (2008) Image objects and geographic objects. In: Blaschke T, Lang S, Hay GJ, (eds) Object-based image analysis. Springer, Berlin, p 91–110
Dorren KKA, Maier B, Seijmonsbergen AC (2003) Improved landsat-based forest mapping in steep mountainous terrain using object-based classification. For Ecol Manage 183:31–46
Drăgut L, Tiede D, Levick SR (2010) Esp: a tool to estimate scale parameter for multiresolution image segmentation of remotely sensed data. Int J Geogr Inf Sci 24:859–871
Drăgut L, Csillick O, Eisank C, Tiede D (2014) Automated parameterization for multi-scale image segmentation on multiple layers. ISPRS J Photogramm Remote Sens 88:119–127
Griffith DA (1987) Spatial autocorrelation. Association of American Geographers, Washington DC
Haralick RM, Shanmugam K, Dinstein IH (1973) Textural features for image classification. IEEE Trans Syst Man Cybern 3(6):610–621
Hay GJ, Castilla G (2008) Geographic object-based image analysis (GEOBIA): a new name for a new discipline. In: Blaschke T, Lang S, Hay GJ (eds) Object-based image analysis. Springer, Berlin, pp. 75–89
Hubert P (2000) The segmentation procedure as a tool for discrete modeling of hydrometeorological regimes. Stoch Env Res Risk Assess 14:297–304
Jappiot M, Philibert-Caillat C, Borgniet L, Dumas E, Alibert N (2003) Analyse spatiale des interfaces agriculture-forêt-urbain. Ingénieries Numéro Spécial, pp 69–81
Kim M, Madden M, Warner T (2008) Estimation of optimal image object size for the segmentation of forest stands with multispectral IKONOS imagery. In: Blaschke T, Lang S, Hay GJ (eds) Object-based image analysis. Springer, Berlin, pp. 291–307
Kim M, Madden M, Warner T (2009) Forest type mapping using object-specific texture measures from multispectral ikonos imagery: segmentation quality and image classification issues. Photogramm Eng Remote Sens 75:819–829
Lefebvre A, Corpetti T, Hubert-Moy L (2008), Object-oriented approach and texture analysis for change detection in very high resolution images. In: IEEE international geoscience and remote sensing symposium, IGARSS 2008, pp IV–663
Marpu PR, Neubert M, Herold H, Niemeyer I (2009) Enhanced evaluation of image segmentation results. J Spat Sci 55:55–68
Meinel G, Neubert M (2004) A comparison of segmentation programs for high resolution remote sensing data. Int Arch Photogramm Remote Sens 35:1097–1105
Neubert M, Herold H (2008), Assessment of remote sensing image segmentation quality. In Proceedings of the GEOBIA international archives of photogrammetry, remote sensing and spatial, information sciences, vol XXXVIII-4/C1
Provencher L, Dubois J-MM (2007) Méthode de photointerprétation et d'interprétation d'image. Précis de télédétection. Presses de l'Université du Québec, Agence universitaire de la Francophonie, Québec, Montréal, p 504
Santilli S (2013) On the fly simplification of topologically defined geometries. Strk's Blog. http://strk.keybit.net/blog/
Schiewe J, Tufte L, Ehlers M (2001) Potential and problems of multi-scale segmentation methods in remote sensing. GeoBIT/GIS 6:34–39
Thomas A (2005) Application de l'approche orientée-objet à l'extraction de fragments forestiers à partir de scènes Spot. DESS SIGMA, 30.

Woodcock CE, Strahler AH (1987) The factor of scale in remote sensing. Remote Sens Environ 21:311–332

Zhang X, Xiao P, Feng X (2012) An unsupervised evaluation method for remotely sensed imagery segmentation. IEEE Geosci Remote Sens Lette 9:156–160

Line Matching for Integration of Photographic and Geographic Databases

Youssef Attia, Thierry Joliveau and Eric Favier

Abstract The aim of this chapter is to describe a new method for assigning a geographical position to an urban picture. The method is based only on the content of the picture. The photograph is compared to a sample of geolocated 3D images generated automatically from a virtual model of the terrain and the buildings. The relation between the picture and the images is built through the matching of detected lines in the photograph and in the image. The lines extraction is based on the Hough transform. This matching is followed by a statistical analysis to propose a probable location of the picture with an estimation of accuracy. The chapter presents and discusses the results of an experiment with data about Saint-Etienne, France and ends with proposals for improving and extending the method.

Keywords GIS · Photo · Matching lines · Hough lines · 3D reconstruction · Labels · City

1 Introduction

The democratization of digital cameras and Web access has completely changed the way people use and share photographs. Nowadays, users publish, share and comment on their photos on the Web on personal sites or, increasingly, on contributing platforms and social networks. Billions of images are stored on the Internet and millions are added every day. In many cases the users annotate their photos with

Y. Attia (✉) · T. Joliveau
ISTHME-EVS, Jean Monnet University, CNRS, 6 Basse des Rives Street, 42023 F-Saint-Etienne Cedex 02, France
e-mail: youssef.attia@univ-st-etienne.fr; josef.attia@gmail.com

E. Favier
ENISE, 58 Jean Parot Street, 42023 F-Saint-Etienne Cedex 02, France

J. Huerta et al. (eds.), *Connecting a Digital Europe Through Location and Place*, Lecture Notes in Geoinformation and Cartography, DOI: 10.1007/978-3-319-03611-3_14,

information about the circumstances, the place, the content or any other element relating to the picture.

Most photo-sharing web platforms, like, for example, Panoramio, Zooomr, http://Loc.alize.us, or Flickr, attach a geographic position to each photo (Torniai et al. 2007). Public photo reshaping and ordering software such as Google's Picasa also provide users with the tools for manually localizing their photos on Google's or Yahoo's virtual globes. The success of these solutions among Internet users, despite the fact that these manual location techniques are tedious and approximate, is a sign of the interest in photo geolocalization.

The current automatic techniques for positioning the picture have limits related to the errors of GPS localization, especially in urban areas, or to the errors of precision in the triangulation of the signal from GSM antennas or Wi-Fi hotspots. But an important number of photographs remain non-localized or approximately localized. In many fields, both for personal use or for professional databases, the need for automatic or semi-automatic photograph location tools is patent.

In this chapter we propose a method for automatically finding the location of the photograph by comparing it with a sample of computer-generated images calculated from a virtual environment of the area. The principle is to extract indicators comparable between the photograph and the synthesis images so as to match and then localize them relying on the coordinates of the best matched synthesis image. The chosen criterion is the number of lines that the photograph shares with the computer-generated image.

In this chapter, we first present an overview of automatic methods existing in the literature for geolocating photographs based on their content. Then we expose our new approach and the validation process we use. We finally give the results and the statistics that allow us to assess the efficiency and limitations of our methodology. We conclude the chapter by some proposals for improving the results.

2 Automatic Geolocalization Methods

In order for them to be automatic, geolocation methods must not have any human intervention in the process of adding geographic information to a photograph. There are a few methods proposed in the bibliography.

A team of researchers at the University of Maryland developed a semi-automatic method of picture annotation (Suh and Bederson 2007). They used a geolocalized photo to identify people in a specific photograph and they apply the same geographic location to all photographs where a person is wearing the same clothes.

Hays and Efros tried to find the geographic position of a photograph by comparing it with all the elements of a geolocalized database of images built to that effect (Hays and Efros 2008). This technique is effective only for touristic pictures because it needs the presence in the photographs of emblematic elements of the landscape. Conversely, Keita Yaegashi and Keiji Yamai used the geographic position of the camera to help detect the picture content (Yaegashi and Yanai 2009).

In his thesis, Moslah (2011) produced an urban model capable of detecting the different elements of the facades such as windows, cornices and balconies. This work deployed a descriptive grammar of the elements constituting an urban landscape.

3 Proposed Approach

We propose realizing a match between the photograph we are trying to locate and a sample of computer-generated images extracted from a georeferenced 3D model of the whole city or the neighbourhood concerned. We are looking to find the position and the exact orientation of the photograph by positioning ourselves in the virtual world so as to get analogous content between the photograph and the calculated image. It is therefore the content of the picture which allows us to deduce the location. An evaluation of the matching quality will then allow us to calculate a position for the photograph from the geographic coordinates of the images that match at different degrees with the picture. It will also be possible to evaluate a distance error in the location.

3.1 The Matching Principles

We use a GIS database that contains the buildings of the considered urban area, the elevation of the ground at the foot of each building and the height of the building. All these data are mobilized to create a 3D virtual world representative of the real one. Therefore, it becomes possible to place a virtual camera in any position of that 3D world to generate an image.

A photograph and a computer-generated image can have an identical content only if both of them share the same position, the same focal length and the same viewing direction. We propose evaluating the degree of resemblance between the computer-generated image and the real photo by using line matching.

Our approach belongs to the data fusion family of techniques. The matching procedure is presented in detail in Fig. 1. It includes five phases:

1. Creation of the virtual world relying on a layer of geographic data containing the buildings and a digital elevation model (DEM).
2. Selection of computer-generated images around the presumed position of the real photograph. The synthesis images are ideally produced at a regular interval of 2 m. Eight images are taken in each position with a 45° rotation step. The choice of that interval is a result of our experimentations. It takes into account the nature of the urban structure.
3. Preparation of the photograph by eliminating the ground.
4. Matching of the photograph with the different computer-generated images selected at phase 2.

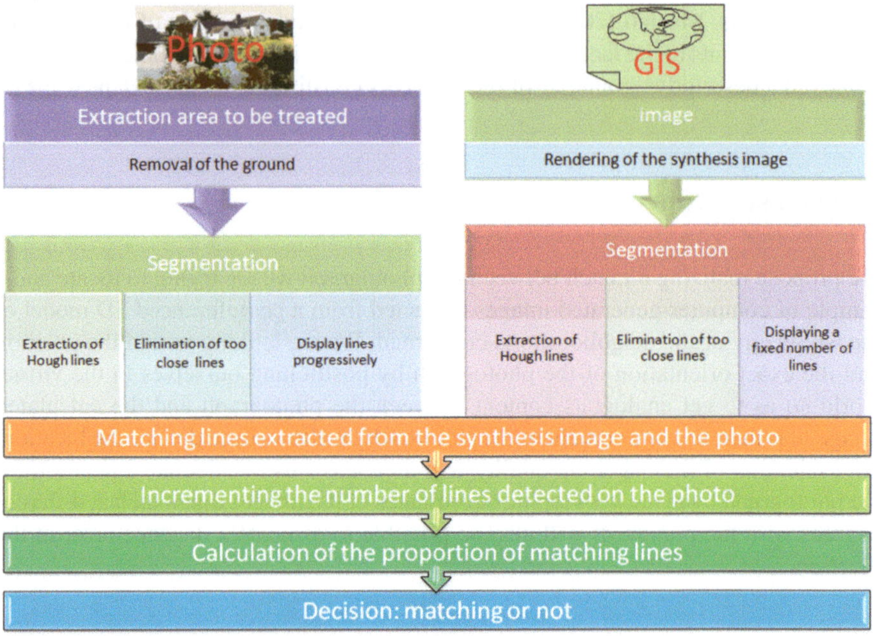

Fig. 1 General approach of location by line matching

5. Applying the position parameters to the photograph. The geographic coordinates are those of the generated image that matches the best. Otherwise we apply weighting functions for calculating an error, according to the value of the matching, we calculate the position of the photograph.

The pairing of the photo and the images consists of moving from a comparison between two images to a comparison between two lists of lines. There is a correspondence (matching) between the lines detected in a computer-generated image and the lines detected in the photograph. We chose to take lines as indicators because they define rather well the building facades in a city. As a preliminary hypothesis, lines can be considered as a signature of a streetscape. The idea is to recognize the streetscape of the photograph in a sample of virtual images.

3.2 Extraction Line Method

The methodology of line extraction that we selected is the Hough transform. Paul V.C. Hough proposed in 1962 a method which makes it possible to detect parametric forms like lines and circles in an image (Hough 1962; Maitre 1985). This technique (Valentin 2009) was re-examined ten years later by other researchers (Duda and Hart 1972; Rosenfeld 1969), who generalized the process from the detection of alignments

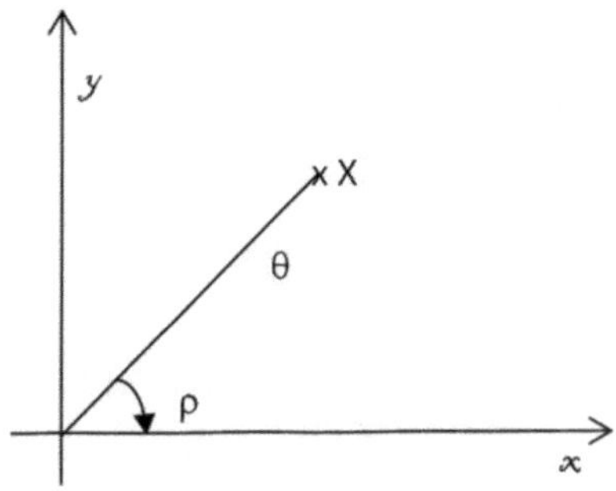

Fig. 2 Line representation

on an oscilloscope to the detection of lines in an image. According to the experiments, the transform of Hough is one of the best methods of the detection of lines for a noisy image (Cohen and Toussaint 1977). It makes it possible, for example, to detect the edges of roads in a dark environment (Kneepkens 2005). The Hough transform is an "optimal" technique according to (Crowley 2010) particularly for detecting lines in very noisy pictures.

This technique does not require the continuity of straight lines to detect them. However, the elements detected by this transformation are not segments, but lines. This transformation makes it possible to switch from a point in an image to a straight line in a space of parameters (Duda and Hart 1972). The space of parameters is defined by the parameters of the representation of the lines in the plan of the image.

Another argument for the use of the transformation of Hough is related to the representation of the lines. Indeed, each detected line is characterized perfectly by only two digital parameters (Fig. 2):

- A distance ρ: this is the distance in pixels between the line and the origin of the coordinate system. The maximum value that can take ρ is the diagonal length of the image.
- An angle θ: this is expressed in radians and ranges from $-\pi$ to $+\pi$. In our case, the simplicity of the representation is important, since we have to record and analyse a significant number of images and lines.

We illustrate in Fig. 3 three lines extracted from a photograph and ten lines in a computer-generated image calculated from the same location and direction. The three lines in common are in red.

3.3 Algorithm

The following algorithm is used for matching lines:

```
Line_matching(Image S ,image P) {
Conversion of the image P and S in greyscale
Detect LS lines of the image S
```

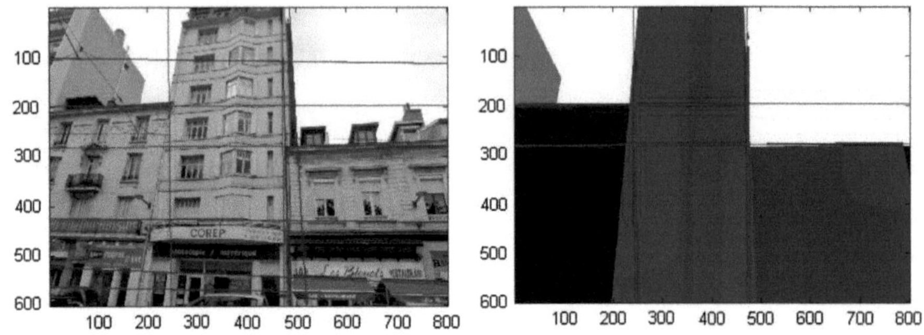

Fig. 3 Example of line extraction

```
Filter lines too close from LS
While C is false do
    Detect LP lines of the image P with the SP threshold
    Filter lines too close from LP
    For all lines ∈ LP do
        similarity ← 0
        For all lines ∈ LS do
            If (ρP ≈ ρS and θP ≈ θS) then
            similarity ← similarity +1
            End if
        End loop
    End loop
    If (similarity ≥ LP × 4/5 and LP ≤ 3 × LS) then
          C ← true (Good matching)
    End if
    If (LP ≥ 3 × LS ) then
          C ← true (No matching)
    End if
    Threshold SP ← SP − 0.01
End loop
}
```

4 Location of Photography

The user can declare a general neighbourhood or an approximate position for the photograph he wants to locate. The system proposes a probable location of the photograph based on the virtual images of the neighbourhood.

Pairing is first launched for virtual images located in this limited area. The location of the photograph is done by using the matching rate, see Sect. 6. The estimated

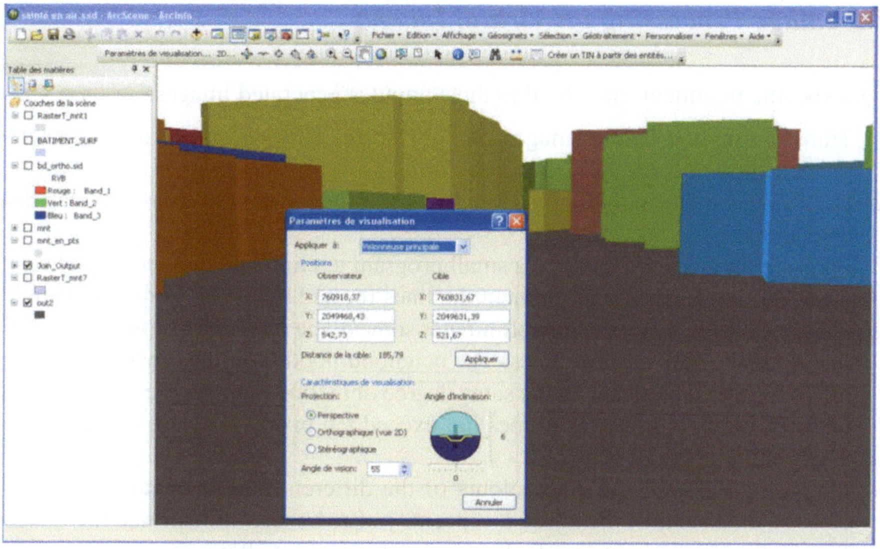

Fig. 4 Parameters of visualization in ArcScene (*Source* screenshot of ESRI ArcScene)

location is done by calculating the barycentre of the locations of images weighted by the matching rates. The coordinates of the weighted barycentre are calculated using the formula (1)

$$X_t = \frac{\Sigma_{t=1}^{n} t_{i x_i}}{\Sigma_{t=1}^{n} t_i} Y_t = \frac{\Sigma_{t=1}^{n} t_{i y_i}}{\Sigma_{t=1}^{n} t_i} \tag{1}$$

The location error is function of the standard-deviation of the distance to the weighted barycentre.

5 Concept Validation

5.1 Technical Choices

We use ESRI ArcScene software to build the 3D model and generate virtual images. It is possible to specify the coordinates X, Y and Z of the camera, the coordinates X, Y and Z of the target, the angle of vision that corresponds to the focal length and orientation angles (Fig. 4).

The application designed to realize the treatment of the tables of lines is running under MATLAB.

5.2 Image Processing on Virtual Images

Four specific treatments are allied to the computer-generated images:

1. Transforming the colour images into greyscale. This step is required to enable the detection of lines by Hough transform.
2. Increasing the threshold for the Hough transform in order to decrease the staircase detection.
3. Filtering the lines. We created a small program to eliminate the lines too close to the other. In an urban environment the lines formed by the buildings are usually well spaced out. This program eliminates some of the lines and keeps only the first line of each group of close lines. The neighbourhood is defined by experimentation. Two parallel lines with the same degree of incline θ and where the difference between the two distances ρ does not exceed ten pixels, they are considered as neighbours.
4. Increasing the contrast. The colours of the different buildings can be close or close to the colour of the ground, especially after having transformed the images from colour to greyscale mode. Increasing the contrast allows an easier detection of contours and lines.

5.3 Image Processing on Photographs

The photographs in an urban environment do not contain only buildings but also trees, advertising signs, pedestrians, cyclists, people... Most of these elements are situated at the foot of the buildings. To remove those elements, we use the method of Hoiem et al. (2005) who proposed an approach for analysing the context of an image. It is based on a heuristic that allows segmenting the image into three parts (horizontal surface or base, vertical area and sky part) by combining four criteria to determine the type of these three areas a pixel belongs to (Hoiem et al. 2007): the location of the treated area, colour, texture and point of view.

5.4 Data Set

We constituted a first set of training data with 20 photographs taken in an urban environment composed of buildings built between the nineteenth century and today. Once the analysis of the photographs was realized and a decision tree created, an evaluation phase of our approach was conducted with 13 more photographs. That makes a total of 33 photographs taken in the city of Saint-Etienne.

We created 46 virtual images in the area where the photographs were taken. On average we took 2.3 images for each position of the photographs. Among these 46 synthesis images, 20 are extremely close to one of the photographs. Another 26 are taken with a different viewing angle, focal length or orientation.

Fig. 5 Sample of photographs and virtual images

The database that we use to validate our approach is made up of 83 images taken in 29 different geographical positions and the 33 photographs of both phases: training and evaluation.

A sample of these photographs is also shown in Fig. 5, with the corresponding virtual images.

The dots on the map of Fig. 6 indicate the location of the camera during the shooting.

The photographs were taken using five different devices: two cameras equipped with an integrated GPS chip, two smartphones, a classic reflex coupled with a GPS to get the location and a compass to indicate the direction of the device.

The data layer used is the version 2 of the BD TOPO of the Institut Géographique National (The French Cartography Agency) in the RGF93 coordinate system. The data used to determine the height of the ground is provided by the digital elevation model (DEM) of the IGN as well.

6 Results

The indicator chosen to evaluate the relationship between the photograph and the computer-generated image is the matching rate of pairing calculated by the formula (2)

$$Matching\ rate = \frac{Number\ of\ de\ common\ lines}{X} \times 100 \quad (2)$$

X is the minimal number of lines detected in the photograph. It corresponds to the possible maximum of matching lines. We call the rate of matching between the photograph and the only virtual image that is in perfect correspondence with it the "self-recognition rate".

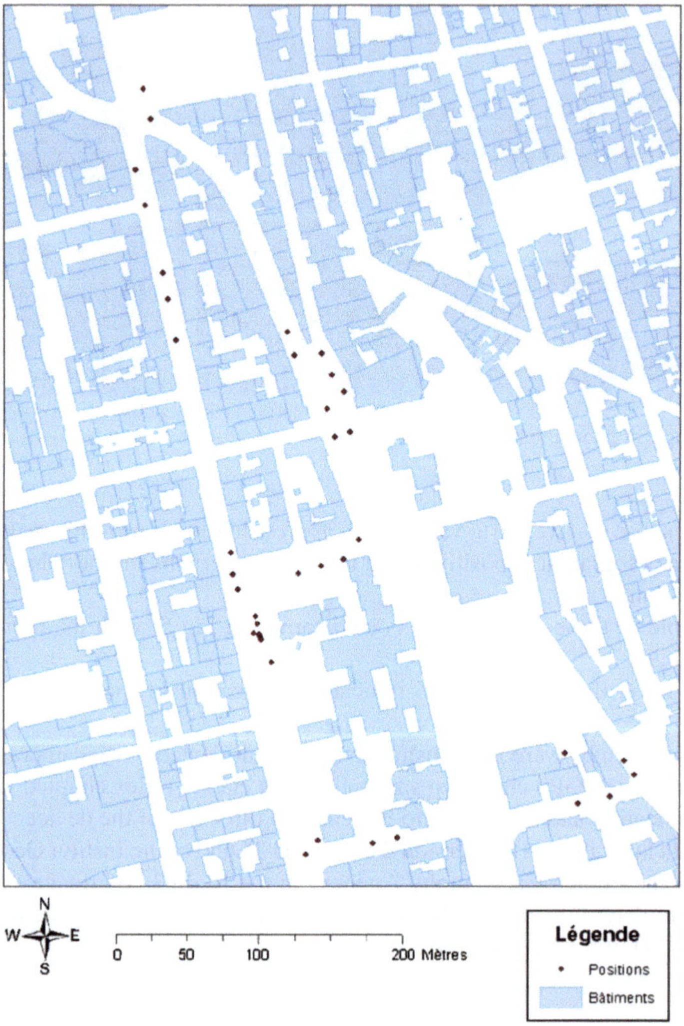

Fig. 6 Position of virtual images

The minimum rate found in our learning database is 21.14 %. The maximum rate is 73.3 %, the median value is 38 % and the standard deviation is 13.56 %. This is the base we use to judge if there is or is not matching between a photograph and a virtual image.

The analysis of the various rates of matching and the rate of self-recognition, allows us to define the decision tree of Fig. 7.

If a matching rate is less than the average of matching rates between a photograph and all synthesis images, we consider this rate not enough to judge the relative synthetic image corresponding to the photograph.

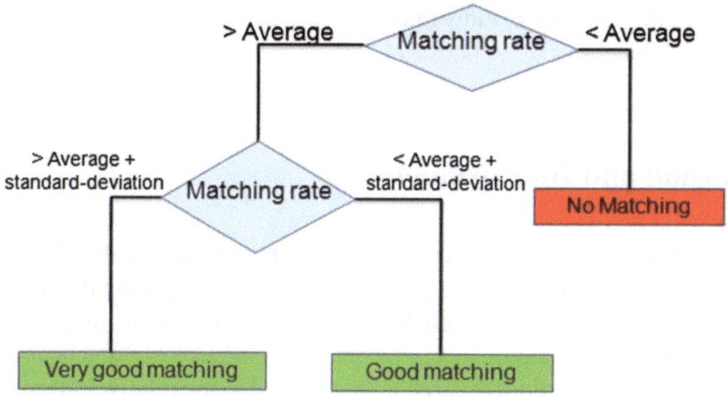

Fig. 7 Decision tree

If this rate is greater than the sum of the average and the standard-deviation, we consider that the synthesis image and the photograph are corresponding. We are comparing the situation in three areas around the photograph: less than 200 m from the photo, less than 100 m and less than 50 m. Only the virtual images located in the area are taken into account to compute the statistics.

In an area of 50 m around a position, using our decision tree, we obtain an average distance of 12 m between the real position of the picture and the virtual image found as the corresponding one to the picture. An area of 100 m generates an average of 29 m between the correct position and the one given by our approach. In a circle of 200 m radius, we can find the average position of a photograph almost 55 m distance from its real position.

If we have no presumed location, we use all synthesis images with a recognition rate higher than the average. The average deviation is then 218 m.

7 Limits

The geographic databases are the first cause of error in the matching process. In the database used, different buildings are often grouped. This reduces the number of lines in the virtual images and disturbs matching with the photograph where all the buildings are actually different.

The second problem is related to the way the picture is taken. The method works when the photographer stands on the ground and points his camera horizontally. Special situations where the photo is taken from the upper floor of a building or the camera is obliquely oriented make the matching harder and the performance drops significantly.

The presence of barriers or obstacles between the camera and the building is also a problem. Festive decorations and snow on the trees accentuate the disturbances and, therefore, the number of unwanted and parasite lines. This drastically lowers the score of the matching.

8 Conclusion and Improvement

We demonstrated that the lines in a picture are representative of an urban streetscape and that it is possible to find the location of an image by comparing the lines extracted from an image with those extracted from computer-generated images of a virtual environment of the same area.

The main limitation of this research is first related to the relatively limited number of virtual images used for this first test. The spatial irregularity of their repartition is also an inconvenience. To calibrate the error rate and fully validate the approach, it would be necessary to have a more complete and regular canvas of virtual images. It would also be necessary to multiply the tests in different and various urban environments in order to verify the performance of the method in areas where buildings' shapes and structures are more uniform than they are in the centre of Saint- Etienne.

Several paths of improvement are possible. The first is an automatic removal of the obstacles located between the camera and the buildings to increase the similarity between the real world and the 3D model. Many new techniques exist for detecting objects located in front of a scene. This should significantly increase the matching rate. A second way of improvement would be to use wire-frame 3D modelling techniques to directly produce the lines of the buildings, without using the Hough transform for the virtual image. Finally, the increasing power of processors and memory in smartphones make it possible to consider processing the picture at the moment it is taken and to use it to get an approximate location on the way.

References

Cohen M, Toussaint G (1977) On the detection of structures in noisy pictures. Pattern Recogn 9:95–98

Crowley JL (2010) Détection et Description de Contraste. Unité de Formation et de Recherche en Informatique et Mathématiques Appliquées (UFR IMAG), de l'Université Joseph Fourier—Grenoble 1 (UJF), Grenoble

Duda RO, Hart PE (1972) Use of the hough transformation to detect lines and curves in pictures. Commun ACM 15:11–15

Hays J, Efros AA (2008) IM2GPS: estimating geographic information from a single image. In: Proceedings of the IEEE conference on computer vision and pattern recognition, 2008, pp 1–8

Hoiem D, Efros AA, Hebert M (2005) Geometric context from a single image. Proceedings of the 10th IEEE international conference on computer vision ICCV05, vol 1. IEEE Computer Society, pp 654–661

Hoiem D, Efros AA, Hebert M (2007) Recovering surface layout from an image. Int J Comput Vis 75:151–172

Hough PVC (1962) Method and means for recognizing complex patterns. US Patent, 3069654, 1962

Kneepkens REJ (2005) Hough-based road detection. Technische Universiteit Eindhoven, PHD thesis

Maitre H (1985) Un panorama de la transformation de hough. Traitement de Signal 2:305–317

Moslah O (2011) Toward large scale urban environment Modeling from images. Dissertation, Cergy-Pontoise

Rosenfeld A (1969) Picture processing by computer. Academic Press, New York and London

Suh B, Bederson BB (2007) Semi-automatic photo annotation strategies using event-based clustering and clothing-based person recognition. Interact Comput 19:524–544. doi:10.1016/j.intcom.2007.02.002

Torniai C, Battle S, Cayzer S (2007) Sharing, discovering and browsing geotagged pictures on the web. Digital Media Systems Laboratory, HP Laboratories Bristol. http://www.hpl.hp.com/techreports/2007/HPL-2007-73.html. Accessed on 15 June 2014

Valentin R (2009) Reconnaissance de formes - Transformée de Hough. ENSEIRB-MATMECA à l'Institut Polytechnique de Bordeaux, Bordeaux

Yaegashi K, Yanai K (2009) Can geotags help image recognition? Advances in image and video technology. Springer, Berlin Heidelberg, Berlin, Heidelberg, pp 361–373

[illegible]

Part IV
Representation, Visualization and Perception

Encoding and Querying Historic Map Content

Simon Scheider, Jim Jones, Alber Sánchez and Carsten Keßler

Abstract Libraries have large collections of map documents with rich spatio-temporal information encoded in the visual representation of the map. Currently, historic map content is covered by the provided metadata only to a very limited degree, and thus is not available in a machine-readable form. A formal representation would support querying for and reasoning over detailed semantic contents of maps, instead of only map documents. From a historian's perspective, this would support search for map resources which contain information that answers very specific questions, such as *maps that show the cities of Prussia in 1830*, without manually searching through maps. A particular challenge lies in the wealth and ambiguity of map content for queries. In this chapter, we propose an approach to describe map contents more explicitly. We suggest ways to formally encode historic map content in an approximate *intensional* manner which still allows useful queries. We discuss tools for georeferencing and enriching historic map descriptions by external sources, such as DBpedia. We demonstrate the use of this approach by content queries on map examples.

S. Scheider · J. Jones · A. Sánchez
Institute for Geoinformatics, University of Münster, Münster, Germany
e-mail: simon.scheider@uni-muenster.de

J. Jones
e-mail: jim.jones@uni-muenster.de

A. Sánchez
e-mail: a.sanchez@uni-muenster.de

C. Keßler (✉)
CARSI, Department of Geography, Hunter College,
City University of New York, New York, USA
e-mail: carsten.kessler@hunter.cuny.edu

J. Huerta et al. (eds.), *Connecting a Digital Europe Through Location and Place*,
Lecture Notes in Geoinformation and Cartography, DOI: 10.1007/978-3-319-03611-3_15,

1 Introduction

Historic maps provide rich knowledge resources that graphically encode information about the state of a fraction of the real world at a certain point in time. As such, libraries and archives with very large collections such as the Library of Congress with its 5.5 million maps[1] offer an invaluable data source for historians and other researchers. However, libraries and historians cannot make full use of the encoded knowledge to date, as it is currently not possible to automatically retrieve maps that contain the answers to specific questions such as *what were the cities of Prussia in 1830*? or *was Posen part of Prussia in 1802*? While the typical metadata for a historical map contain the title, author, year of production, year represented, and a number of standardized keywords, it often remains unclear whether or not a map is able to help answering such a specific question. If a historian wants to find an answer, it depends on her background knowledge and a fair bit of luck in picking the right keywords for the query to find maps that potentially contain the answer. If those maps have not been digitized yet, it may require searching through a large number of actual paper maps. This is even the case for the fairly small number of maps in the University and State Library (about 2,000) or the Institute for Comparative Urban History's library[2] (about 20,000) at the University of Münster, Germany, that we have dealt with in this work.

A particular problem concerns the way how such detailed map contents should be encoded in order to be machine-readable and in order to allow such detailed queries. Furthermore, the *wealth* as well as the *ambiguity* of the content, even of a single map, poses a challenge for libraries. Extending the metadata for a map to cover all potential keywords about content is clearly not feasible, especially for very large collections.

To overcome this problem, we propose an approach for encoding and querying historic map content based on Semantic Web technology. It makes use of the fact that all map content is *semantically describable in an approximate manner*, *spatially* arranged on the map, and *temporally* referenced through metadata. In a nutshell, we discuss how to make visually encoded map information semantically explicit and queryable to different degrees. We demonstrate how to draw on external knowledge resources, such as DBpedia, for *named explicit content*, and discuss how *implicit contents* can be described in an *intensional* manner which is still useful for search. We base our discussion on a number of competency questions and query scenarios with three map examples in Sect. 2. Section 3 gives an overview of relevant related work. In Sect. 4, we discuss formal ways of encoding map contents, corresponding vocabularies, and review existing software tools for semantic enrichment. Section 5 applies this to the map examples and Sect. 6 demonstrates queries which correspond to the competency questions.

---

[1] http://www.loc.gov/rr/geogmap/guides.html

[2] http://www.uni-muenster.de/Staedtegeschichte/portal/datenbanken

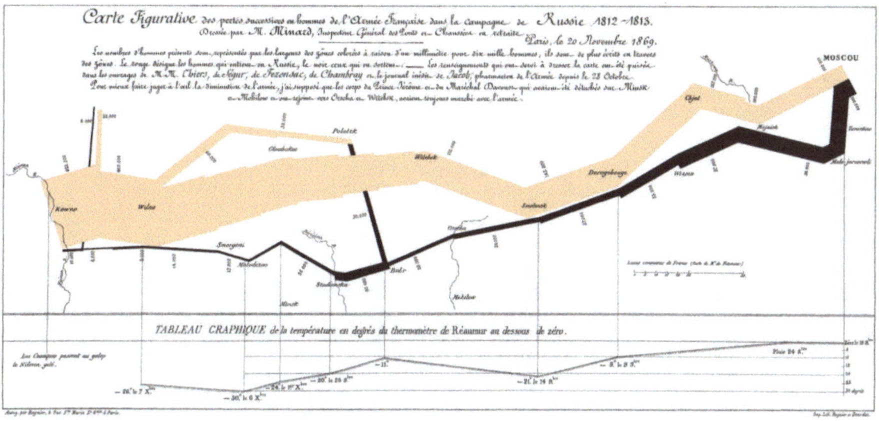

Fig. 1 Charles Minards map from 1861 about Napoleon's 1812 march on Russia

2 Motivating Examples

Following the methodology proposed by Gangemi and Presutti (2009), we start with discussing three map examples together with some *competency questions* that may be asked about them. These questions make some of the semantic content of the maps explicit, and they inform ontology engineers how such contents might be represented. We are aware that our selection does not in any way cover many relevant types of historic maps and possible questions that one might have. Yet, we believe that it already shows considerable variety and depth in content, and therefore serves as a good starting point for research. We focus on three challenging historic maps whose contents are not straightforward to describe.

The first example we discuss is the famous map of Napoleon's 1812 march on Russia by Charles Joseph Minard (see Fig. 1). It depicts the losses of Napoleon's army during his Russian campaign, showing the advance and retreat paths with cartographic signs depicting the number of people of the campaign, the visited places, as well as the temperature. Note that the map is not accurate in terms of the locations and the shape of the paths. The details of content and alternative ways of visualizing it are discussed in Kraak (2003). We focus here on some of the informative questions that might be asked about this map:

1. Where did Napoleon's 1812 campaign to Russia happen?
2. How many people did Napoleon's army have when soldiers arrived in Smolensk during his 1812 campaign?
3. What were the lowest temperatures during Napoleon's campaign?
4. Which places did Napoleon's army come across during the 1812 campaign?

Our second example is a political map which depicts the administrative parts of Prussia in the 17th century (see Fig. 2). It shows the different sub-territories and the Prussian dukes involved in obtaining them in the course of this century.

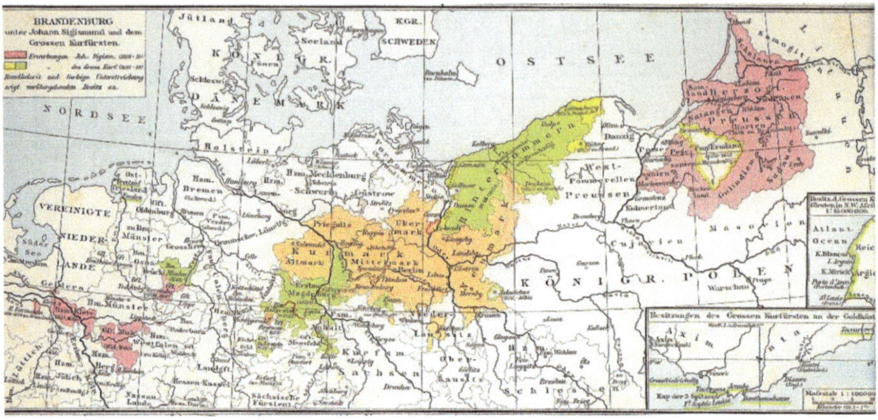

Fig. 2 Prussia (Brandenburg) in the 17th century. *Source* (public domain) G. Droysens Historischer Handatlas, 1886

1. Where was Prussia in 1688?
2. Which territories were part of Prussia in 1688?
3. Which Prussian territories were acquired by Friedrich-Wilhelm of Brandenburg, the great elector?

Our third example is a topographic map of Hildesheim from 1839 (see Fig. 3). This map is very rich in detail and in thematic content, including roads, landscape features, landcover, as well as the built environment. However only few of these depicted kinds of things are actually named entities.

1. Where was Hildesheim in 1839?
2. What were the types of landcover around Hildesheim in 1839?

The problem we address in this chapter is how an explicit semantic representation of the content should look like in order to answer these questions in terms of queries. Based on our examples, we explore a number of related research challenges: Which language and which expressivity is needed for encoding? In particular, how can we encode *complex spatio-temporal map contents*, such as the story behind Minard's map? How can we describe contents involving well known *named entities*, such as the Prussian territories and their associations with kings? How can we encode contents involving *nameless entities*, such as the landscape around Hildesheim or city blocks? And, since encoding each and every detail of a map's content is clearly not feasible: how can map content be encoded in an *intensional form*, i.e., such that only aspects of the content are made explicit?

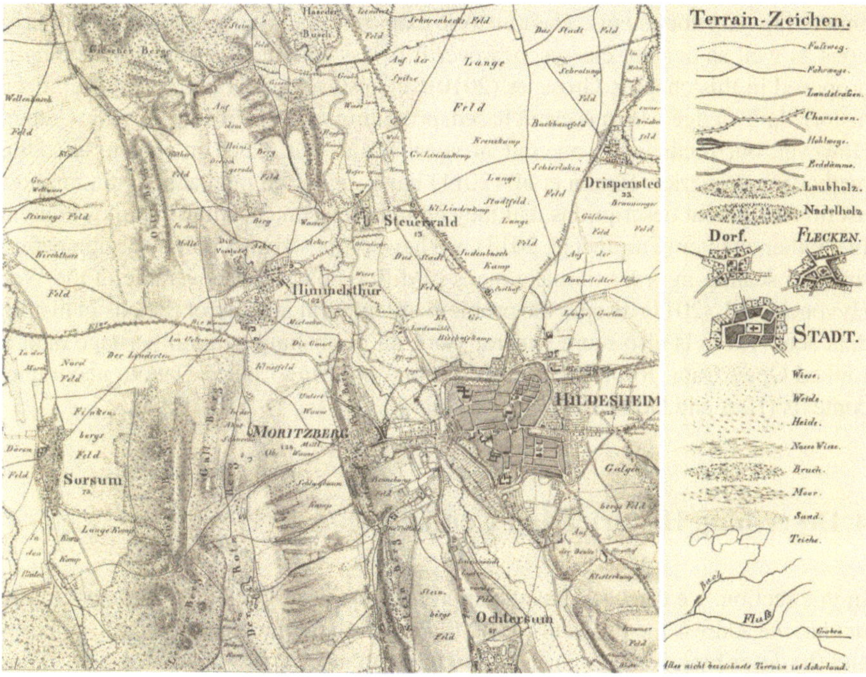

Fig. 3 Excerpt of a map of Hildesheim of the "Gaußsche Landesaufnahme" from 1839. *Source* Historische Kommission für Niedersachsen, Hannover 1963

3 Related Work

Making the content of historic maps and libraries available to the public is an ongoing topic of research as well as development. Library records are commonly encoded in the MARC21 standard or similar, however, more flexible and precise ways of representing document contents are needed. One focus is on combining historical databases with geographical information systems (GIS) (see Grossner (2010) for an overview) and on novel approaches of visual map representation Kraak (2003). A recent line of developments is concerned with *historical gazetteers*, such as CHGIS[3] and Simon et al. (2012). However, only few authors have focused on using Semantic Web technology for map descriptions so far. In the context of digital libraries, there is recent work on tools for annotating historic maps (Simon et al. 2011; Haslhofer et al. 2013), extracting content (Arteaga 2013) and on semantic vocabularies and ontologies (Gkadolou and Stefanakis 2013; Gkadolou et al. 2013; Grossner 2010). Gkadolou and Stefanakis (2013) propose an extension of CIDOC-CRM[4] for

[3] http://www.fas.harvard.edu/~chgis/

[4] The International Committee for Museum Documentation's conceptual reference model for cultural heritage documentation, see http://cidoc.ics.forth.gr.

annotating map documents. Their historic maps ontology[5] covers document descriptions as well as content classes, however it is less suitable for content encoding as proposed in this chapter. Grossner (2010) suggests a general ontology to represent historic knowledge which is event based, providing relevant insights for this chapter. However, our focus is not on content vocabularies but rather on content encoding methods and queries. Simon et al. (2011) have suggested map annotation methods useful for content descriptions, but they do not focus on encoding complex content. In another chapter (Carral et al. 2013), we have proposed an ontology design pattern for map scaling in order to make maps of different scale discoverable in the Web. Hyvönen et al. (2011) discuss the problem of temporal identity of regions in historic data sets. There is also more general work on publishing cultural heritage data as Linked Open Data (Ruotsalo et al. 2013) and on using Linked Data in geographic contexts (Hart and Dolbear 2013).

4 Describing Historic Map Contents

In this section, we discuss semantic approaches to describing map contents. In particular, we address the problems of (a) finding a syntax for encoding these contents; and (b) describing contents in an approximate manner.

4.1 Formally Encoding Map Contents

What exactly is the content of a map? This question, at least in this form, might be too big to yield an answer, given the contested nature of maps as complex signs (MacEachren 2004). It may be as difficult to answer as the question about the semantic content of texts. However, from a pragmatic viewpoint and without being overly reductionist, one could say that the content of a map as a document is the *set of assertions* which can be extracted by looking at it (see Fig. 4).

Such *map interpretation* has a long tradition in cartography and geography, and it consists in drawing explicit conclusions from looking at a map. It is well known that conclusions can be different depending on who is looking at a map. Furthermore, an interpreter may have difficulties in actually writing down *all* contained assertions in an exhaustive manner. Still, it makes sense to think about map content in terms of a set of assertions which *can* be made by some interpreter. For example, the assertions extractable from the Minard map may include one statement saying that upon Napoleon's arrival in Moscow, the temperature was 0 degree on the Reaumur scale. And the map content of the Prussia map may contain assertions saying that Hinterpommern is a territory, that was part of Prussia 1806, and another one saying that it was acquired by elector Friedrich-Wilhelm. All of these assertions are visually

[5] http://gaia.gge.unb.ca/eg/HistoricalMap.owl

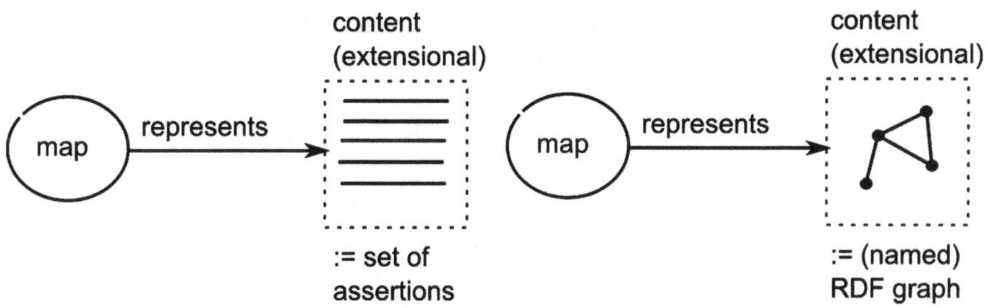

Fig. 4 Map contents as sets of assertions. A useful form of encoding such assertions are RDF graphs. Encoding sets of such assertions can be done in terms of "named graphs"

encoded in the corresponding map and can thus be extracted by human beings without drawing on further knowledge sources. In the following paragraphs, we discuss formal encoding schemes for such content.

Encoding map content as a named graph In order to formally encode these assertions and for making them machine-readable, however, we need a formal language and a vocabulary which allows us to talk about *individual represented phenomena*. That is, we need names for represented individual entities, like "Prussia" and "Hinterpommern", and we need also logical constants for classes and relations, such as "territory" or "state" and "is a", "part of" or "temperature was". Furthermore, we need a way of linking maps with their content.

One simple form of encoding such assertions are *RDF*[6] *Triples*, i.e., logical assertions with *subject*, *predicate*, and *object*. They form edges of a labeled graph in which subject and object are nodes and predicates denote types of edges. RDF assertions written down in this manner[7] look like this:

$$dbp{:}Hinterpommern \quad rdf{:}type \quad phen{:}Territory. \tag{1}$$

$$dbp{:}Hinterpommern \quad phen{:}partOfObject \quad dbp{:}Prussia. \tag{2}$$

where *dbp:*, *rdf:* and *phen:* specify namespaces for shared vocabularies, such as DBpedia[8] and RDF.[9] In the following, we will use the abbreviation *a* for *rdf:type*. RDF triples of this form do not contain any variables, only constant names. Names can either be Web addresses (URIs) or strings (called *literals*). The resulting assertions then form a *Linked Data graph* of explicit map contents which can be published on the Web (Bizer et al. 2009).

How should this content description be linked to a map? Maps can easily be encoded as single node which denotes some document. We then use a predicate

[6] http://www.w3.org/RDF/

[7] Throughout the chapter, we use the Turtle syntax to write down triples; see http://www.w3.org/TeamSubmission/turtle/.

[8] http://dbpedia.org

[9] http://www.w3.org/1999/02/22-rdf-syntax-ns

`maps:represents` as a semiotic short hand for the fact that some part of the map image, namely a certain *map sign*, represents something. Note that this "something" needs to be a graph in our case: *Each map represents one graph which encodes its content*, see Fig. 4. This requires to link maps to whole graphs, not to single nodes, which is only possible using a *named graph*,[10] i.e., a graph of assertions which has been assigned a URI for identification. The latter can be linked via `maps:represents` to a map and then serves to answer corresponding map queries. However, while triple stores such as OWLIM[11] support named graphs, they actually break the language barriers of RDF. They require, in effect, quadruples instead of triples.

Encoding map content by direct links If one wants to stay in RDF (or in some other Semantic Web logic) and if one is willing to sacrifice content relations for content nodes, it is possible to link contents without a named graph:

$$:map \quad maps{:}represents \quad dbp{:}Prussia. \tag{3}$$

It is furthermore possible and indeed makes sense to use both of these approaches for encoding map contents. In the following, we assume that all nodes of a content graph are also directly linked to the corresponding map.

The problem of encoding content implicitly The decision which names, classes and relations are needed to encode contents is not easy to take. Furthermore, it is almost unfeasible to write down map content assertions in an exhaustive manner. The reason is not only that it is too much of an effort to write them down, but also that for many assertions visually encoded in a map, we lack corresponding unambiguous names and constants. A look at the Hildesheim map (see Fig. 3) makes this clear: Most assertions depicted in this map talk about things that are not explicitly named, and about relations for which we may not know any adequate established constant names. For example, there are landscape objects like hills without names, and the question is open which geometrical relations we should choose.

Intensional description of map contents by graph patterns This makes it very hard to turn map content into a machine-readable form. Using formal logic, however, there is still a way of describing map contents in an approximate manner. If for some reason, we are unable to list content assertions, we may still write down some logical description of these assertions useful for search. For example, it is not required to explicitly encode all names and content assertions in terms of ground sentences. We can instead use variables that range over potential entities in order to avoid talking explicitly about some content. This allows us to encode map content in an *intensional* instead of an *extensional form*.

A useful form of such an intensional map content description is a *constraint network* or a *graph pattern*. This is a graph with variables. Assume each variable is implicitly existentially quantified (see Fig. 5). Such a graph is like a successful query that would deliver results if fired against an exhaustive set of RDF statements which

[10] http://en.wikipedia.org/wiki/Named_graph

[11] http://www.ontotext.com/owlim

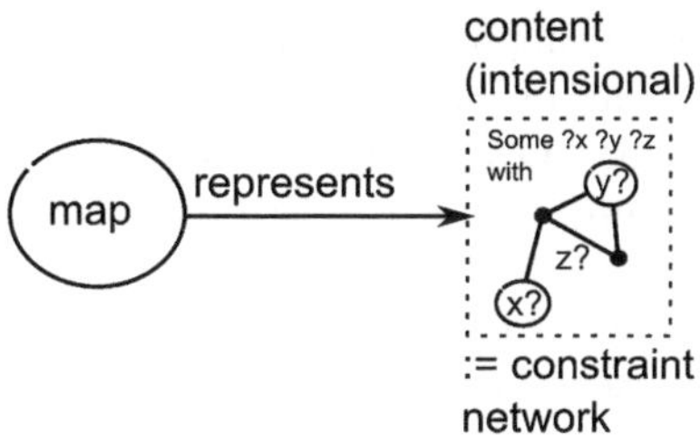

Fig. 5 Intensional description of map content using a constraint network. *Black nodes* are constants and question marks indicate variables

constitutes a map's content (or that of a large map collection). And just like such a query, it says something about the map's content. Graph patterns are therefore also used as bodies in the SPARQL query language.[12]

For example, without any knowledge of the historic names, one could encode the unnamed content of the Prussian map simply as the following query, where question marks indicate variables:

$$
\begin{aligned}
?x\ :partOfObject\ ?y.\\
?y\ a\ :State.\\
?x\ :wasAcquiredBy\ ?z.\\
?w\ :isSettingForPerson\ ?z.\\
?w\ :isSettingForRole\ :King.\\
?w\ :rulesOver\ ?y.\\
?w\ :isSettingatTime\ ?q.
\end{aligned}
\tag{4}
$$

The query says that some person *?z*, such as Friedrich the Great, who took the role of king of some state *?y*, such as Prussia, during time interval *?q*, such as 1740–1786, acquired some *?x*, such as Silesia, which was part of Prussia.[13]

A useful way of encoding such content graph patterns is to write them down as *graphs with blank nodes* instead of variables. Blank nodes, denoted by `_:id`, where *id* can be any string, can be interpreted as existentially quantified variables, because they can be interpreted into any other node satisfying a query. In our implementation, we encoded intensional map contents in terms of *named graphs with blank nodes*, as this allows to use the query mechanism of a triple store.

A simple intensional map content graph For further illustration, we discuss an example which corresponds to a very simple but useful constraint network. Because of

[12] http://www.w3.org/TR/rdf-sparql-query/

[13] In order to stay historically correct, one would need to say that Silesia was part of Prussia only after its conquest in 1763. This would require to introduce time-indexed *partOf* relations. Similarly, *wasAcquiredby* reflects some event. However, as a matter of fact, such kind of information is actually not contained in the map, and thus should not be represented by the content graph. Moreover, representing such time-indexed relationships presents a challenge of its own (Trame et al. 2013) which goes beyond the scope of this chapter.

their simplicity, graphs of this form can be encoded also in terms of constructs supported by the Web Ontology Language (OWL)[14] instead of named graphs with blank nodes. OWL encodes a fragment of Description Logic (DL) (Krötzsch et al. 2012), a subset of First-order Logic (FOL) which allows quantifying variables to a limited extent.

The simplest but still useful case is when we ask whether an instance of a certain class is contained in a map. People who catalog contents may only want to say that "some phenomenon of a class" is depicted in a map. For example, one may want to say that a map contains some administrative units, some roads, rivers and cities, without telling exactly which units, roads and cities are present. This directly corresponds to the following pattern:

$$?x\ a\ :River. \tag{5}$$

Without making use of a named graph, this content description can also be directly linked to a map by some Description Logic (DL) existential:

$$:map\ a\ \exists :represents.:River. \tag{6}$$

In this expression, the class $\exists represents.River$ stands for an OWL class restriction (where square brackets stand for a blank node), which is encoded in RDF as:

$$\begin{aligned}&[a\ owl{:}Restriction;\\ &owl{:}onProperty\ :represents;\\ &owl{:}someValuesFrom\ :River].\end{aligned} \tag{7}$$

The expression literally means "the class of maps which represent something which is a river". A future task is to identify similar intensional content patterns which can be encoded directly into some DL fragment and which can then be used for reasoning.

4.2 Vocabularies for Historic Map Content

Map descriptions need to make use of content vocabularies as well as vocabularies for describing documents. As these are two fundamentally different matters, they need to be distinguished. We first discuss vocabularies for geographic phenomena, and then propose vocabularies for describing semantic, spatial and temporal reference of maps as documents.

Geographic phenomena A geographic phenomenon (denoted by the class *GeoPhenomenon*) is any phenomenon on a geographic scale (Montello 1993) which can be

[14] http://www.w3.org/TR/owl-ref/

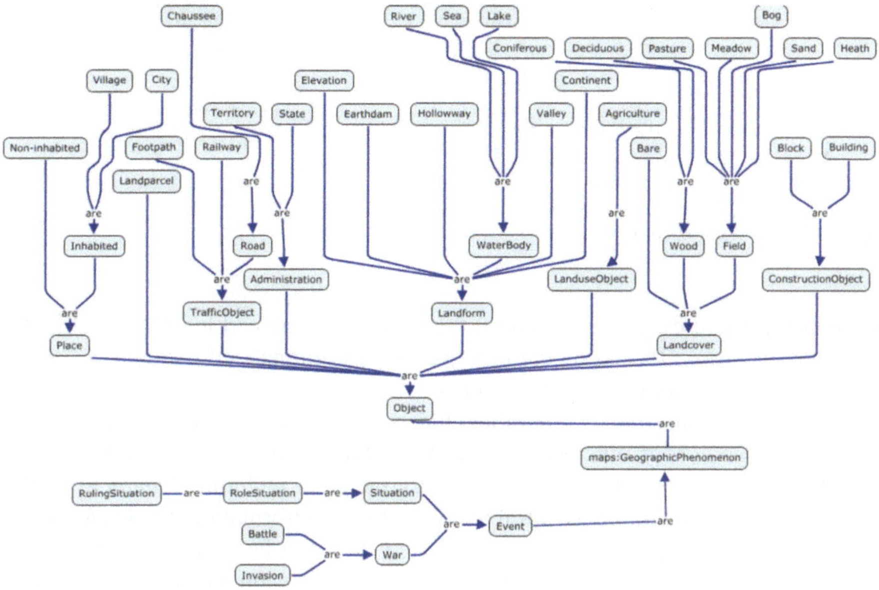

Fig. 6 Geographic phenomenon classes relevant for describing the content of historic maps

mapped. It roughly corresponds to what the Open Geospatial Consortium (OGC) calls a "geographic feature" (Kottmann and Reed 2009). For the domain of historic maps, the specific cartographic techniques used constrain the range of types of phenomena which are depicted in such maps. Grossner (2010) argues for an event-centric approach and suggests a corresponding spatial history ontology. A semi-structured overview of phenomenon classes is often available in the form *map legends*. We chose concepts needed based on our examples (see Fig. 6) and put them into the *historicmapsphen* ontology.[15] We thereby reused concepts from other geographic ontologies.[16] However, the question of ontological coverage is beyond the scope of this paper and we expect that the phenomenon ontology needs to be extended for each particular collection of maps under consideration.

Different classes of geographic phenomena correspond to conventional *cartographic perspectives* on geographic space, and thus to conventionalized domains of experience. These include geographic objects such as *landforms* (based on the shape and texture or the ground surface), *landcover* (based on the form of vegetation on the ground surface), *landuse* (based on cultivation of the ground surface), *construction* (based on the presence of buildings), *traffic* (based on the presence of

[15] Available at http://geographicknowledge.de/vocab/historicmapsphen [.rdf/.jpg], denoted by prefix *phen*.

[16] In fact, our classes cover the classes of geographic kinds suggested by Smith and Mark (2001), which was based on an empirical study. We furthermore imported the ontology for Linking Open Descriptions of Events (LODE) http://linkedevents.org/ontology/.

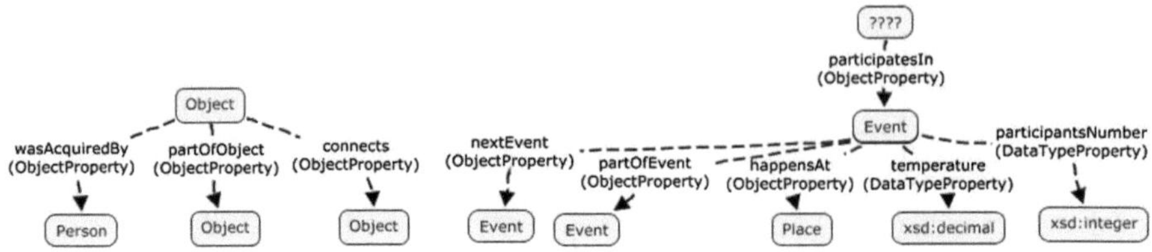

Fig. 7 Relations between phenomena

traffic infrastructure), *place* (based on the presence of named locales, inhabited or not), *administration* (based on the presence of institutionalized entities), as well as *landparcel* objects (objects of ownership).

These objects can *connect*[17] to other objects (e.g. a road connects to some village) and they can be *part of* other objects (such as administrative units) (see Fig. 7).

Furthermore, *events* and their *participants*, the latter including *human* as well as *institutional agents*, take on a central role in historical information. Events *happenAt* places, they can be *part of* other events, and they can be a *member of an event sequence*. Furthermore, all phenomena may have observable properties. A particularly relevant kind of historic event is a *role situation* in which some person takes on some role, e.g., of being a king. The complex relation between a person, his role as a king, his country, and the time of reign can be encoded using Gangemi's *time indexed person role* (Gangemi and Presutti 2009).[18] We included a simplified version of this pattern in our ontology which follows the example given in Sect. 4.1.

We assume that an individual phenomenon may be instantiated by more than one phenomenon class, depending on perspective. For example, a landuse instance such as an agricultural area may at the same time be considered a landcover area (bare), a landform (hill), as well as an object of ownership (landparcel). This corresponds to the common habit of intermixing geographic categories in a map legend, giving rise, e.g., to forest as a type of landuse. We therefore do not assume that the phenomena classes are mutually exclusive.

Space and time of map contents A spatial geometry, denoted by the class *Geometry*, denotes a region, i.e., a subset of point locations in some spatial reference system. We use the *GeoSPARQL ontology*[19] (Battle and Kolas 2012) in order to encode geometries together with their reference system. A geometry has one or more RDF literals which encode its region. We express the geometry as a WKT literal[20] using the GeoSPARQL predicate *asWKT*, which maps from geometries to WKT literals:

[17] With this predicate, we express that phenomena are visually connected in the map image, without making any further implications.

[18] http://ontologydesignpatterns.org/wiki/Submissions:Time_indexed_person_role

[19] Available at http://www.opengis.net/ont/geosparql/1.0, prefix *geo*.

[20] A serialization of geometry based on OGC's *simple feature* standard.

$$geo{:}asWKT\ rdfs{:}domain\ geo{:}Geometry. \tag{8}$$

$$geo{:}asWKT\ rdfs{:}range\ geo{:}WKT(Literal). \tag{9}$$

Geographic phenomena have locations. In order to express that a phenomenon is located at a geospatial geometry, we use a sub predicate *where* of the GeoSPARQL predicate *hasGeometry*, which maps from geographic phenomena to their geometric representation:

$$phen{:}where\ rdfs{:}domain\ phen{:}GeoPhenomenon. \tag{10}$$

$$phen{:}where\ rdfs{:}range\ geo{:}Geometry. \tag{11}$$

Maps *always* (per definitionem) represent the location of geographic phenomena. This can be added as a corresponding DL axiom (where we omitted namespaces):

$$GeoPhenomenon \sqcap \exists represents^{-}.Map \sqsubseteq \exists where.Geometry \tag{12}$$

This axiom says that if some geographic phenomenon is represented by a map, then the location of that phenomenon is also represented. Such an axiom can be expressed in OWL and automatically adds blank nodes as geometries of encoded geographic phenomena.[21]

For historic maps, the temporal coverage is as important as the spatial coverage. We denote temporal entities based on *OWL time*,[22] which encodes intervals simply as entities with start and end literals encoded as *xsd:dateTime*.

Maps as documents In the *maps ontology*,[23] we describe maps as image documents which represent windows of space, time as well as spatio-temporally referenced content. For this purpose, we subclass the class *Map* of the *Bibo* ontology,[24] which denotes image documents, add properties such as scale, and reuse *Dublin Core Terms*[25] predicates, such as title, author, medium, size (see Fig. 8). We also linked to *HistoricalMap* in http://gaia.gge.unb.ca/eg/HistoricalMap.owl. Maps represent assertions about geographic phenomena (encoded in terms of named graphs with blank nodes), the phenomena themselves (encoded by direct links), as well as regions in spatial and temporal reference systems. For this purpose, we use the semiotic predicate `maps:represents` and add corresponding sub-properties.

For example, maps directly represent the *spatial area of their map window*, a particular geometry which covers the spatial extent of the map, and which is denoted by a sub-predicate *mapsArea*:

[21] Alternatively, one may add corresponding blank nodes by some SPARQL construct.

[22] Available at http://www.w3.org/2006/time; prefix *time*.

[23] Available at http://geographicknowledge.de/vocab/maps [.rdf/.jpg], prefix *maps*.

[24] http://purl.org/ontology/bibo/

[25] http://purl.org/dc/terms/

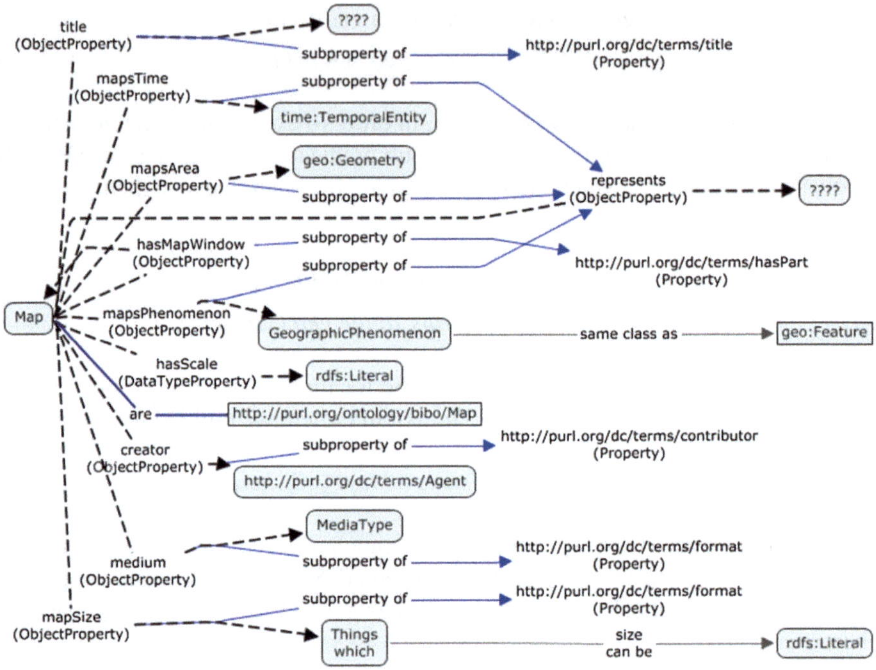

Fig. 8 Ontology of maps as documents

$$maps{:}mapsArea\ rdfs{:}subPropertyOf\ maps{:}represents. \tag{13}$$

$$maps{:}mapsArea\ rdfs{:}range\ geo{:}Geometry. \tag{14}$$

Correspondingly, maps have a *temporal window*, which corresponds to the time which is represented in the map:

$$maps{:}mapsTime\ rdfs{:}subPropertyOf\ maps{:}represents. \tag{15}$$

$$maps{:}mapsTime\ rdfs{:}range\ time{:}TemporalEntity. \tag{16}$$

4.3 Georeferencing Historic Maps and Finding Contents

The traditional way to find a historic map, for example in a library, relies on tagging maps with key words and expecting them to match user queries. In contrast to this, our approach consists in spatio-temporal and semantic content descriptions. The combination of semantic descriptions and georeferencing enables the use of explicit historical background knowledge published in the Web, such as DBpedia.[26] We

---

[26] http://dbpedia.org

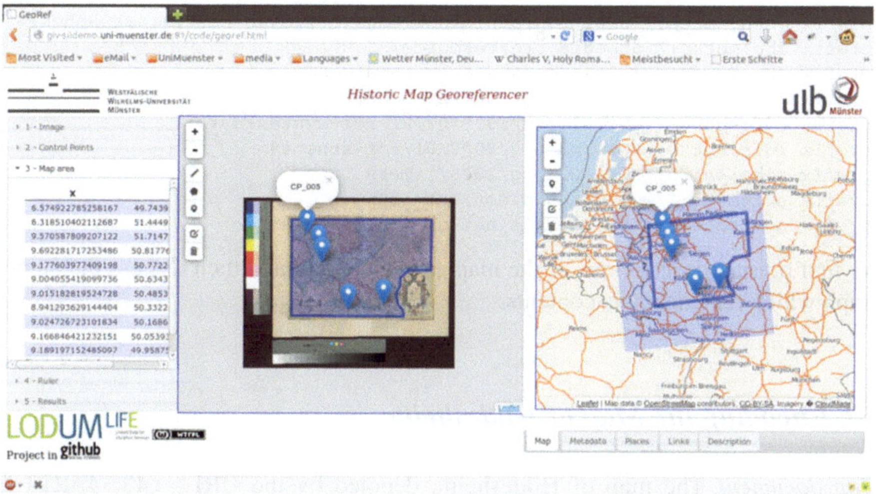

Fig. 9 A georeferencing tool which allows combined spatial, temporal and semantic descriptions of historic maps

have developed a *georeferencing tool* which allows to combine complex content descriptions with georeferencing.[27] The map area and time are determined by the users through *georeferencing* a map image. This information allows the tool to search for entries located inside the map area or space-time window in DBpedia. Since DBpedia does not support GeoSPARQL yet, the spatial component is formulated using the bounding box of the map area. The result is a list of resources potentially related to the map (i.e. modern locations and historical events overlapping the historic map), which is presented to the user, who chooses those resources to be included in the map's description. Furthermore, it allows to encode intensional content in a simple way, as nameless typed entities (see Sect. 4.1). The application is still under development, but it may be turned into a more comprehensive map annotation tool, which allows to extend and reuse content ontologies and to encode arbitrary content in an intensional form (Fig. 9).

A remaining question is how to let users extract content in an automated fashion. One possibility is to use an image vectorization approach. This was employed in the *Map polygon and feature extractor* tool (Arteaga 2013).[28]

5 Encoding the Example Maps

In this section, we present the content encodings of the three example maps using the solutions discussed in Sect. 4. The following vocabularies are used:

```
@prefix maps:<http://www.geographicknowledge.de/vocab/maps#> .
```

[27] https://github.com/lodum/georef

[28] https://github.com/NYPL/map-vectorizer

```
@prefix phen:<http://www.geographicknowledge.de/vocab/historicmapsphen#> .
@prefix dbp:<http://dbpedia.org/resource/> .
@prefix dbp-de:<http://de.dbpedia.org/resource/> .
@prefix xsd:<http://www.w3.org/2001/XMLSchema#> .
@prefix rdf:<http://www.w3.org/1999/02/22-rdf-syntax-ns#> .
@prefix rdfs:<http://www.w3.org/2000/01/rdf-schema#> .
@prefix time:<http://www.w3.org/2006/time#> .
@prefix sf:<http://www.opengis.net/ont/sf#> .
@prefix geo:<http://www.opengis.net/ont/geosparql#> .
```

We first present descriptions of the map as a document and then discuss the named graph which represents its contents.

5.1 Encoding the Map of Hildesheim

Map document The map of Hildesheim, denoted by the URI `+:4354_Hildesheim+`, is georeferenced and described as a map document in the following listing:

```
:4354_Hildesheim a maps:Map;
    maps:digitalImageVersion :4354_Hildesheim.jpeg;
    maps:represents :hildesheim;
    maps:hasScale "1:28526.1"^^xsd:string;
    maps:mapSize"62.4 * 55.5 cm"^^xsd:string;
    maps:medium maps:Paper;
    maps:mapsTime"1840"^^xsd:gYear;
    maps:mapsArea _:4354_Hildesheim_geom.
    _:4354_Hildesheim_geom a geo:Geometry
    geo:asWKT"<http://www.opengis.net/def/crs/EPSG/0/4326>
POLYGON((9.874690102339652 52.25156096729222, 9.874324681594004
52.126487663211606, 10.07547489355107 52.1268449901813,
10.073392224324136 52.252405987705664, 9.874690102339652
52.25156096729222))"^^sf:wktLiteral.
```

Content graph (:hildesheim) Our encoding of the Hildesheim map content makes heavy use of blank nodes for nameless and approximate contents. In this way, it becomes possible to say that villages and buildings and city blocks are shown in the map, the latter being part of the city of Hildesheim, that Hildesheim is connected to some road as well as the river "Innerste", and that the map also contains wood, pasture, elevations and non-inhabited named places, such as "Steinbergsfeld".

```
_:someroad a phen:Road ;
    phen:connects _:somevillage ;
    phen:connects dbp:Hildesheim .
_:someroad2 a phen:Footpath .
dbp:Hildesheim a phen:City .
_:somevillage a phen:Village .
_:someblock a phen:Block ;
    phen:partOfObject dbp:Hildesheim .
_:somebuilding a phen:Building .
_:somehill  a phen:Elevation .
```

```
 _:someforest a phen:Wood .
 _:somepasture a phen:Pasture .
 _:someplace a phen:Non-inhabited .
 dbp:Innerste a phen:River ;
     phen:connects dbp:Hildesheim .
...
```

5.2 Encoding Minard's Map

Map document The following triples describe the map and link it to the named graph `:french_invasion+`. We omit further document descriptions as they are equivalent to our first example.

```
:minard_map a maps:Map;
    maps:represents :french_invasion ;
    maps:mapsTime "1812"^^xsd:gYear .
...
```

Content graph (*:french_invasion*) In this map, a logical sequence of nameless stationary war events is visually encoded in terms of a spatio-temporal flow band. Correspondingly, these events are encoded as blank nodes. The map itself tells us nothing about the type of event (e.g. whether it is a battle or just a campaign arrival). The map is also vague about the path taken by the army in between these stationary events. The only information that we can get from the map is the logical event sequence and some of the event properties, such as the temperature during an event, the (remaining) number of people in the campaign, as well as the geographic place (and the time) of the event. The following listing shows an excerpt of this event sequence:

```
_:invasionOfRussia a phen:Invasion .
_:EVENT_13 a phen:War ;
    phen:partOfEvent _:invasionOfRussia ;
    phen:happensAt dbp:Smolensk ;
    phen:participantsNumber "37000"^^xsd:decimal ;
    phen:temperature"-21"^^dbp-de:Raumur-Skala ;
    phen:nextEvent _:EVENT_14 .
_:EVENT_14 a phen:War .
    phen:partOfEvent _:invasionOfRussia ;
    phen:happensAt dbp:Orscha ;
    phen:participantsNumber"24000"^^xsd:decimal ;
    phen:nextEvent _:EVENT_15 .
...
```

5.3 Encoding the Map of Prussia

Map document.

```
:Prussia_map a maps:Map;
    maps:represents :prussia ;
    maps:mapsTime "1688"^^xsd:gYear .
...
```

Content graph (:prussia) The following listing encodes the information that Johann Sigismund, a duke of Prussia from 1618–1619, acquired the "Grafschaft Mark", which is a subterritory of Prussia. Note that the situation of Johann Sigismund being the duke of Prussia is not given any name (and thus represented by a blank node, i.e., a variable), and similarly, the time interval of his ruling.

```
dbp:Grafschaft_Mark phen:wasAcquiredBy dbp:Johann_Sigismund ;
    phen:partOfObject dbp:Prussia .
_:EVENT_DUCHY_JS_PREUSSEN a phen:Event ;
    phen:isSettingForPerson dbp:Johann_Sigismund ;
    phen:isSettingForRole dbp:Duke ;
    phen:rulesOver dbp:Prussia ;
    phen:isSettingAtTime _:TIME_INTERVAL_JS_PREUSSEN .
_:TIME_INTERVAL_JS_PREUSSEN a time:TemporalInterval ;
    time:hasBeginning _:INSTANT_BEGINNING_JS ;
    time:hasEnd _:INSTANT_END_JS .
_:INSTANT_BEGINNING_JS a time:Instant ;
    time:inXSDDateTime "1618"^^xsd:gYear .
_:INSTANT_END_JS a time:Instant ;
    time:inXSDDateTime"1619"^^xsd:gYear .
...
```

This is continued for all subterritories shown in the map.

6 Querying Historic Map Contents

If historians search in map collections, they are primarily interested in *finding maps which help answer their question*. While this is currently done through searching for maps which have certain properties, acting as a proxy, the former is a much more general problem. We translate this question into the following *map content query*, expressed in SPARQL syntax and based on our encoding scheme:

```
SELECT ?map ... WHERE {
  ?map maps:represents ?g .
  GRAPH ?g { ...   }
}
```

This query can be translated as: *which maps represent content which contains statements of the following form* (i.e., *answering the graph pattern*)...? The historian's question to be answered by the map is posed in terms of the graph pattern. The encoded content (as far as it was made explicit) can also be retrieved along with the map. Note that map content graphs may be intensional, and thus do not have to contain an answer. Still, a triple store can deliver a map which contains an answer if the map's intensional description (including blank nodes) satisfies the pattern. That is, even though the answer might not be explicitly encoded, knowledge about whether the map contains the answer can be used to automate map selection. We illustrate this mechanism in the following by our examples. All of these queries were tested on a standard triple store using a standard reasoner.[29]

Where was Hildesheim in 1840?

```
SELECT DISTINCT ?map ?where WHERE {
    ?map maps:represents ?g ;
         maps:mapsTime "1840"^^xsd:gYear .
    GRAPH ?g {{dbp:Hildesheim ?p ?o}UNION{?a ?d dbp:Hildesheim}
    }
    dbp:Hildesheim phen:where ?where .
}
```

This query simply retrieves maps which represent the location of Hildesheim in 1840. The `UNION` keyword, a logical or, allows detecting Hildesheim in a map's content regardless of whether it is subject or object of an assertion. As mentioned in Sect. 4.1, all maps which represent Hildesheim link to it directly, and by Eq. 12, Hildesheim therefore must have some location. Thus, the query delivers a meaningful result, even though the geometry of Hildesheim was never explicitly encoded.

What were the types of landcover around Hildesheim in 1840?

```
SELECT DISTINCT ?map ?class WHERE {
   ?map maps:represents ?g ;
        maps:mapsTime "1840"^^xsd:gYear .
   GRAPH ?g {{dbp:Hildesheim ?p ?o}UNION{?a ?d dbp:Hildesheim}
      ?instance a ?cl .
   }
    ?instance a ?class .
    ?class rdfs:subClassOf phen:Landcover.
}
```

In this query, the intended types of landcover (`?class`) are those subclasses of `phen: Landcover` for which we know that there is some instance of this class depicted in a map which shows Hildesheim in 1840. We are not interested in these instances as such, only in their classes. And in fact, for this query to deliver results, the content graph does not have to contain any actual instances. In fact, they are all

---

[29] OWLIM-Lite 4.0 with OWL-Horst reasoner (optimized) and OpenRDF workbench version 2.7.7. The endpoint is available at http://data.uni-muenster.de:8080/openrdf-sesame/repositories/mapcontents.

blank nodes. Note that since reasoning is done based on the ontology, the landcover classification pattern needs to be located outside the content graph. In this way, an RDFS reasoner will be able to automatically classify all instances which are landcover, regardless of whether this was made explicit in the content graph.

How many people did Napoleon's army have when soldiers arrived in Smolensk during his 1812 campaign?

```
SELECT ?map ?soldiers WHERE {
   ?map maps:represents ?g .
   GRAPH ?g {
      ?event phen:partOfEvent dbp:French_invasion_of_Russia ;
             phen:happensAt dbp:Smolensk ;
             phen:participantsNumber ?soldiers .
   }
}
```

In order to answer this question, we need to find the respective subevent. Only if this subevent and the corresponding property is depicted in the map, we can be sure that the map serves to answer that question.

Very similarly, this query allows to retrieve a minimal property of all subevents. *What were the lowest temperatures during Napoleon's campaign?*[30]

```
SELECT (MIN(?temperature) AS ?temp) WHERE {
   ?map maps:represents ?g .
   GRAPH ?g {
      ?event phen:partOfEvent dbp:French_invasion_of_Russia ;
             phen:temperature ?temperature .
   }
}
```

Which places did Napoleon's army come across during the 1812 campaign?

```
SELECT DISTINCT ?map ?places WHERE {
   ?map maps:represents ?g .
   GRAPH ?g {
      ?event phen:partOfEvent dbp:French_invasion_of_Russia ;
             phen:happensAt ?places. }
}
```

For the Prussia map, we need to make use of object parthood relations in order find answers:

Which territories were part of Prussia in 1783?

```
SELECT ?map ?parts WHERE {
   ?map maps:represents ?g ;
        maps:mapsTime "1783"^^xsd:gYear .
```

[30] Note: Due the fact that the temperature uses a literal dbp-de:Reaumur-Skala, the SPARQL function MIN does not work without additional effort.

```
    GRAPH ?g {
        ?parts phen:partOfObject dbp:Prussia .
    }
}
```

Which Prussian territories were acquired by Friedrich Wilhelm, the great elector?

```
SELECT ?map ?parts WHERE {
   ?map maps:represents ?g .
   GRAPH ?g {
      ?parts phen:partOfObject dbp:Prussia .
      ?parts phen:wasAcquiredBy dbp:Friedrich_of_Brandenburg .
   }
}.
```

7 Conclusion

In this chapter, we argued that the contents of historic maps can be encoded and efficiently queried in terms of named RDF graphs with blank nodes. This allows to formally represent map content in terms of a set of triple assertions and to link the description of the map as a document with its content in an explicit way. It furthermore makes possible intensional descriptions of map content, in which parts of the content are not made explicit while still being useful to inform about the types and relations depicted in a map. This solution allows to logically express explicit contents to different degrees: Complex spatio-temporal content can be encoded in great detail, going well beyond simple links from maps to historic entities. Named phenomena can be linked to external knowledge sources on the Web, while nameless or implicit content can be encoded in an intensional manner using blank nodes. All this may help libraries in encoding map content, and it allows to retrieve historic maps based on specific content-related questions that historians have when searching for maps, instead of keywords or simple map properties.

We discussed corresponding encoding schemes, vocabularies and tools. We furthermore tested encodings as well as map content queries on a standard triple store, based on a list of competency questions about three historic map examples.

Future work should focus on the following aspects: Which tools would support users in encoding complex map contents without deeper knowledge of Semantic Web technology? The georeferencing tool presented in this chapter needs to be extended in order to generate complex content descriptions such as named graphs. And correspondingly, which tools would allow library users to formulate arbitrary content queries and hence go beyond browsing and exploring (Simon et al. 2012)? Annotation quality and usability of these tools needs to be evaluated in a library context. It will require a community of interested volunteers to generate content descriptions for a significant portion of a given map collection. Furthermore, the proposed solution for encoding and reasoning needs to be further developed, based on actual information needs of historians. It would, e.g., be useful to hide blank nodes from users and to provide frequent query patterns as templates. It remains also unclear which sort of reasoning is needed to deal with intensional map contents in general.

Acknowledgments This work has been funded by the German Research Foundation (DFG) through the *Linked Data for eScience Services* (LIFE) project (KU 1368/11-1). We also would like to thank our project partners, the Münster University Library (ULB) and the Institute for comparative urban history (ISTG) for their constant support of this work.

References

Arteaga MG,(2013) Historical map polygon and feature extractor. In: MAPINTERACT'13, Orlando, FL, USA. ACM, New York, NY, 05–08 Nov 2013

Battle R, Kolas D (2012) Enabling the geospatial semantic web with parliament and GeoSPARQL. Semantic Web 3(4):355–370

Bizer C, Heath T, Berners-Lee T (2009) Linked data: the story so far. Int J Semantic Web Inf Syst 5(3):1–22

Carral D, Scheider S, Janowicz K, Vardeman C, Krisnadhi A, Hitzler P (2013) An ontology design pattern for cartographic map scaling. In: Cimiano P, Corcho O, Presutti V, Hollink L, Rudolph S (eds) The semantic web: semantics and big data. Lecture notes in computer science, vol 7882. Springer, Berlin, pp 76–93

Gangemi A, Presutti V (2009) Ontology design patterns. In: Staab S, Studer R (eds) Handbook on ontologies, international handbooks on information systems. Springer, Berlin, pp 221–243. doi:10.1007/978-3-540-92673-3_10

Gkadolou E, Stefanakis E (2013) A formal ontology for historical maps. In: 26th international cartographic conference

Gkadolou E, Tomai E, Stefanakis E, Kritikos G (2013) Ontological standardization for historical map collections: studying the Greek borderlines of 1881. In: ISPRS annals of the photogrammetry, remote sensing and spatial, information sciences, vol I-2

Grossner K (2010) Representing historical knowledge in geographic information systems. Ph.D. thesis, University of California, Santa Barbara

Hart G, Dolbear C (2013) Linked data: a geographic perspective. CRC Press, Boca Raton

Haslhofer B, Robitza W, Guimbretiere F, Lagoze C (2013) Semantic tagging on historical maps. In: Proceedings of the 5th annual ACM web science conference, WebSci '13. ACM, New York, NY, USA, pp 148–157

Hyvönen E, Tuominen J, Kauppinen T, Väätäinen J (2011) Representing and utilizing changing historical places as an ontology time series. In: Ashish N, Sheth AP (eds) Geospatial semantics and the semantic web, semantic web and beyond, vol 12. Springer, New York, pp 1–25

Kottmann C, Reed C (2009) The OpenGIS abstract specification. Topic 5: features

Kraak MJ (2003) Geovisualization illustrated. ISPRS J Photogrammetry Remote Sens 57(56): 390–399

Krötzsch M, Simancik F, Horrocks I (2012) A description logic primer. CoRR abs/1201.4089

MacEachren AM (2004) How maps work: representation, visualization, and design, 2nd edn. The Guilford Press, New York

Montello D (1993) Scale and multiple psychologies of space. In: Frank AU, Campari I (eds) Spatial information theory a theoretical basis for GIS. Lecture notes in computer science, vol 716. Springer, Berlin, pp 312–321

Ruotsalo T, Haav K, Stoyanov A, Roche S, Fani E, Deliai R, Mäkelä E, Kauppinen T, Hyvönen E (2013) SMARTMUSEUM: a mobile recommender system for the web of data. Web Semant Sci Serv Agents World Wide Web 20:50–67

Simon R, Haslhofer B, Robitza W, Momeni E (2011) Semantically augmented annotations in digitized map collections. In: Proceedings of the 11th annual international ACM/IEEE joint conference on digital libraries, JCDL '11. ACM, New York, NY, USA, pp 199–202

Simon R, Barker E, Isaksen L (2012) Exploring Pelagios: a visual browser for geo-tagged datasets. In: International workshop on supporting users' exploration of digital libraries. Paphos, Cyprus, 23–27 Sept 2012

Smith B, Mark D (2001) Geographic categories: an ontological investigation. Int J Geog Inf Sci 15(7):591–612

Trame J, Keßler C, Kuhn W (2013) Linked data and time: modeling researcher life lines by events. In: Tenbrink T, Stell J, Galton A, Wood Z (eds) Spatial information theory. Lecture notes in computer science, vol 8116. Springer International Publishing, pp 205–223. doi:10.1007/978-3-319-01790-7_12

An Area Merge Operation for Smooth Zooming

Radan Šuba, Martijn Meijers, Lina Huang and Peter van Oosterom

Abstract When zooming a digital map it is often necessary that two or more area features must be merged. If this is done abruptly, it leads to big changes in geometry, perceived by the user as a "jump" on the screen. To obtain a gradual merge of two input objects to one output object this chapter presents three algorithms to construct a corresponding 3D geometry that may be used for the user's smooth zooming operations. This is based on the assumption that every feature in the map is represented in 3D, where the 2D coordinates are the original representation, and 1D represents the scale as a Z value. Smooth zooming in or out is thus equivalent to the vertical movement of a horizontal slice plane (downwards or upwards).

1 Introduction

Cartographic generalization is the process of transforming a map from a detailed scale to a less detailed scale. Only the end result of such a process is usually considered. However, for some applications the visual process of continuously changing the Level of Detail (LoD) is important. For example what is visible when zooming the map.

R. Šuba (✉) · M. Meijers · L. Huang · P. van Oosterom
Section GIS Technology, OTB—Research for the Built Environment, Delft University of Technology, Delft, The Netherlands
e-mail: R.Suba@TUDelft.nl

M. Meijers
e-mail: B.M.Meijers@TUDelft.nl

P. van Oosterom
e-mail: P.J.M.vanOosterom@TUDelft.nl

L. Huang
School of Resource and Environmental Science, Wuhan University, Wuhan 430079, China
e-mail: L.Huang-1@tudelft.nl

J. Huerta et al. (eds.), *Connecting a Digital Europe Through Location and Place*, Lecture Notes in Geoinformation and Cartography, DOI: 10.1007/978-3-319-03611-3_16,

Instead of discretely switching from one scale to another, a continuous transformation from source to target scale is preferable.

An essential factor in map usage is how the user perceives a transition between map scales at different levels of detail in the course of zooming in or out. From the experiment carried out by Midtbø and Nordvik (2007) it is evident that sudden changes during zooming are distracting to the user, and may also result in losing track of the objects the user is interested in.

The classical solutions for zooming are based on a multi-scale approach, where every scale is stored separately and the zooming is effectively "switching" from one map to another. This drawback is offset by tricks such as graphic enlargement of the map layer before "switching", blurring the map at the time of switching or map morphing i.e. an animated translation from one map to another (Reilly and Inkpen 2004, 2007).

The visualization of the continuous scale change has been investigated. van Kreveld (2001) focuses on analysing the different ways of visually continuously changing a map, defining a number of operators that can be used. His work is based on *transitional* maps (maps that connect different predefined scales) and techniques of *cartographic animation* (Robinson et al. 1995; Maceachren and Kraak 1997), such as morphing and fading.

Sester and Brenner (2005) demonstrate the gradual change of objects as a decomposition into a sequence of elementary steps. They call this *continuous generalization* and so far it has been applied only for buildings.

Danciger et al. (2009) introduce deformation of the shapes of regions in a map during a continuous scale change. They define mathematical functions for objects. However, the geometry forming a complete subdivision of space (a planar partition), which is important for vector map data, is not considered in this work.

Alternatively, van Oosterom and Meijers (2011a) introduced an approach termed *smooth zooming* of vector data where the geometry gradually changes, but it has not been implemented. This approach for smooth zooming is based on a 3D space, two dimensions representing the map at a level of detail and one the scale representation, allowing smooth merging of features as will be demonstrated in this chapter. Map features are represented as volumetric data, which are "sliced" to produce a representation. It is based on the vario-scale concept and it offers a gradual change of map objects in the process of zooming, without switching between discrete map layers.

We focus on the transformation of vector data in a representation convenient for supporting smooth zooming and aim to find the requirements for such an operation.

The remainder of this chapter is organised as follows: the part entitled 'Related work' describes the principles and applications of the vario-scale approach. The implementation details are explained in the 'Methods' part. '3D Storage of small dataset' describes the storage efficiency and this is followed by 'Analysis'. Finally, the 'Conclusion and future work' are presented.

2 Related Work

There are two ways of managing a dataset on different levels of detail: the multi-scale approach and the variable scale (vario-scale) approach. Each of these approaches deals differently with scale transition.

The classical multi-scale approach makes use of several versions of the map, explicitly stored at different scales in the multiple representation databases (MRDB). Every time a map of specific scale is required, the most appropriate scale from pre-defined scales is selected and displayed. The zooming only switches from one pre-defined scale version to another pre-defined scale of the map, which is closer to the target scale.

The second method is the vario-scale approach, which creates an index structure on the base map that enables one to extract a map on exactly the right scale whenever one requires. If you want a map on a specific scale it is constructed for you on-the-fly. Vario-scale needs only one dataset to be managed, while data can be displayed on any scale. These aspects strongly motivated our research and in our implementation we only make use of the vario-scale approach.

2.1 Smooth Topological Generalized Area Partition

One of the structures used for the vario-scale approach is known as tGAP (topological Generalized Area Partition) (van Oosterom 2005; van Oosterom and Meijers 2011b). The principle of the tGAP structure can lead to smoother user interaction, for which the concept has been introduced by Meijers and van Oosterom (2011); van Oosterom and Meijers (2011a). It is presented as a space-scale partition. The 3D structure is termed the Space-Scale Cube (SSC). It allows a single real world feature to have a single database representation, by contrast with the discrete scales approach which not only has different representations, but that these are often separately maintained. Example: The same river is named at one scale, and unnamed at a larger scale. Two versions exist: classic and smooth (tGAP/SSC).

The classic SSC stores prism based representations for objects; see Fig. 1a. This gives an idea of how map generalization can be seen as an extrusion of the original data into an additional dimension connecting the discrete scales (without topology errors). Map scale is seen as an additional geometric dimension. The resulting vario-scale representation for 2D map objects is a 3D structure. The 2D area object is then presented as a 3D volume. However, a merge operation still causes a sudden local map change ("shock"). A small change in the map scale does not automatically result in a small geometrical change. This situation will not, however, occur in the smooth SSC representation of tGAP using polyhedra without horizontal boundary faces; see Fig. 1b. All generalization actions must lead to a smoothly changing map. One can imagine this as making a gradual shift of a slice plane (where a horizontal slice is taken) from the top of the cube downwards, there will not be any object suddenly

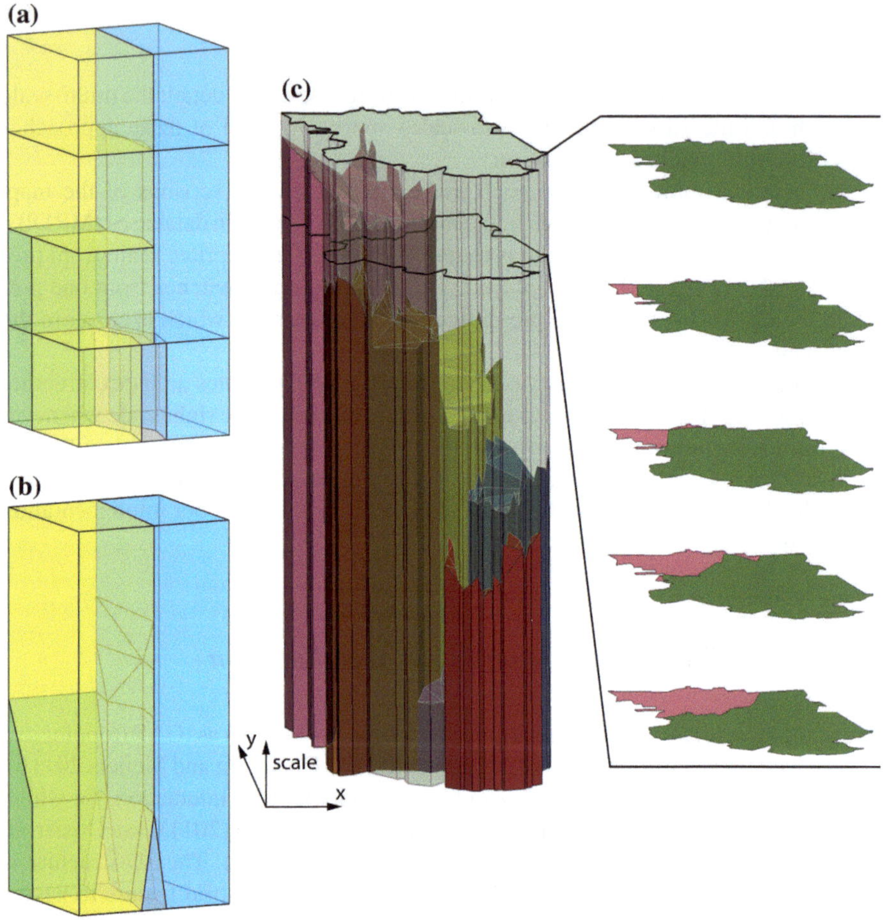

Fig. 1 The space–scale cube of the tGAP structure: **a** SSC for classic tGAP, **b** SSC for smooth tGAP, **c** A small subset of real data dataset with arbitrary slices. The *colors* are randomly assigned. Figures (**a**) and (**b**) *source* (van Oosterom and Meijers 2011a). (**c**) Shows seven area objects (at the most detailed level). The horizontal *slices* illustrate the last step of the tGAP structure where the *pink* object is merging with *green* one

appearing or vanishing. All changes result in a smoothly changing 2D map: a small change in the map scale means a small change in the geometry of the resulting map, see Fig. 1b. Figure 1c provides an example of smooth representation for small subset of CORINE Land Cover dataset. The arbitrary slices in Fig. 1c demostrate the final impresion from a vertical shift of the slice plane.

This chapter investigates requirements of merge operation (aggregation) for smooth zooming on the map and it furthermore presents three algorithms which create 3D object representations for this transition. In other words, the merge operation is converted in a real volumetric representation of the vario-scale data structure,

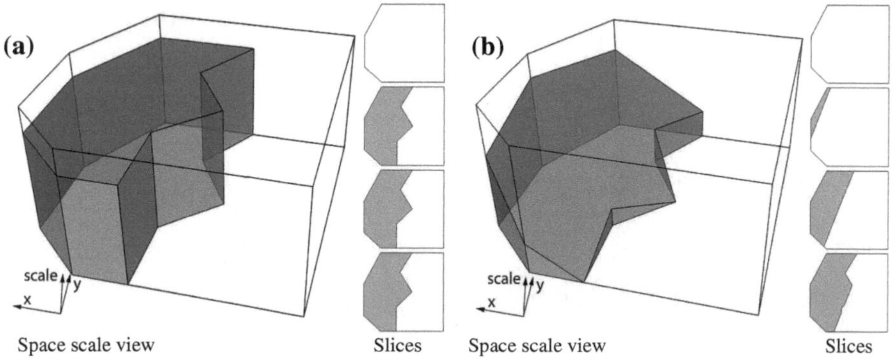

Fig. 2 The merge operation in classic tGAP (**a**) and smooth tGAP (**b**). The *white* winner takes space from the *gray* loser. *Slices* are shown at four levels, from *top* to *bottom* in the image: from *high*, to *low*

where vario-scale 2D objects are represented as 3D geometric objects (2D geometry + 1D scale).

It is important to underline that this chapter is focused on development of merge operation for smooth zooming with the tGAP structure only and thus inherits the advantages and drawbacks of this structure.

2.2 The Principle of Smooth-Merge Operation

The creation of the tGAP structure is based on the merge operation of the least important object which is called call the loser, with its most compatible neighbour, called the winner. Figure 2a presents such an operation in a classic tGAP structure, where the white winner merges with the grey loser and creates the white object. Figure 2b presents the same process in the smooth tGAP, which will be termed the smooth-merge operation. Figure 2b shows that any arbitrary horizontal slice leads to a new 2D map. If the slice plane is moved up, then the white winner grows and the grey loser shrinks. All the geometry changes gradually.

During the transition to the smooth tGAP structure, some objects can be deformed or misrepresented and the resulting map (slice of the smooth tGAP representation) can be confusing. For instance, a single object representation can break into multiple parts—see the white object in Fig. 3. Such spurious multiple parts cause a transitory increase in the number of objects in the map and result in a degradation of the quality of the map. Therefore our intention is to ensure that area features do not break into discontinuous parts.

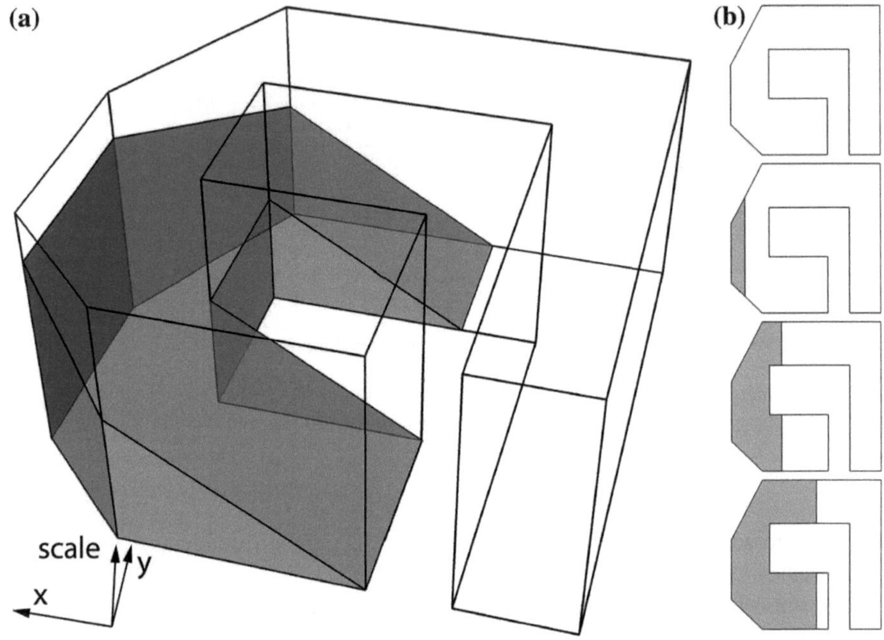

Fig. 3 More complicated shapes with arbitrary slices in smooth tGAP. The *middle* and *low slices* in (**b**) shows two parts of the same object (*white* winner). (**a**) Space scale view. Only share boundary between objects is colored, (**b**) Slices

2.3 Requirements

The main goal of our research was to transfer objects from classic tGAP to the smooth tGAP represented in SSC where horizontal faces, causing abrupt changes, do not exist. A smooth-merge operation was developed. This should compute in 3D a set of tilted faces representing the boundary surface between the loser and the winner. We defined the following requirements for good smooth-merge operations (each one supported by its own motivation):

1. Topologically correct in 3D—The input 2D map is topologically correct and forms a 2D partition of space, the resulting 3D representation should also be topologically correct and not have gaps, overlaps or intersections.
2. No new points, that is no new x, y coordinates—The construction of tilted faces make use of the already existing geometry. It follows one of the main aspects of the vario-scale approach, which is to minimize the redundancy of data. Only the new edges and faces are created using existing geometry.
3. No horizontal faces—This follows from the definition of the smooth tGAP structure mentioned above as horizontal faces cause sudden changes.
4. No vertical faces between winner and loser—The winner should gradually take over the area of loser, but vertical face means nothing happens here for a while.

5. No multi-parts—Spurious multi-part objects (which should be single part), confusion to the map.
6. Constant steepness of tilted face/faces—The shared tilted boundary composed from multiple faces with different steepness will result in the effect that some parts of the loser merge much faster than others during smooth zoom. The steepness of the faces should be as constant as possible.
7. Optimal shape of shared boundary—The shared 3D surface boundary can consist of multiple faces. There often exist more configurations of faces, and each defines a 3D surface of the shared boundary. The configuration that gives the best merge impression to user should be selected. That is, a natural looking boundary during the smooth zoom (slicing operator).

The list above contains three "hard" and four "soft" requirements. The first three requirements will always be guaranteed by the algorithms. It is not possible to guarantee the all other requirements, as they are sometime contradicting and they are therefore classified as soft requirements. It will be tried to optimized these in a well balanced manner. Note that the different "soft" constraints may be competing; e.g. constraint 6 has preference to single flat plane (equal steepness), but may result in multi-parts; see Fig. 3. The hard requirements guarantee functionality and they are crucial to finding the solution. The soft requirements are more aesthetic requirements which are more like recommendations. We would like to fulfill them. However, they are not to be strictly satisfied. Example In some cases a loser composed from multi-parts can result.

3 Methods

Three different algorithms for smooth-merge operation have been created and implemented to satisfy the hard requirements (and optimize the soft requirements): the "Single flat plane", the "Zipper" and the "Eater". The difficulty of implementation with these algorithms ranges from trivial, with the "Single flat plane"; to more complex, with the "Zipper"; to rather complex, with the "Eater". Based on the various types of map objects—the thematic classification and shape, which may result in different emphasis on the soft requirements, it should be possible to select the most suitable algorithm.

3.1 Single Flat Plane

The first implemented algorithm was "single flat plane". It originated from idea that smooth-merge can be represented by simple single flat plane of constant steepness as illustrated in Fig. 2b. In principle, the loser (the face that has to be removed) can always be removed with a single flat plane. The plane is defined in 2D by the shared

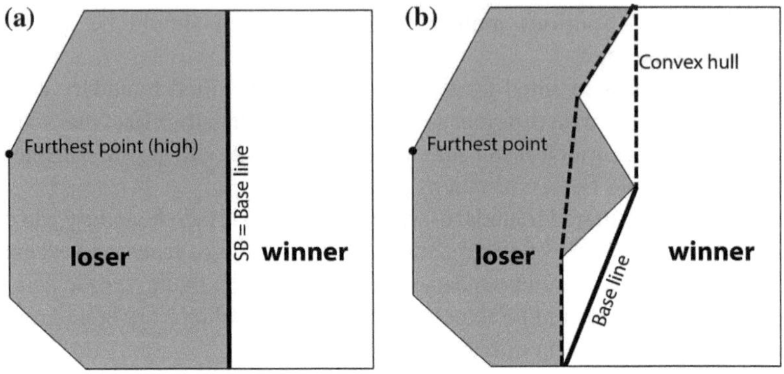

Fig. 4 Definition of the base line for the "Single flat plane". (**a**) A case where the base line is same as the shared boundary, (**b**) A case where the shared boundary is not a *straight line*. The base line is the longest site of the convex hull (*dashed line*) of the shared boundary located in the winner

boundary, the edge(s) between the winner and the loser, and the furthest point of the loser from the shared boundary; see Fig. 4a. Then in the 3D the plane is lowest at the shared boundary and highest at the furthest point. As it is a single flat plane, linear interpolation can be applied for calculating the height at any other point on the share boundary surface between the winner and the loser. Figure 2b shows the final 3D representation of smooth merge using the "Single flat plane" algorithm.

In context of the smooth-merge operation the shared boundary is used, where the merging starts and the distance from the shared boundary to every point of the loser is calculated to find the point most far away. If the shared boundary is not a straight line; see Fig. 4b, then the concept of a base line is introduced. The base line is an "approximation" of the shared boundary and it is used for measuring distances and finding the furthest point. Three ways of finding the base line are investigated:

- The base line is the longest site convex hull of the shared boundary located in (or on the boundary of) the winner; see Fig. 4b.
- The base line is the edge of the smallest rectangle around the loser which has biggest overlap with the face of winner (and has loser completely at other side).
- The base line is result of a best-fit line technique known as Eigenvector line fitting to obtain the orientation of the base line (van Oosterom 1990). There are two options: use all points of the loser or just the points of the shared boundary. Then this Eigenvector line is translated to make sure the loser is completely at one side (and touching the line) with the winner on both sides, but preferably with the largest part opposite the loser.

The first approach, the longest edge of convex hull of the shared boundary has been used for our implementation.

The most serious limitation of the "Single flat plane" approach is orientation. There always exists only one direction where the plane can be oriented. However, in some cases the winner and the loser can have more shared boundaries, e.g. the boundary between the winner and the loser is broken by another polygon. In such a case, the decision where the tilted plane should be placed must be made. The situation is even worse when the loser lies inside the winner, where no good quality solution exists. Also in case of a very long and facetted boundary, the base line will not be able to represent this well. Some vertical faces are needed to make the single flat plane fit into the 3D SSC. Note this is also the case when base line fits rather well, only then the vertical faces will not need to be very high. The vertical faces result in the effect that at their locations for a while, nothing changes while other parts the winner is already taking space from the loser.

Despite the above mentioned limitations of the "Single flat plane" algorithm, it can be very effective for simple convex polygons where a simple shared boundary exists. The computation of the base line, together with a definition of the tilted face, is trivial. From the smooth zooming point of view, the "Single flat plane" has the advantage of the winner face growing with constant speed. These aspects make the "Single flat plane" algorithm a good candidate for processing simple convex faces, such as rectangular buildings.

3.2 Zipper

The second approach is based on the decomposition of the loser polygon into triangles. These triangles will then in 3D become the tilted surface of the shared boundary. The segments of the boundary of the loser are classified into two types: 1. The Shared Boundary (SB), which is the boundary between the winner and the loser and 2. The Opposite Boundary (OB), which is the remaining part of the boundary of the loser, holes included. Figure 5 shows the final 3D representation. It is assumed that the tilted faces should be tilted from SB (low) to OB (high). Before we designed the "Zipper" algorithm, we first investigated another method for finding a good solution satisfying as many requirements as possible:

- "Delaunay triangulation and flipping"—which starts with a Delaunay triangulation of the loser, giving the fattest triangles possible (satisfying the soft requirements). Then it traverses the triangulation and flips edges which might provide a better configuration of the triangles. This method is computationally expensive and does not guarantee the best solution, because a local flip of the triangles may not lead to the global optimal solution.

Therefore, another method, the "Zipper", was implemented. The hard requirements remain valid during the whole process: topology is correct, no new vertices (x, y coordinates) are created and no horizontal faces are created. Besides, the other optimization rules have been met:

- Every triangle connects SB and OB, which means that at least one vertex lies on the SB and at least one other vertex lies on the OB;

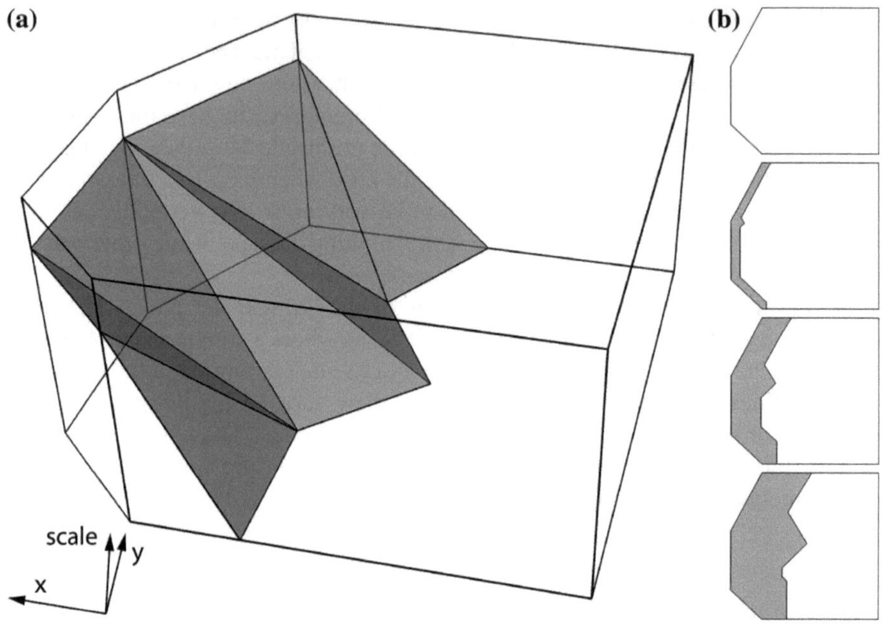

Fig. 5 The zipper algorithm. The *white* winner taking over space from the *gray* loser. (**a**) Space scale view, (**b**) *Slices* at four levels, from *top* to *bottom* in the image: from *high* to *low*.

- The triangles should have a low aspect ratio—the aspect ratio is the longest side divided by the shortest altitude of triangle (Shewchuk 1996). It guarantees that the triangles have no sharp angles;

These additional rules guarantee that we get the best solution according to the hard and soft requirements. Another optimization rule, not further explored could be: the two edges that run from SB to OB should be as much as possible of equal length, making sure that the speed of transition is as nearly as possible equal.

Figure 6 describes the principle of the "Zipper" algorithm. One can imagine the process as a walk along the SB and OB simultaneously, creating of a connection from one vertex at SB to another vertex at OB, if possible. When the connection is created, a new triangle is defined. The process starts at the junction of SB and OB (vertex I in Fig. 6) and ends in another junction (vertex II in Fig. 6). For every edge, connecting SB and OB, the aspect ratio is computed. Then the process can start from the other side of the SB. The optimal solution is defined as one where the whole area of the loser is processed and where the sum of aspect ratios is minimal.

Figure 7 presents an alternative visualization of the process. The process can be represented as a graph where every connecting edge is a node in the graph. The vertical axis corresponds to the index of SB. The horizontal axis corresponds to the index of OB. Only moving to the right or moving downwards in the graph is possible as the algorithm to create triangles either takes a step on SO or OB, but

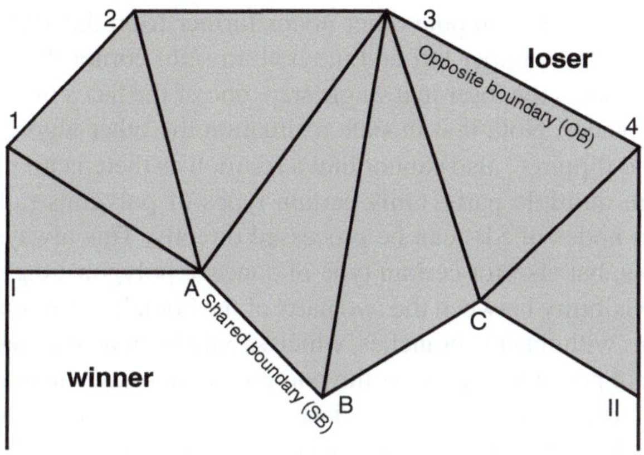

Fig. 6 Principle of the "Zipper". The optimal solution, only the final connections are present

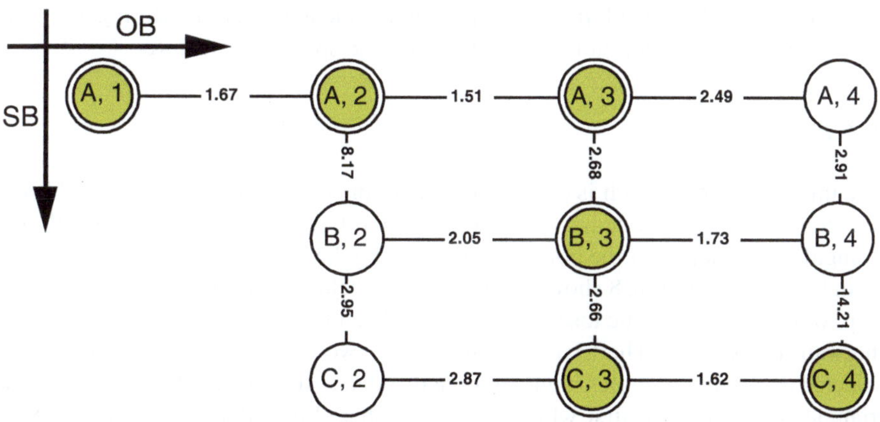

Fig. 7 *Graph* representation of the "Zipper" approach for configuration in Fig. 6. The "zipping" starts at *triangle* (I, A, 1) towards (II, 4, C). *Note* that the process can run in the opposite direction. The *yellow* presents

not both as this would result in a quadrangle. The weight of a graph edge is defined as the aspect ratio of the triangle associated with the graph edge connects.

The process starts in the upper left corner and finishes (if a solution exists) in the bottom right corner. If a solution exists, then a connection between the upper left and the bottom right corner exists.

The graph representation can be used for optimization. The best solution is the solution with the lowest total sum of aspect ratios of triangles (corresponding to graph edges). Basically, it is the cheapest path through the graph. As it can be observed, the edges with the lowest ratio tend to lie mostly on the diagonal of the graph, see Fig. 7. This means that for an optimal solution it is sufficient compute the diagonal connection from the upper left to the bottom right corner and then if such a connection does not exist one can start to compute other nodes further from diagonal. If there is no connection between the upper left and the bottom right corner this means that it is not possible to process the loser in a single step, one of the faces of the loser must be split into more pieces. Note that in such a situation the other algorithm "Delaunay triangulation and flipping" also cannot find a solution as there is no solution without splitting loser in multiple parts. Only certain types of polygons (all nodes on OB are visible from nodes of SB) can be processed directly. This always the case with convex polygons, but also for certain type of concave polygons; e.g. relatively long polygons with visibility between the two parts of the boundary SB and OB (e.g. road or river polygons without side branches, which should be treated as separate objects).

Where it is not possible to process the polygon in this way, the graph can suggest where a split can take place. The associated vertex of the graph nodes, where the connection fails in most cases, can be a candidate for splitting. Unfortunately, with splitting a number of possible solutions arise and for the overall optimal solution many or all of them should be checked. Therefore another algorithm, which is called the "Eater" was developed. It will be presented in the next section offering a general solution for arbitrary polygons without the need to split either feature.

3.3 Eater

The third approach, which is called the "Eater" provides a solution to any arbitrary polygon (with holes, concave, or multiple shared boundary parts). It is based on a triangular tessellation of the polygon and ordering the triangle into a gradual change from low to high. Figure 8 shows the result. The Delaunay triangulation takes initial step, where the face of the loser is tessellated. Then finding the starting triangles for the walk takes place. The triangles which have two edges of shared boundary are selected as starting triangles if any exist, and added to the so-called *active set*. These triangles are processed first which makes the shared boundary less "curvy". Note that when processing such a triangle in 3D there are two options how to set relative heights: 1. keep both edges of SB completely at the lowest start height, or 2. keep only the shared node of these two edges at the lowest start and other two nodes at the first height step. The first option will result in a horizontal triangle and therefore violates a hard requirement. Therefore the second option is selected, which has only the drawback of introducing some vertical triangles, but this is 'just' violating a soft constraint and not a hard one. If there are no triangles with two edges in SB, then the triangles which have one edge in SB are selected as starting triangles and these add into the *active set* (and both nodes of involved edge get fixed height at start level). The process starts with all triangles set on *not-visited*, some triangles in the *active set*, the *relative height* set to 1 and empty *next active set*. The elements in square brackets refer to the *active set*, the elements in curly brackets refer to the *next active set* for initial iteration. The steps to process and sort the triangles are as follows:

1. start with starting triangles as the *active set*, [A] { };
2. until the *active set* becomes empty do:

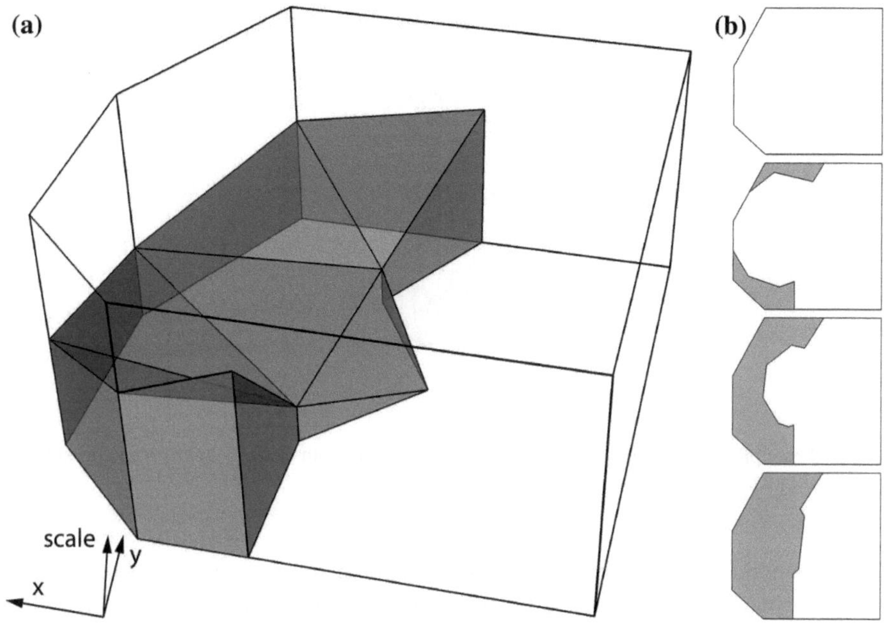

Fig. 8 *Tilted* faces of loser processed by the "Eater". (**a**) Space scale view, (**b**) *Slices* at four levels, from *top* to *bottom* in the image: from *high* to *low*

(1) select a triangle from the *active set*, A;
(2) if a node has fixed height use it, otherwise use the *relative height* for a node (defining the triangle in 3D space, fitting into the SSC topology);
(3) add all non-visited neighbours of triangle into the *next active set,* [A] {B};
(4) remove the triangle from the *active set* and set to *visited,* [ ] {B};
(5) if there are no more triangles in the *active set*, then increase the *relative height* and move the triangles from the *next active set* into the *active set,* [B] { };

For next iterations we start with [B] { } and other steps are as follows:

	2. Iteration	3. Iteration	4. Iteration	5. Iteration	6. Iteration
(1)	B	C_1	C_2	E_1	E_2
(2)					
(3)	[B] {C_1,C_2}	[C_1,C_2] {E_1}	[C_2] {E_1,E_2}	[E_1, E_2] {F}	[E_2] {F}
(4)	[] {C_1,C_2}	[C_2] {E_1}	[] {E_1,E_2}	[E_2] {F}	[] {F}
(5)	[C_1,C_2] { }	[C_2] {E_1}	[E_1,E_2] {}	[E_2] {F}	[F] {}

Figure 9 demonstrates the principle and shows the sequence of the traversal of the triangles. The "Eater" ends with the triangles ordered, and a relative height assigned to every node.

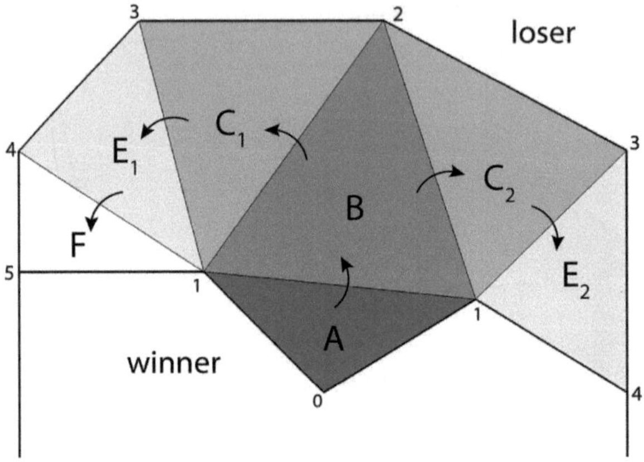

Fig. 9 Principle of the "Eater". The *triangles* are walked from *dark* to *light grey*. The numbers present relative Z value

In general, this algorithm guarantees a solution for every object, the face with multiple shared boundaries included. With this algorithm it is always possible to convert a dataset in classic tGAP representation into a smooth tGAP representation, where no horizontal faces exist. The price for that is some vertical faces and a significant chance of multi-parts. The cooperation with other approaches and conclusion will follow in next sections.

4 3D Storage of Small Dataset

The number of resulting elements can give us some idea how efficient the storage is. The small subset of CORINE Land Cover dataset (7 faces) is used as an input; see Fig. 1c. It contains 676 vertices and 15 records of faces in classic tGAP structure. Scale attributes will be used as z values in the conversion to the 3D representation.

Because of its general applicability the "Eater" algorithm was used for converting data stored in classic tGAP into the smooth representation. The whole dataset in smooth tGAP has been stored explicitly as polyhedral volumes in a 3D topological structure containing:

- A list of vertices where x, y, z coordinates are stored for every vertex.
- Faces, stored as a list of indices pointing to the vertex list. Shared faces between objects are always represented just once, not duplicated. On the other hand, the edges between two faces are stored twice.

This small example is represented by 989 vertices (which is more than the original 676 vertices as some of these have multiple counterparts at different height levels; however, there are no new x, y coordinates introduced) and 684 faces, consisting of: 289 triangles—the tilted faces of shared 3D boundaries between winners and losers,

Table 1 Summary table, where − = bad, + = neutral and + + = good

	Single flat plane	Zipper	Eater
Always has solution	−	+	++
Type of polygon	−	+	
	(One SB only, convex, rectangular)	(Convex, some concave with visible SB-OB)	++ (All)
Computational efficiency	+	−	+
Multi-parts	−	+	−
Optimal shape of faces	−	++	+
Number of extra elements	+	+	−
Constant steepness	++	+	−
No vertical faces at SB	−	++	−

The first three rows evaluate the algorithms in general. The remaining rows give an overview of fulfilment of the mainly aesthetic soft requirements

and 395 vertical rectangular faces (normal boundaries, which are not the result of a smooth merge). Finally there are 7 horizontal bottom faces and 1 horizontal top face.

Alternatively, the smooth tGAP dataset has been converted to tetrahedrons (a tetrahedron is a geometric object composed of four vertices and four triangular faces) for visualization and slicing purposes (Šuba et al. 2013). 1734 tetrahedrons have been stored explicitly in a 3D topological structure comprised of:

- A list of vertices;
- Tetrahedrons, stored as a list of indices pointing to the vertex list.

The slicing can be done very efficiently because to make a slice of set of the tetrahedrons is much easier than making a slice of arbitrary 3D polyhedral objects. On the other hand the number of elements is in the thousands for even a small example.

From the simple example presented above we can see that making the 3D representation explicit can be space consuming, but more important is when and where the conversion is performed. Once a *subset* of the tGAP structure is transferred from a server to a client, the conversion can be performed very efficiently. Then the 3D reconstruction and visualization can use the full potential of graphic hardware such as computing slices of tetrahedrons or rendering.

5 Analysis

This research investigated the possibility of smooth-merge operation and implemented three algorithms, the "Single flat plane", the "Zipper" and the "Eater". These approaches are summarized in Table 1. Some of the criteria on the left side of the table come up from soft requirements. Each of these criteria can be quantified and based on this the different algorithms are compared. In addition not all criteria are of equal weight when choosing best solution for specific situation.

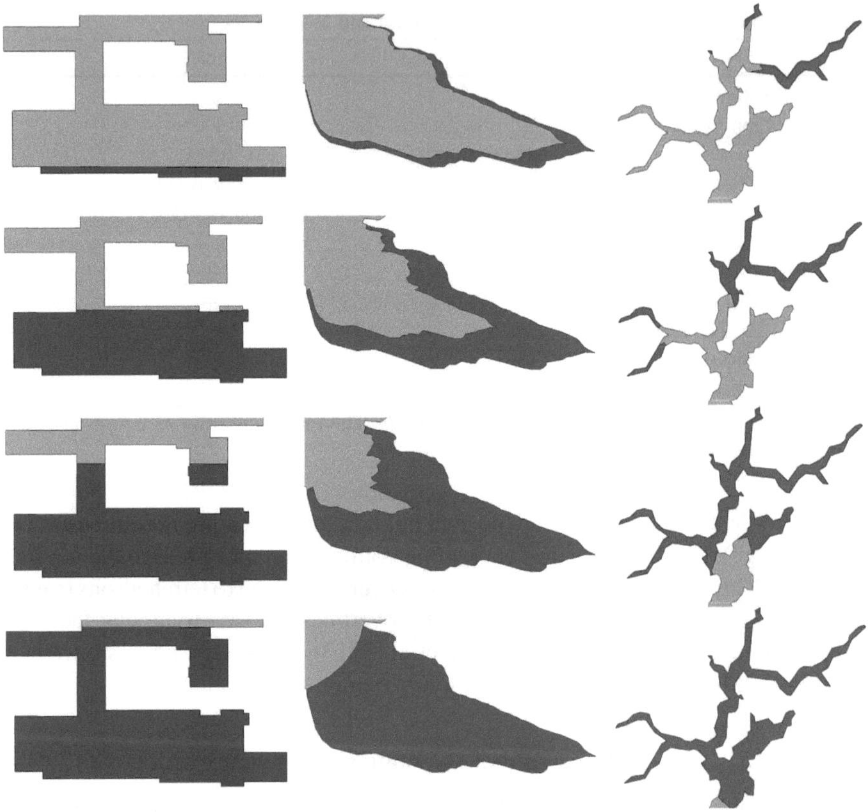

The "Single flat plane" – example of complex building The "Zipper" – example of a dam The "Eater" – example of a river.

Fig. 10 Example of three approaches with real data, the "single flat plane" on the *left*, the "Zipper" in the *middle* and the "Eater" on the *right*. The arbitrary slides simulate gradual merge (from less merged (*bottom*) to more merged (*top*)). The *light grey face* is the winner, the *dark grey face* is the loser

First, the "Single flat plane" offers easy computation, constant speed of merging and it always creates just one tilted plane. The main disadvantages are defining the orientation of the tilted plane where the shared boundary is rather complex or more shared boundaries exist. The algorithm also does not try to avoid a multi-parts and the final impression of merging looks artificial—a single line "sweeping on the screen". A low score has been given to "Always has solution", because it in some cases the solution is not meaningful, e.g. where the winner lies inside the loser. Despite those limits the "Single flat plane" algorithm can be very effective for simple convex rectangular polygons where only one shared boundary exists, such as building see Fig. 10.

Second, the "Zipper" offers more generic solution. It works for all convex and even for some concave polygons, this depends on visibility between SB and OB. It fulfils soft requirements and finds the optimal solution, if it exists; which results in good impression of merging and minimal risk of multi-parts. On the other hand, the quadratic computational complexity can be overcome by graph representation together with finding optimal solution. If there is no single part solution, the graph also gives an indication of how to split the polygon into multiple parts.

Third, the "Eater" offers a solution for any arbitrary polygon. It always processes one triangle at the time and the whole polygon is processed in one operation; which makes the algorithm more effective and robust. The Delaunay triangulation with $O = (n \log n)$ is the most expensive part of the algorithm. When using the 3D structure and creating slices, the risk of multi-parts is high and it cannot guarantee constant steepness. However, steepness can be improved by global information about configuration of triangles.

Figure 10 shows process of merging in few steps (slices of the 3D representation). It gives an impression of smooth merge operation. The reader can see that delta zoom means delta change in geometry. As can be seen, the "Single flat plane" approach simulate merging by sweeping straight line trough area of looser which (from our experience) gives an artificial impression while the zipper and the eater looks more "natural".

6 Conclusion and Future Work

This research investigated various algorithms for the smooth-merge operation, of which three were implemented (the "Single flat plane", the "Zipper" and the "Eater") and tested with real world data. Each method has its own strong and weak points. However, various improvements are possible in future work.

First of all, to minimize the chance of multi-parts. The "Eater" as a promising algorithm can be further improved in the future. Using other information involved, such as predecessor of processing triangle, the negative effect of multi-parts can be reduced. However, some cases need to be treated more globally. Example: Where the loser has many holes and "branches", and where some "branches" go faster than others.

Additionally, more user testing can be involved. Right now, there are no fixed rules how the process of merging should be done and how the final impression should look. In this research the hard and soft requirements have been identified, based on analysing the properties of the smooth merge operation. However, it is not completely clear which process of merging is capable of delivering the best quality; i.e. balancing the various soft requirements.

Another concern is the storage aspect of the 3D data structure. Minimum redundancy is one of the main principles of vario-scale, but our current encoding of Space-Scale Cube takes significant storage (and especially the number of tetrahedrons is quite big). How can redundancy be minimized also for explicit 3D storage and is explicit 3D storage really needed? Storing the data in a 3D topology data structure could be one option. Another option could be to derive what is needed from the tGAP data structures that store more or less separately the 2D geometry and 1D scale range, creating the 3D representation when needed, e.g. at client side, because the current tGAP data structure implicitly contains 3D data, but does not store it as such.

Currently, all smooth merge operations are sequenced. First one operation is finished and then the next merge takes place. So, only at one location at the time something in the maps is changing. It will look more natural if several, non-interfering, smooth merge operations take in parallel.

Furthermore, investigating the effects of other generalization operators in the smooth tGAP structure. Besides merging, also simplification and collapsing/splitting are supported in current tGAP structure. Apart from making a smooth version for these operation an open question is how to store the result in the tGAP structure. For example, after the collapse of an area object to a line (or point) object, the line (or point) object lives on. In the SSC this object is therefore represented by a polyhedral volume to which a vertical surface (or vertical line) is connected at the top. All attributes are attached to the same object, which is represented in the SSC by connected multiple parts of respectively dimension 3 and 2 in case of collapse to line and 1 in case of collapse to point.

Acknowledgments This research is supported by the Dutch Technology Foundation STW (project number 11185), which is part of the Netherlands Organisation for Scientific Research (NWO), and which is partly funded by the Ministry of Economic Affairs. Furthermore, we would like to thank the anonymous AGILE reviewers and Rod Thompson for their constructive comments, suggestions and questions. Extra thanks to Rod for correcting our English.

References

Danciger J, Devadoss SL, Mugno J, Sheehy D, Ward R (2009) Shape deformation in continuous map generalization. GeoInformatica 13(2):203–221

Maceachren AM, Kraak M-J (1997) Exploratory cartographic visualization: advancing the agenda. Comput Geosci 23(4):335–343. doi:http://dx.doi.org/10.1016/S0098-3004(97)00018-6

Meijers BM, van Oosterom PJM (2011) The space-scale cube: an integrated model for 2D polygonal areas and scale. Int Arch Photogram Remote Sens Spatial Inf Sci XXXVIII-4/C21:95–102. doi:10.5194/isprsarchives-XXXVIII-4-C21-95-2011

Midtbø T, Nordvik T, (2007) Effects of animations in zooming and panning pperations on web maps: a web-based experiment. Cartogr J 44(4):292–303. doi:10.1179/000870407X241845

Reilly DF, Inkpen KM (2004) Map morphing: making sense of incongruent maps. Paper presented at the proceedings of graphics interface 2004, London, Ontario, Canada

Reilly DF, Inkpen KM (2007) White rooms and morphing don't mix: setting and the evaluation of visualization techniques. Paper presented at the proceedings of the SIGCHI conference on human factors in computing systems, San Jose, California, USA

Robinson AH, Morrison JL, Muehrcke PC, Kimerling AJ, Guptill SC (1995) Dynamic/interactive mapping, Chap. 29. In: Robinson AH, Morrison JL, Muehrcke PC, Kimerling AJ, Guptill SC (eds) Elements of cartography, 6th edn. Wiley, New York

Sester M, Brenner C (2005) Continuous generalization for visualization on small mobile devices. In: Peter Fisher (ed) Developments in spatial data handling. Springer, Berlin, pp 355–368. doi:10.1007/3-540-26772-7_27

Shewchuk JR (1996) Triangle: engineering a 2D quality mesh generator and delaunay triangulator. Paper presented at the selected papers from the workshop on applied computational geormetry, Towards Geometric Engineering

Šuba R, Meijers M, van Oosterom P (2013) 2D vario-scale representations based on real 3D structure. In: Proceedings of the 16th ICA generalization workshop, pp 1–11

van Kreveld M (2001) Smooth generalization for continuous zooming. In: Proceedings of the 20th international cartographic conference, pp 2180–2185

van Oosterom P (1990) Reactive data structures for geographic information systems. Ph.D. Theses, Department of Computer Science, Leiden University, The Netherlands

van Oosterom P (2005) Variable-scale topological data structures suitable for progressive data transfer: the gap-face tree and gap-edge forest. Cartography and geographic information science 32(4):331–346. doi:10.1559/152304005775194782

van Oosterom P, Meijers M (2011a) Towards a true vario-scale structure supporting smooth-zoom. In: Proceedings of 14th ICA/ISPRS workshop on generalisation and multiple representation, pp 1–19

van Oosterom P, Meijers M (2011b) Vario-scale data structures supporting smooth zoom and progressive transfer of 2D and 3D data. NCG Jaarverslag, pp 21–42

Point Labeling with Sliding Labels in Interactive Maps

Nadine Schwartges, Jan-Henrik Haunert, Alexander Wolff and Dennis Zwiebler

Abstract We consider the problem of labeling point objects in interactive maps where the user can pan and zoom continuously. We allow labels to slide along the point they label. We assume that each point comes with a priority; the higher the priority the more important it is to label the point. Given a dynamic scenario with user interactions, our objective is to maintain an occlusion-free labeling such that, on average over time, the sum of the priorities of the labeled points is maximized. Even the static version of the problem is known to be NP-hard. We present an efficient and effective heuristic that labels points with sliding labels in real time. Our heuristic proceeds incrementally; it tries to insert one label at a time, possibly pushing away labels that have already been placed. To quickly predict which labels have to be pushed away, we use a geometric data structure that partitions screen space. With this data structure we were able to double the frame rate when rendering maps with many labels.

Keywords Dynamic maps · Interactive maps · Automated map labeling · Sliding labels · Point labeling

1 Introduction

In navigation systems and online map services, map objects (such as cities) are typically annotated by labels (such as city names) in order to convey information about the map objects. While it is desirable to place many labels, it is difficult to do

N. Schwartges (✉) · A. Wolff · D. Zwiebler
Chair of Computer Science I, University of Würzburg, Würzburg, Germany
e-mail: nadine.schwartges@uni-wuerzburg.de

D. Zwiebler
e-mail: dennis.zwiebler@freenet.de

J.-H. Haunert
Institut für Geoinformatik und Fernerkundung, University of Osnabrück, Osnabrück, Germany

J. Huerta et al. (eds.), *Connecting a Digital Europe Through Location and Place*, Lecture Notes in Geoinformation and Cartography, DOI: 10.1007/978-3-319-03611-3_17,

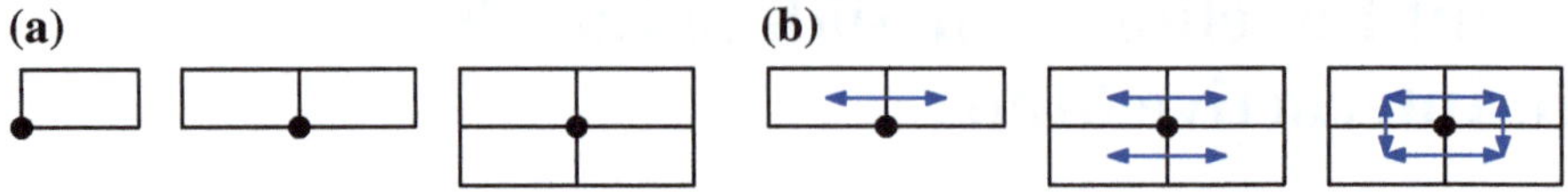

Fig. 1 Examples of common labeling models: fixed-position models (**a**) and slider models (**b**). **a** One-position model (*1P*), two-position model (*2P*), and four-position model (*4P*), **b** one-slider model (*1S*), two-slider model (*2S*), and four-slider model (*4S*)

so since labels must not overlap each other. Many interactive map products are not satisfactory in terms of label placement; they block large areas around labels in order to avoid that labels overlap when the user interacts with the map.

Map labeling is a classical problem in cartography. Cartographers such as Alinhac (1962) or Imhof (1975) have given rules for good label placement. Partially based on these rules, computer scientists have suggested many map-labeling algorithms in the 1980s and 1990s, especially for point objects. For practical purposes, the static point-labeling problem can be considered solved. Point labeling requires a *labeling model* that defines possible label positions. There are two types of such models. In *fixed-position models*, each label is restricted to a discrete set of positions relative to the point it labels; see Fig. 1a. In *slider models* (Van Kreveld et al. 1999), each label can be placed at any position such that (a certain part) of its boundary touches the corresponding point; see Fig. 1b. Each feasible placement of a label is called a label *candidate*. Usually, every point comes with a *weight* (or priority); the higher the weight the more important it is to label the point. Then, the aim is to maximize the sum of the weights of the labeled points.

This leads to the following static weighted point-labeling problem StatPointLab (for a fixed labeling model). Given a set P of points in the plane and, for each point p in P, a weight $w(p)$ and a set $L(p)$ of label candidates, find a subset P' of P and, for each point p in P', a label $\ell(p) \in L(p)$ such that no two labels overlap and the sum $\sum_{p \in P'} w(p)$ is maximized. The case of axis-aligned rectangular labels has been studied from a theoretical point of view. For fixed-position models, this problem is known as *maximum independent set* in weighted rectangle intersection graphs, which is known to be NP-hard (Fowler et al. 1981). There are, however, some approximation algorithms for the unweighted (Agarwal et al. 1998; Chalermsook and Chuzhoy 2009) and for the weighted case (Adamaszek and Wiese 2013; Erlebach et al. 2005). For slider models, too, the problem is known to be NP-hard (Poon et al. 2003), even for the most restricted slider model, the *one-slider model* (1S), where the bottom edge of the label must touch the corresponding point; see Fig. 1b. For the weighted case and rectangular labels of equal height, Erlebach et al. (2009) have given a *polynomial-time approximation scheme* (PTAS), that is, a $(1 + \varepsilon)$-approximation algorithm (for any $\varepsilon > 0$).

Our Model. In this chapter, we are interested in a dynamic setting where the user can interact with the map by panning and zooming continuously. We consider a time interval $[0, T]$ in which the user interacts with the map. Current screens are redrawn repeatedly; the content of the screen between two updates is called a *frame*. Accordingly, we discretize the time interval into a sequence $t_1, \ldots, t_n$ (with $t_1 = 0$ and $t_n = T$) of points in time, which correspond to frames. At any time t_i, the user can see a rectangular region R_i of the plane, the *view*. When the user pans, R_i is translated; when the user zooms in or out, R_i is scaled accordingly.

Now we can define the dynamic point-labeling problem DYNAPOINTLAB. For each $i = 1, \ldots, n$, let P_i' be the subset of points in the view R_i that are labeled at time t_i. We insist that all labels must lie completely within R_i. As in the static case, the quality of the current labeling is $W_i = \sum_{p \in P_i'} w(p)$. Then we define the overall quality of a dynamic label placement to be the quality, averaged over all frames: $\sum_{i=1}^{n} W_i / n$. Note that DYNAPOINTLAB generalizes STATPOINTLAB, which corresponds to the restriction to a single frame ($n = 1$) and a large enough view R_1.

There are, however, two further requirements for interactive maps, which were introduced by Been et al. (2006). They argued that in a *consistent* dynamic map labeling, labels should neither jump nor flicker (pop). In order to guarantee consistency, they disallowed labels to move at all and, when zooming, they insisted that a label is visible in at most one scale interval, the label's *active range*. Been et al. (2010) adopted the same rules and gave approximation algorithms for various special cases of the resulting (unweighted) optimization problem where the sum of the lengths of the active ranges is to be maximized. This is a continuous version of the objective function that we adopted above.

In this chapter, we take a somewhat more pragmatic standpoint. We do allow labels to move. Still, our labels do not jump since we assume the one-slider model and our frame rates are high enough to ensure a smooth-looking movement when labels "slide". We do not, however, guarantee that labels don't flicker. We mitigate the problem for the map user by introducing a simple *waiting function* that suppresses labels for about 30 frames (that is, between 0.5 and 4 s) after they disappear.

As in most previous work, we assume that labels are axis-parallel rectangles. We decided to adopt the one-slider model due to a result of van Kreveld et al. (1999) who compared various fixed-position and slider models using the same simple greedy algorithm for static point labeling. They found that a slider model yields about 15 % more labels than the corresponding fixed-position model (on real-world data).

With our algorithm we mainly target applications in which very large sets of points are to be labeled and thus time is critical, for example, the train radar for regional (RB/RE) trains of Deutsche Bahn[1] (German Railways) or browsers for large images of crowds that can be tagged.[2] If efficiency is less important, however, we suggest extending our algorithm to improve the quality of the labeling. For example, like Harrie et al. (2005) and Zhang and Harrie (2006), we could rule out positions where labels obscure other map objects or, following a rule of Imhof (1975), we could define preferences for certain label positions.

Similar to the algorithm of Zhang and Harrie (2006), our heuristic proceeds incrementally. It repeatedly goes through all points in the view and tries to label each

[1] http://bahn.de/zugradar, accessed Feb. 6, 2014.

[2] http://www.u2.com/gigapixelfancam/, accessed Feb. 7, 2014.

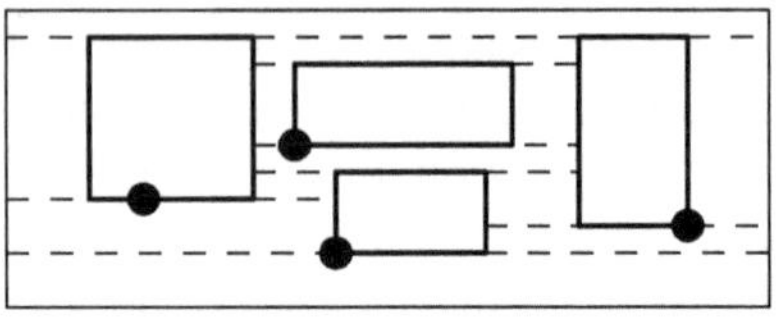

Fig. 2 A rectangulation of a set of labels within a bounding *rectangle R* (the view) is a subdivision of *R* into labels and *empty rectangles*

unlabeled point, one at a time. Other than the algorithm of Zhang and Harrie, however, our algorithm may push away labels that have already been placed in order to make space for a new label. We use a geometric data structure that allows us to efficiently predict collisions when pushing labels. We may intuitively think of the labels as vessels drifting in water. At any time and for any label we need to know the neighbors of that label since these neighbors are the labels that are the possible counterparts for a collision. Exactly that problem, maintaining the adjacency relationships of moving vessels for collision avoidance, can be solved with a kinetic Voronoi diagram (Goralski et al. 2007). In our application, however, every vessel (that is, label) can slide only horizontally and thus can collide only with vessels to its left or right. Therefore, a Voronoi diagram does not reflect the adjacency relationship that is relevant in our application. Instead, we show how to use a *rectangulation* (see Fig. 2) to access the relationships that matter and how to maintain the rectangulation when adding labels. A rectangulation is the special case of a trapezoidal map (de Berg et al. 2008) where all trapezoids are rectangles. A rectangulation of the labels can be obtained by shooting horizontal rays from the top and bottom edges of the labels.

More Related Work. Labeling interactive maps or 3d scenes is a relatively new research topic. When a user interacts with a map, the labeling has to be updated frequently. A naive approach is to perform each update by running a map labeling algorithm for static maps, not regarding the labeling that was visible before the update. Due to the recomputation of the labeling in each frame, however, labels flicker. Maass and Döllner (2006) presented such a "memoryless" algorithm. Their labeling model doesn't insist that a label touches its point. To help the user understand the label–object association, they connect labels and objects with line segments, so-called *leaders*. Their approach runs in real-time. Mote (2007) introduced an algorithm for labeling points in interactive maps using the 4P labeling model. The algorithm requires labels of uniform size. With a little workaround and loss of quality, the algorithm can also deal with labels of arbitrary size. The author shows that his algorithm labels 5,000 points in 50 ms and 75,000 points in less than a second. For this reason, he recomputes the labeling in each frame. More recently, Luboschik et al. (2008) gave a heuristic for the problem of maximizing the number of placed labels using the 4S labeling model as well as distant labels with leaders. According to their experiments, their approach is fully real-time capable although it computes the labeling in each frame. Due to the (additional) use of leaders, they often manage to label all points within the view. They do not, however, ensure that the leaders are crossing-free. This makes it hard to quickly decipher the labeling.

Gemsa et al. (2011b) considered the problem of maximizing the total length of the active ranges when labels are allowed to slide horizontally and the points are restricted to lie on the x-axis. The authors have presented an efficient PTAS for this problem. In order to support consistent labeling when users interactively rotate a map, Gemsa et al. (2011a) recently extended the idea of active ranges of scales to active ranges of rotation angles. Similarly, Gemsa et al. (2013) introduced active ranges of time, assuming that the user follows a pre-computed trajectory and that the viewport is centered on the user and oriented in the direction of movement.

The existing approaches based on active ranges allow one degree of freedom, that is, scale, rotation angle, *or* time. It may be possible to deal with two-dimensional active ranges, but, since current map viewers allow for zooming, rotating, panning, *and* tilting, we doubt that interactive labeling can be solved with the help of pre-computed active ranges alone. On the other hand, current algorithms that do not use preprocessing accept labels that flicker. Our approach with sliding labels, a waiting function, and a geometric data structure in the background can be seen as a compromise between these two worlds.

Our Contribution. We use the dynamic rectangulation data structure mentioned above to design an interactive algorithm for DYNAPOINTLAB (see Sect. 2) and suggest ways to speed up this algorithm (see Sect. 3). Finally, we present some experiments with real-world data (see Sect. 4) and conclude the chapter (see Sect. 5). A video that shows a result of our method can be found online.[3]

2 Incremental Algorithm

In interactive maps, new labels can appear whenever the user interacts with the map. To avoid that labels flicker, we build and maintain our labeling and the corresponding rectangulation *incrementally* and use a waiting function (see Sect. 3.1). One incremental step roughly works as follows (also see Algorithm 1). First, we locate the new point in the rectangulation. Next, we try to place its label such that it does not overlap other labels. This may imply that some labels may have to be pushed away or to be deleted. If the cost for these operations is too high, we do not execute them and instead reject the new label. Otherwise we update the rectangulation accordingly. In the following, we go through each of these steps in more detail.

2.1 Algorithm for Point Location

In computational geometry, point location in subdivisions is a well-known and well-solved problem. For trapezoidal maps, point-location data structures with logarithmic

[3] http://lamut.informatik.uni-wuerzburg.de/dynapointlab.html

Algorithm 1: IncrementalAlgorithm

foreach *point p to be labeled* **do**
- determine the rectangle that contains p in order to quickly find elements that are involved when placing the label $\ell(p)$ of p
- slide $\ell(p)$ from its leftmost to its rightmost position
- slide $\ell(p)$ from its rightmost to its leftmost position
- combine the two sliding directions in order to determine a good position $\ell^*(p)$ for $\ell(p)$
- **if** *labeling p increases the total weight of the labeling* **then**
  - place $\ell(p)$ at $\ell^*(p)$
  - fix the rectangulation

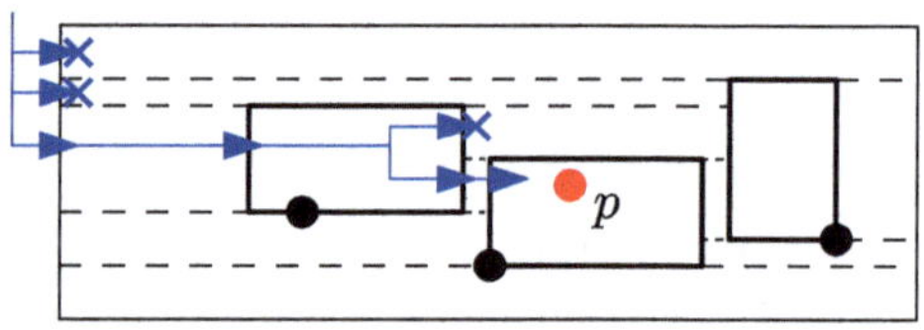

Fig. 3 Illustration of the point-location algorithm. The point p is the reference point of the label to place

query time exist (De Berg et al. 2008). Since we did not want to invest too much time into implementing such a data structure without knowing whether point location was the bottleneck in our algorithm, we settled for a much simpler (though slower) approach.

Our search algorithm is a type of target-oriented breadth-first search; see Fig. 3. Let p be the reference point to be labeled and let $y(p)$ be the y-coordinate of p. We start the search at the top left corner of the map. The left boundary of the map corresponds to a list of empty rectangles that is ordered by y-coordinate. We go through this list until we find the rectangle r whose y-interval contains $y(p)$. Then we test whether r contains p. If yes, we are done. Otherwise, we go right. As each rectangle knows its (unique) right neighbor label ℓ, we can easily test whether ℓ contains p. If not, we continue the search from ℓ in the same manner as searching from the left boundary of the map until we find the element that contains p. Under the assumption that our rectangulation is roughly grid-like, the query time is $O(\sqrt{n})$, where n is the current number of labels in the view.

2.2 Algorithm for Sliding Labels

With the help of the point-location algorithm, we know the element of the rectangulation that contains the point p to label. We next determine the final label position $\ell^*(p)$ of the label $\ell(p)$. In order to save running time, we only label the current view. We require that labels rather vanish than overlap the view boundary. Normally, we have to make space for placing $\ell^*(p)$ by sliding and removing labels. Thus, we search for a position such that the sum over the priorities of all removed labels is as small as possible; the priority of a label $\ell(p)$ is the same as the priority of its point p. We first compute labelings at which labels can only slide to the left or to the right. We use the rectangulation to quickly query potential collision counterparts. While sliding, chains of labels form. Usually, there will be a label that finally prevents that we move the entire label chain further. Out of this chain, we remove a label that touches the view boundary or has reached its uttermost position and that has the lowest priority among those. At last, we compute a labeling at which labels slide in both directions by combining the two sliding directions. In the following, we describe this algorithm in more detail. For a better understanding, see Fig. 4. Only the final decision is visible to the user.

First, we set the label $\ell(p)$ to its leftmost position. We ignore all labels whose reference points lie to the left of p (we will correct this error by combining the two sliding directions). Next, we determine *clusters* of labels. To this end, we use a directed *contact graph* whose vertices are the labels that are currently visible. There is an edge between the vertex $\ell(p)$ and each vertex whose label overlaps $\ell(p)$ as well as between two vertices if the boundaries of their labels touch sideways. We direct an edge $(\ell(u), \ell(v))$ such that $x(u) < x(w)$. To complete, a cluster $c(s)$ is the set of vertices that can be reached by a (source) vertex $\ell(s)$; see Fig. 5.

Assume that $\ell(p)$ is overlapped. By removing $\ell(p)$ from the contact graph, we obtain vertices without ingoing edges. Let $\ell(s)$ be such a vertex so that $\ell(s)$ additionally overlaps $\ell(p)$. We now slide the cluster $c(s)$ until it does not overlap $\ell(p)$ anymore, it touches another label, it touches the view boundary, or one of its labels reaches its rightmost position. We repeat rebuilding the conflict graph and sliding clusters until $\ell(p)$ is occlusion-free or there is no cluster that we can slide further. If $\ell(p)$ is still overlapped, we determine a label $\ell(q)$ with a lowest priority that lies between $\ell(p)$ (excluding) and a *blocking* label (including), that is, a label that we cannot slide further as it has reached its rightmost position or as it touches the view boundary. If the priority $w(p)$ of p is too small, that is, if $w(p) \leq \sum_{d \in D} w(d) + w(q)$, where D is the set of removed labels, we reject $\ell(p)$; otherwise we remove $\ell(q)$. Then, labels that were clustered with $\ell(q)$ and whose reference points lie to the right of q slide back until they touch another label or reach the position they had before they were slided. We repeat building and sliding clusters and removing blocking labels until $\ell(p)$ is occlusion-free.

As soon as $\ell(p)$ is occlusion-free, we repeat the entire process with the objective that $\ell(p)$ reaches its right-most position, that is, we slide $\ell(p)$ within its cluster $c(p)$. To this end, we modify the process as follows: we use $c(p)$ instead of $c(s)$; we use a cost function and stop sliding to the right instead of rejecting $\ell(p)$ due to priorities. Whenever we remove a label $\ell(q)$, we store the priority of q and the current position of p at $\ell(p)$, this is, the *amplitude*, in a cost function. If $w(p) \leq \sum_{d \in D} w(d) + w(q)$, we stop sliding $\ell(p)$ to the right and set the cost function from this amplitude to the rightmost position to $-\infty$.

With the costs and amplitudes that we have stored, we finally obtain a step function for sliding $\ell(p)$ to the right. We repeat the entire process for sliding $\ell(p)$ from its

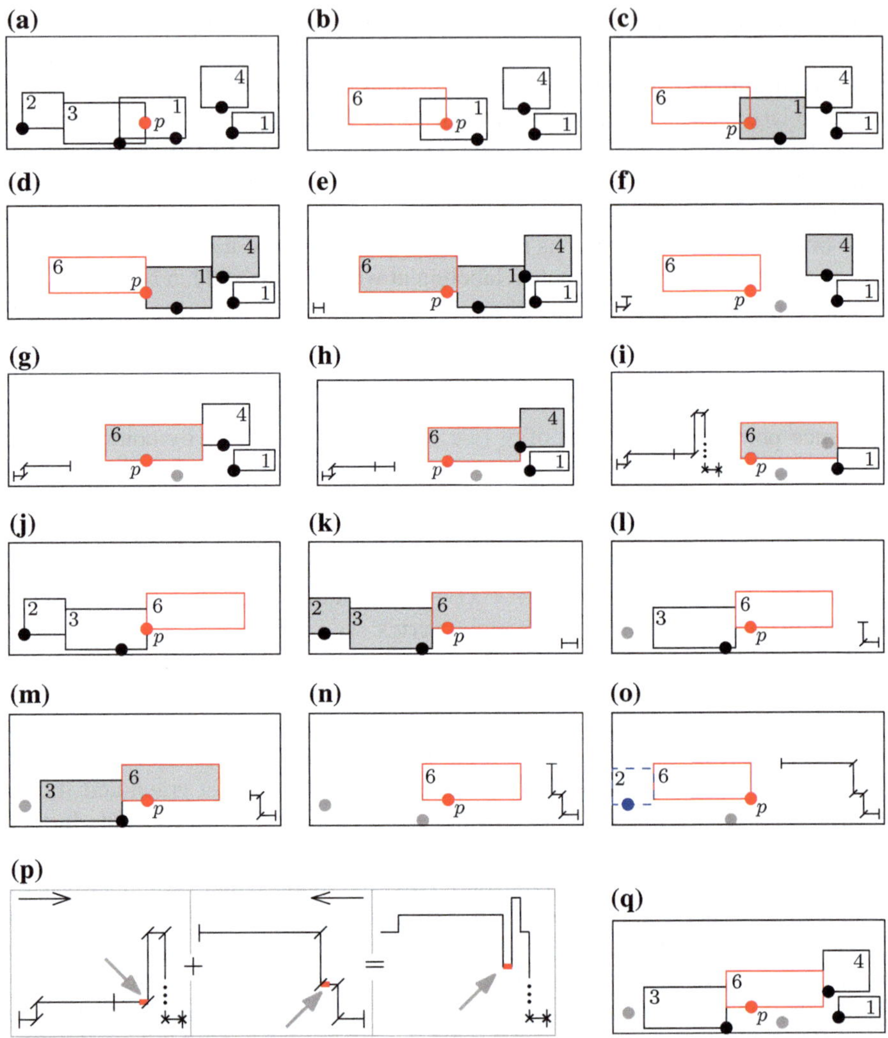

Fig. 4 Illustration of several steps of the algorithm for sliding labels. The point to be labeled is p. We annotated every label with its priority. The rectangulation is not shown. **a** The point p to be labeled appears, **b** set $\ell(p)$ to its leftmost position; neglect labels with reference points *left* to p, **c** slide clusters to the *right* to make $\ell(p)$ occlusion-free, **d** slide entire cluster, **e** $\ell(p)$ is occlusion free; slide cluster; record amplitude (*left*), **f** raise blockade (*uttermost position*); some labels slide back; record costs, **g** slide further, **h** slide further; next, raise blockade and slide further, **i** set cost function negative due to priorities; we are done, **j** from *right* to *left*: set $\ell(p)$ to its *rightmost position*, **k** slide, **l** raise blockade (view boundary), **m** slide further, **n** raise blockade (uttermost position), **o** $\ell(p)$ reached leftmost position; *no* re-insertion of labels, **p** cost functions, aggregated function, and minimum costs (*arrowed*), **q** final configuration; here, we decided for the largest amplitude

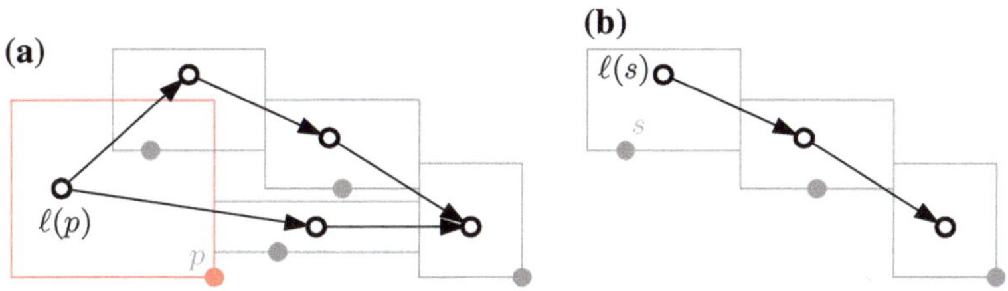

Fig. 5 A contact graph and one possible cluster. **a** A labeling and its corresponding contact graph, **b** a cluster $c(s)$ with source $\ell(p)$

rightmost to its leftmost position. We sum up the cost functions. These aggregated costs represent the costs for sliding some labels to the right and some to the left. Next, we extract the minimum of the aggregated function. Remark that the minimum (normally) is bounded by two amplitudes. Indeed, each label position for $\ell^*(p)$ between these two amplitudes yields low costs. There are several criteria to decide for one position. In our implementation, we choose a low-cost amplitude that causes the fewest labels to slide. Now, we make our final decision visible for the user. To this end, with the help of the cost function, we once more slide some labels to the right and some to the left—this time simultaneously—in order to make space to place $\ell^*(p)$.

Note that our algorithm is a heuristic. In Fig. 4o we could re-insert the label on the left. So, we sometimes overestimate costs. This can finally result in the choice of a non-optimal amplitude, this is, we place fewer labels than possible. If we (try to) label unlabeled points in each frame, this error is quickly fixed.

2.3 Algorithm for Fixing the Data Structure

We now discuss how to update the rectangulation after sliding, removing, or placing a label.

Sliding a label $\ell(q)$ is the easiest operation since it does not change the topology of the rectangulation. We only have to update the widths of the empty rectangles to the left and right of $\ell(q)$, their amplitudes, as well as the amplitude of $\ell(q)$.

Removing a label $\ell(q)$, however, is slightly more complicated; see Fig. 6. By means of the rectangulation, we directly know all the left and right neighbor rectangles of $\ell(q)$. To find the neighbor rectangles above and below $\ell(q)$, we perform a search, originating from $\ell(q)$, that is similar to the point-location algorithm; see Fig. 6a. Now, the set of neighbors of $\ell(q)$ is complete; see Fig. 6b. We remove $\ell(q)$ and extend the horizontal edges of its neighbor rectangles to close the gap left by $\ell(q)$; see Fig. 6c. As the number of empty rectangles influences the running time deeply, we finally merge rectangles that are vertically adjacent to each other and have the same left and right neighbor.

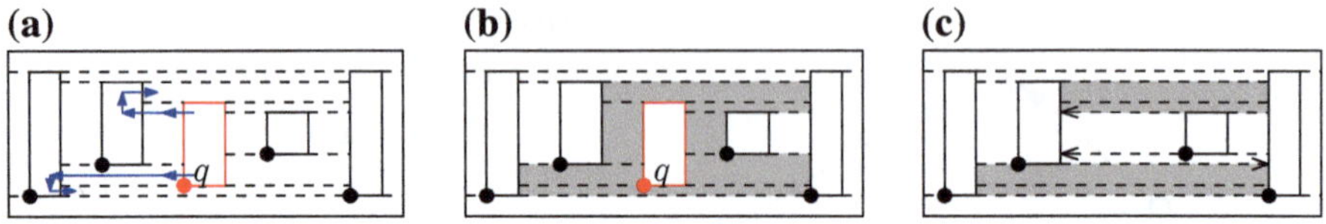

Fig. 6 Illustration of several steps of the algorithm for updating the rectangulation. The label to remove is $\ell(q)$. **a** Search *rectangles above* and *below* $\ell(q)$, **b** *rectangles* to update are *shaded*, **c** *lengthen lines*; *rectangles* to merge *shaded*

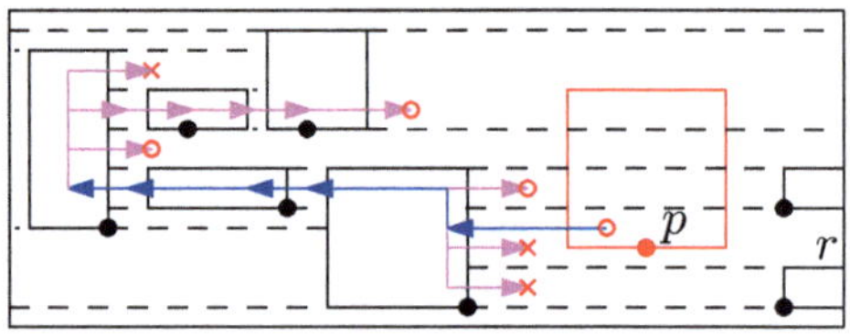

Fig. 7 Illustration of the search originating from r for detecting *overlapped rectangles*; a *circle* indicates an overlap with the label $\ell^*(p)$; a *cross* indicates the end of a search path

We add a new label $\ell^*(p)$ to the rectangulation after we have eliminated and slided existing labels to make space for $\ell^*(p)$. Therefore, we must not care about label–label overlaps. Still, we need to update the rectangulation. For this purpose, we first detect all empty rectangles that $\ell^*(p)$ overlaps. Again, we use a search similar to the point-location algorithm; see Fig. 7. Starting from the rectangle r that contains p we go to the left neighbor of r. Now, we repeatedly move from the topmost left neighbor rectangle to the next label until we reach a label whose top edge lies at a higher y-coordinate than the top edge of $\ell^*(p)$. From every label we passed while going left, we start to go right. We stop if we find a rectangle that lies completely above or below $\ell^*(p)$, that overlaps $\ell^*(p)$, or that we have visited before. During this search we collect all rectangles that overlap $\ell^*(p)$. Next, we split each of these rectangles into at most three new rectangles, that is, the part above $\ell^*(p)$, the part below $\ell^*(p)$, and the remaining middle part. This middle part again needs to be split into at most three parts, that is, the part left of $\ell^*(p)$, the part right of $\ell^*(p)$, and the part covered by $\ell^*(p)$. After splitting $\ell^*(p)$ into its parts (see Fig. 8 for the result) we need to merge rectangles that are vertically adjacent to each other and have the same left and right neighbor.

3 Running Time Improvements

The incremental algorithm is quite fast. Triggering it in each frame for testing if we can place a new label or updating label sizes due to zooming operations is time consuming, though. Therefore, we present two concepts to speed up the algorithm. First, we introduce a *waiting function*; this is, we wait several frames until we try to label a certain reference point again. Furthermore, we discuss how to predict the point in time at which we have to trigger an update of the rectangulation.

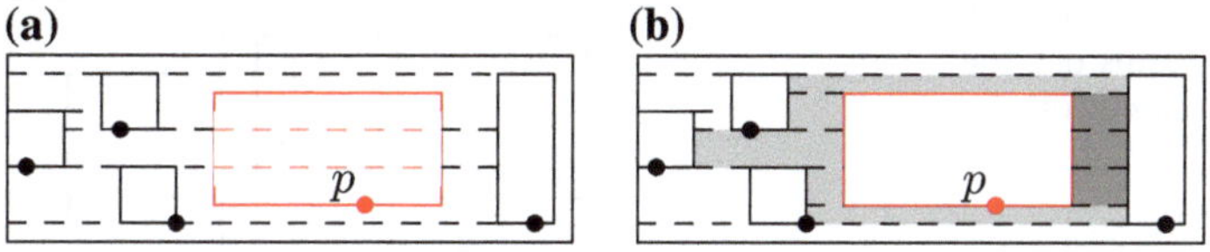

Fig. 8 How to update the rectangulation if we place the label $\ell^*(p)$. **a** Situation before placing $\ell^*(p)$, **b** *Shaded rectangles* were built; *dark shaded rectangles* must be merged

3.1 Waiting Function

Certainly, in a view, there can be many reference points without labels. It is rather unlikely that we can place a label that we could not place in the preceding frame. Additionally, it does not disturb the user if we place a label with a small delay. Due to these considerations, we introduced a waiting function.

We always try to label all reference points that just appeared. Let p be a reference point that we unsuccessfully tested for placing its label. For this, we add p to the list W of waiting reference points. Now, we wait at least for f frames until we test p again. (We only count frames with interactions, though.) For load balancing, we just test a certain number M of labels. Currently, M is the minimum of $|W|/f$ and all labels whose last test lies at least f frames in the past. Thereby $|W|$ is the number of labels in W and $|W|/f$ is an empirical value.

Recall that the algorithm for sliding labels does not re-insert labels; see Fig. 4o. Applying the waiting function, it lasts some frames until a label appears again. Sometimes, it also can happen that an unimportant label instead of an important one is placed awhile. Indeed, the waiting function can cause quickly changing labels. Consider a labeling to which we can place $\ell(p)$ in frame f. In $f+1$ a more important label makes $\ell(p)$ disappear. We can continue this. Thus, each of these labels is only visible for a single frame. There are several possibilities to solve this problem. We could state that we must not remove a just-placed label $\ell(p)$ for several frames. We also could increase the priority of $\ell(p)$ and decrease it little by little. This approach has the advantage that we place labels with a much higher priority than $\ell(p)$ earlier than labels that are only slightly more important.

3.2 Predicting Changes of the Rectangulation

When a user pans or zooms the map, we need to update the rectangulation.

For panning operations, it is easy to predict the *event points* at which changes will be necessary. The labels in the map will not intersect unless a new label appears at the view boundary or a label is blocked by the view boundary and thus needs to slide. This allows us to compute the distance that the user can pan to the right, left, bottom, and top without any event. If a reference point enters the view, we can apply

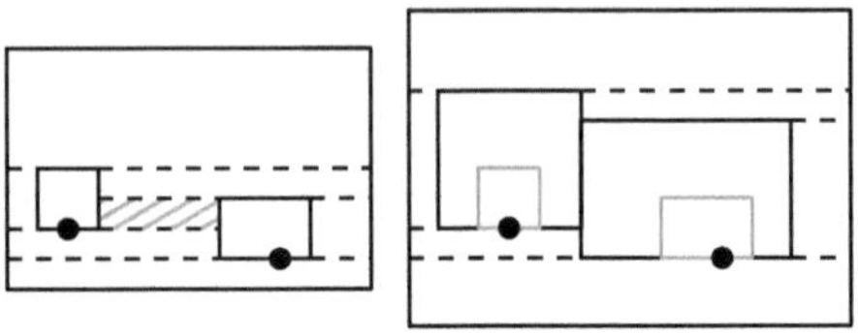

Fig. 9 From *left* to *right*: If the user zooms out, the hatched *rectangle* collapses *horizontally*

the incremental algorithm of Sect. 2 in just the same way as for any other point. In the case that a label touches the view boundary, we can treat the boundary as a big label that must not be moved. Thus, the touching label slides (or finally vanishes) rather than crosses the boundary. Again, we can apply the algorithm of Sect. 2. After each update of the rectangulation, we compute new event points.

While the user zooms the map, we require that each label keeps its size on the screen. More precisely, labels have to shrink with respect to real-world coordinates while the user zooms in and labels have to grow while the user zooms out. Certainly, while zooming the map, empty rectangles can collapse and the y-order of the top edge of a label and the bottom edge of another label (nearby) can change, see Fig. 9. This makes the prediction of event points and a local update while zooming rather difficult. Therefore, in our current implementation, we recompute the rectangulation in each frame if the user zooms the map. An interesting question for future research is whether we can speed up our method by predicting changes of the rectangulation that are caused by a zooming operation.

4 Experiments

We have implemented the incremental algorithm from Sect. 2 using a rectangulation and a waiting function (Sect. 3.1). To estimate the value of our algorithm, we compared it to a naive approach. The naive approach differs from the rectangulation-based approach in how it detects overlapping labels and potential collision counterparts. Instead of using a geometric data structure, the naive approach repeatedly checks *all* pairs of visible labels. The naive approach yields the same labeling as the rectangulation-based approach.

Both approaches have in common that we can (i) apply the waiting function and (ii) replace the slider model by a fixed-position model where the center of the label's bottom edge touches the reference point.

For our implementations, we used C++ with OpenSceneGraph 3.0.[4] We executed our experiments on a Windows 7 system with a 3.3-GHz AMD triple-core processor and 8 GB of RAM, applying the Microsoft Visual Studio 2010 Ultimate compiler in 32-bit release mode. The complete code has about 12,300 lines. For our tests,

[4] http://www.openscenegraph.org/, accessed Nov. 24, 2013.

Fig. 10 Point set of world map that we used in our experiments

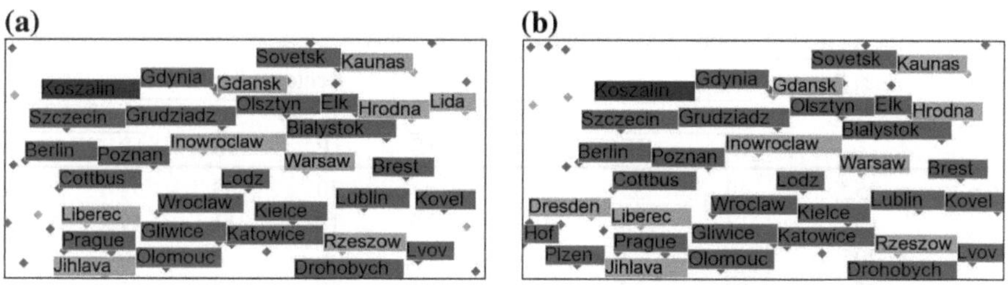

Fig. 11 A labeling before (**a**) and after (**b**) panning the map. New labels appeared at the *left boundary* of the view; at the *right boundary*, a label vanished. On the *lower right*, Lvov pushed Rzeszow, Rzeszow pushed Katowice, and so on

we used a world map from Natural Earth[5] providing 7,322 cities with priorities; see Fig. 10. We used priorities 1 (unimportant) to 4 (most important). We implemented several different single-interaction camera paths, this is, paths for only panning and for only zooming. Additionally, we defined multi-interaction paths that pan and zoom in and zoom out. Each of these paths executes its interactions for 42 s. For all single and multi-interaction paths, on average, either 35, 105, or 205 labels are visible. For each of these numbers, we implemented three different paths. We taped one of the multi-interaction paths and made the resulting video available online, from the url referenced at the end of Sect. 1. Figure 11 shows two snapshots of a panning interaction.

To the resulting 27 different paths, we applied the naive approach as well as the rectangulation-based approach with and without sliding and with and without a waiting function of $f = 30$ and $f = 60$ frames.

[5] http://www.naturalearthdata.com/, accessed Nov. 28, 2013.

Table 1 Quality: average value of the objective function (sum of the priorities over the labeled points on the screen) for the rectangulation-based approach

	$\varnothing\lvert P'\rvert$	Rectangulation			
		$f=0$		$f=30$	
		1P	1S	1P	1S
Pan	35	71	102	69	97
	105	201	302	196	277
	205	417	605	404	568
Zoom	35	54	81	53	79
	105	178	259	162	225
	205	375	547	318	452
Both	35	71	107	68	99
	105	197	294	183	258
	205	394	582	344	479

f denotes the waiting interval (in frames); $\varnothing\lvert P'\rvert$ is the average number of labeled points on the screen; 1P and 1S are our labeling models

Table 2 Speed: average frame rates (in FPS) of camera paths

	$\varnothing\lvert P'\rvert$	Naive				Rectangulation			
		$f=0$		$f=30$		$f=0$		$f=30$	
		1P	1S	1P	1S	1P	1S	1P	1S
Pan	35	49	33	51	37	50	33	52	38
	105	13	8	14	10	18	11	19	14
	205	8	4	8	6	9	6	10	7
Zoom	35	60	37	60	41	60	38	61	42
	105	19	12	21	15	19	12	21	16
	205	9	5	11	8	9	6	11	8
Both	35	46	28	48	34	45	28	48	34
	105	17	10	18	13	17	11	19	14
	205	7	4	8	6	9	6	10	8

f denotes the waiting interval (in frames); $\varnothing\lvert P'\rvert$ is the average number of labeled points on the screen; 1P and 1S are our labeling models; *rectangulation* and *naive* are our algorithms

For determining the width of a rectangle, we counted the number of the letters in the city name and scaled it with an empirical value that depends on the desired width of a letter and the priority of the label. As the drawing routine of OpenSceneGraph for Windows is rather time consuming, we "only" drew reference points and labels in the view.

For each frame, we recorded the sum of weights over all labeled points. We summed up the weights over the three paths with the same interaction type and the same average number of labels in the view. Finally, we averaged the weights over the total number of frames in order to compute the averaged quality; see Table 1. Additionally, we averaged the total number of frames over the processing time in order to compute the *frame rate* in frames per second (FPS); see Table 2.

We observed that in many cases, the frame rate is rather low when we start a camera path as well as while zooming. Recall that, in these cases, we compute the rectangulation from scratch. We also observed that our algorithms yield different results with regard to the averaged weight and FPS for each pass of the *same* camera path. This is because the current load factor influences our measurements. As a result, also the average quality of our algorithm and the naive approach differ slightly. Since the difference is not noteworthy, Table 1 shows only the quality results for the original algorithm. Moreover, the results for zooming in only slightly differ from the results for zooming out (the inverted path). Thus, we averaged the results for zooming out.

We conclude that, using the slider model, our algorithm yielded an improvement of 30–50 % in the labeling quality with respect to the algorithm using the fixed-position model; see Table 1. Second, we point out that the rectangulation-based approach increased the frame rate by up to 40 % if the screen contained a large number of labels; see Table 2. If we additionally applied a waiting function of 30 frames, the frame rate for small point sets increased by about 15 %. For large point sets, it sometimes doubled. The maximum loss in quality was 18 %. When we applied a waiting function of 60 frames, to our surprise, the frame rates increased by at most 2 FPS whereas the quality dropped by up to 30 %. Therefore, we do not show the details concerning the longer waiting function in Tables 1 and 2.

5 Conclusion and Future Work

In this work, we have described an algorithm that labels points in interactive maps using a slider model. To speed up our algorithm, we used a rectangulation data structure and a waiting function. We conclude that sliding labels improve the labeling quality (in terms of our objective function) by up to 50 %. Compared to a naive approach, our heuristic significantly improved the frame rate, that is, in some cases, it doubled.

In the future, we plan to implement the prediction of changes in the rectangulation. Further, we want to analyse the cost of our current simplistic point-location strategy. Will it be worthwhile replacing it with a dedicated dynamic point-location data structure? It would also be interesting to deal with rotations and 3D environments where the view can be tilted or to study how users cope with the additional cognitive load of sliding labels. Ooms et al. (2009) showed that when panning horizontally, users did not react significantly to certain differences in the labeling.

References

Adamaszek A, Wiese A (2013) Approximation schemes for maximum weight independent set of rectangles. In: Proceedings of 54th annual IEEE symposium on foundations of computer science (FOCS'13), pp 400–409

Agarwal PK, van Kreveld M, Suri S (1998) Label placement by maximum independent set in rectangles. Comput Geom Theory Appl 11:209–218

Alinhac G (1962) Cartographie Théorique et Technique, chapter IV. Institut Géographique National, Paris
Been K, Daiches E, Yap C (2006) Dynamic map labeling. IEEE Trans Visual Comput Graphics 12(5):773–780
Been K, Nöllenburg M, Poon SH, Wolff A (2010) Optimizing active ranges for consistent dynamic map labeling. Comput Geom Theory Appl 43(3):312–328. http://dx.doi.org/10.1016/j.comgeo.2009.03.006
Chalermsook P, Chuzhoy J (2009) Maximum independent set of rectangles. In: Proceedings of 20th annual ACM-SIAM symposium on discrete algorithms (SODA'09), pp 892–901
de Berg M, Cheong O, van Kreveld M, Overmars M (2008) Computational geometry: algorithms and applications, chapter 6, 3rd edn. Springer, Berlin
Erlebach T, Jansen K, Seidel E (2005) Polynomial-time approximation schemes for geometric intersection graphs. SIAM J Comput 34(6):1302–1323
Erlebach T, Hagerup T, Jansen K, Minzlaff M, Wolff A (2009) Trimming of graphs, with application to point labeling. Theory Comput Syst 47(3):613–636. http://dx.doi.org/10.1007/s00224-009-9184-8
Fowler RJ, Paterson MS, Tanimoto SL (1981) Optimal packing and covering in the plane are NP-complete. Inform Process Lett 12(3):133–137
Gemsa A, Niedermann B, Nöllenburg M (2013) Trajectory-based dynamic map labeling. In: Cai L, Cheng SW, Lam TW (eds) Proceedings of 24th annual international symposium on algorithms computation (ISAAC'13). Lecture notes in computer science, vol 8283. Springer, pp 413–423. http://dx.doi.org/10.1007/978-3-642-45030-3_39
Gemsa A, Nöllenburg M, Rutter I (2011a) Consistent labeling of rotating maps. In: Dehne F, Iacono J, Sack JR (eds) Proceedings of 12th international symposium on algorithms and data structures (WADS'11). Lecture notes in computer science, vol 6844. Springer, pp 451–462. http://dx.doi.org/10.1007/978-3-642-22300-6_38
Gemsa A, Nöllenburg M, Rutter I (2011b) Sliding labels for dynamic point labeling. In: Proceedings of 23th Canadian conference on computational geometry (CCCG'11), pp 205–210
Goralski R, Gold CM, Dakowicz M (2007) Application of the kinetic Voronoi diagram to the real-time navigation of marine vessels. In: Proceedings of 6th international conference on computer information systems and industrial management applications (CISIM'07), pp 129–134
Harrie L, Stigmar H, Koivula T, Lehto L (2005) An algorithm for icon labelling on a real-time map. In: Fisher PF (ed) Proceedings of 11th international symposium on spatial data handling (SDH'05), pp 493–507
Imhof E (1975) Positioning names on maps. Am Cartogr 2(2):128–144
Luboschik M, Schumann H, Cords H (2008) Particle-based labeling: fast point-feature labeling without obscuring other visual features. IEEE Trans Visual Comput Graphics 14(6):1237–1244. http://dx.doi.org/10.1109/TVCG.2008.152
Maass S, Döllner J (2006) Efficient view management for dynamic annotation placement in virtual landscapes. In: Butz A, Fischer B, Krüger A, Oliver P (eds) Proceedings of 6th international symposium on smart graphics (SG'06). Lecture notes in computer science, vol 4073. Springer, Berlin, pp 1–12
Mote KD (2007) Fast point-feature label placement for dynamic visualizations. Inf Visual 6(4): 249–260. http://dx.doi.org/10.1057/palgrave.ivs.9500163
Ooms K, Kellens W, Fack V (2009) Dynamic map labelling for users. In: Cartwright W, Lopez P (eds) Proceedings of the 24th international cartographic conference (ICC'09)
Poon SH, Shin CS, Strijk T, Uno T, Wolff A (2003) Labeling points with weights. Algorithmica 38(2):341–362. http://dx.doi.org/10.1007/s00453-003-1063-0
van Kreveld M, Strijk T, Wolff A (1999) Point labeling with sliding labels. Comput Geom Theory Appl 13:21–47. http://dx.doi.org/10.1016/S0925-7721(99)00005-X
Zhang Q, Harrie L (2006) Real-time map labelling for mobile applications. Comput Environ Urban Syst 30(6):773–783

Routes to Remember: Comparing Verbal Instructions and Sketch Maps

Vanessa Joy A. Anacta, Jia Wang and Angela Schwering

Abstract Sketch maps of routes have been widely used to externalize human spatial knowledge and to study wayfinding behavior. However, specific studies on what information and how people recall route information they obtain from verbal instructions by drawing sketch maps are limited. This chapter aims to know how much information, especially landmarks and streets, people recall after following a wayfinding task. We conducted an experiment and asked participants to draw a sketch map of the route they travelled. Landmarks were classified based on their locations on the route. Sketch maps were compared with verbal instructions to analyze what specific landmarks and street information people recalled as well as what other information was added. Our study showed that (1) landmarks along the route were sketched as often as landmarks located at decision points; (2) participants added landmarks and streets which were not mentioned in the verbal instructions. This chapter provides a better understanding of wayfinding strategies and spatial learning.

Keywords Route · Sketch maps · Verbal instructions · Landmarks · Streets · Wayfinding

1 Introduction

In wayfinding, it is of great interest to know what a wayfinder recalls after following a route instruction from a direction provider. Usually, not all information from the route instruction is remembered. The amount of recollected information

V. J. A. Anacta (✉) · J. Wang · A. Schwering
Institute for Geoinformatics, University of Muenster, Muenster, Germany
e-mail: v.anacta@uni-muenster.de

J. Wang
e-mail: j_wang05@uni-muenster.de

A. Schwering
e-mail: schwering@uni-muenster.de

J. Huerta et al. (eds.), *Connecting a Digital Europe Through Location and Place*,
Lecture Notes in Geoinformation and Cartography, DOI: 10.1007/978-3-319-03611-3_18,

depends on characteristics of spatial features such as the saliency of landmarks (Raubal and Winter 2002) or how vivid the descriptions are of streets and landmarks (Tom and Tversky 2012). Except for the incompleteness, there are also cases when the wayfinder remembers other information that is not mentioned in the route instruction but rather from his or her travel experiences (Albert et al. 1999). Landmark attributes such as visibility (Denis 1997), prominence (Heft 1979), and location (Denis 1997; Lovelace et al. 1999; Michon and Denis 2001) contribute how quickly people recall information. Even the color of buildings (Jansen-Osmann and Wiedenbauer 2004) has been considered a factor which aids in wayfinding and spatial learning. In this chapter, we focused on the types of landmarks recalled based on their locations on the route. The recall of landmarks of different locations can be more beneficial in studying human spatial orientation and in learning how wayfinders relate objects in space. We also looked at types of landmarks people recalled, and what other sketching information was added which was not mentioned in the verbal instructions. We believe our study contributes to a better understanding of wayfinding strategy and spatial learning.

We used sketch maps to externalize the recollected route information which wayfinders learned from the verbal instructions and their direct travel experiences. We provided wayfinders with a set of verbal route instructions and asked them to perform a wayfinding task in an unfamiliar area. After completing the wayfinding task, we immediately asked wayfinders to draw sketch maps of the routes they had travelled. The sketch mapping activity was video recorded. Landmarks sketched were examined and classified into three groups based on their locations on the route: *along the route*, *at decision point,* and *off the route*. Sketch maps were also compared with verbal instructions to study added and missing landmarks and streets. Analysis on the characteristics of landmarks is beyond the scope of this chapter, but we consider it an interesting aspect for future work.

2 Related Work

Most people can remember the routes they travelled by following verbal instructions. Landmarks and streets play an important role for people to learn and memorize routes and environments. Garling et al. (1982) claimed that people could already learn a path almost perfectly on the first trial while taken on tour by car in an unfamiliar environment. Denis and Zimmer (1992) studied the role of verbal instructions in constructing cognitive maps. The results suggest that people are able to transform linguistic descriptions of configurations into mental representations based on the descriptions in the same way as from a perceptual experience. This also highlighted the fact that a person is able to create good visuo-spatial representations of his or her environment through verbal descriptions. Magliano et al. (1995) stated that a wayfinder was able to learn the environment based on a specific goal. For example, if the task is on landmark-based activity, people are able to follow it but learning is only concentrated in this specific task. Our study also investigated whether people could learn an unfamiliar environment after following a wayfinding task.

Lynch (1960) emphasized the importance of paths, edges, districts, nodes, and landmarks in helping a person to understand the structural image of a place that guides wayfinding. Some studies considered paths as the skeletal structure in any map drawn first and landmarks encoded afterwards (Appleyard 1970; Garling et al. 1982). With first mapping the paths (or routes) people are able to define where specific landmarks are located (MacEachren 1992). On the other hand, Siegel and White (1975) claimed that people first learn the environment through landmark-based. Related to this claim is the anchor point theory (Golledge and Stimson 1997; Couclelis et al. 1987) stating that landmarks, nodes and regions are linked together creating subregions of organized space. Thus, one is able to create a hierarchical structure of spatial features by relating one landmark to other landmarks which helps in building cognitive map.

Lovelace et al. (1999) looked at elements in route instructions that were relevant for effective wayfinding. The study showed that not only landmarks on choice points were considered important quality of route directions. Michon and Denis (2001) found that participants often referred to landmarks for reorientation. These referred landmarks were located at "critical nodes" where a change of direction happens. The absence of such landmarks makes it difficult for people to progress in a route where reorientation is needed or when there are possible choices of directions. (Raubal and Winter 2002) highlighted the importance of saliency of local landmarks in wayfinding instructions. Landmark saliency has been further analyzed through structurally integrating landmark position on specific route instructions (Klippel and Winter 2005) and in the attempt of modelling salient landmarks in route directions looking at both visual and semantic characteristics (Nothegger et al. 2009). Waller and Lippa (2007) investigated the effectiveness of landmarks both as beacons and associative cues in learning the routes. Our study looked at which types of landmarks people remembered based on its location on the route as investigated by previous studies.

Sketch mapping has been used to externalize and study route knowledge as well as wayfinding performance (Rovine and Weisman 1989; Walmsley and Jenkins 1992; Young 1999) to evaluate how sketch maps can draw individual's environmental knowledge. Similarly, we used sketch maps to study the routes which were recalled and learned from verbal descriptions and direct travel experiences.

3 Wayfinding and Sketch Mapping

This chapter is a follow up study on a previous research on analyzing the wayfinding performance of different reference frames (Anacta and Schwering 2010). The previous wayfinding task design was modeled on the methodology of Ishikawa and Kiyomoto (2008) wherein participants were given route instructions in a shifting reference frame of directions: absolute or relative. The switching reference frame means if the participant is given an "absolute" instruction for one route then he or

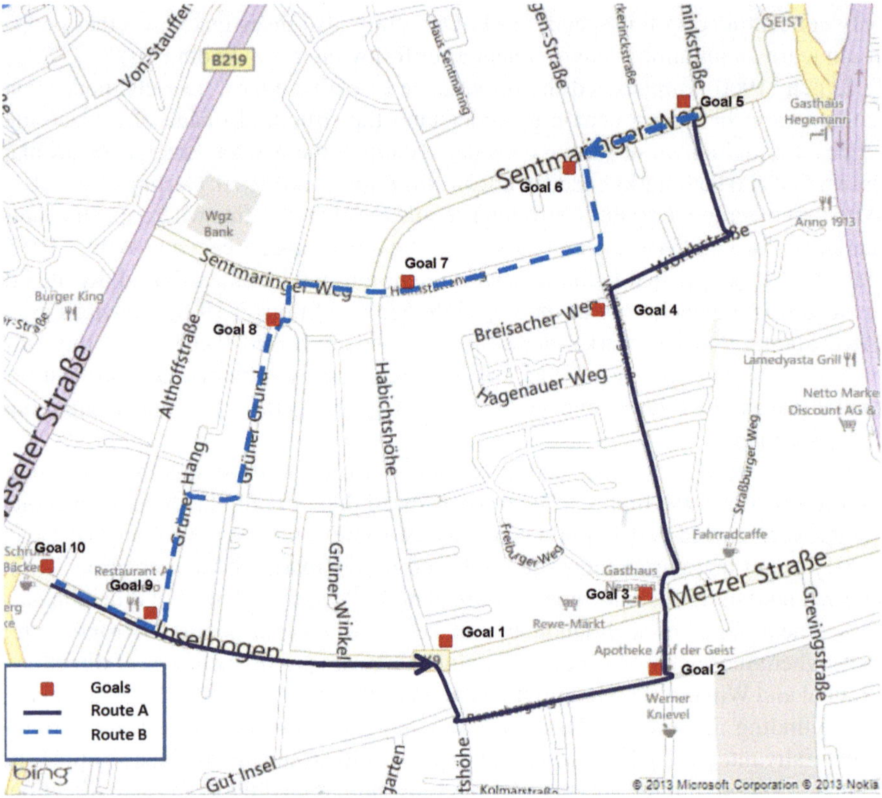

Fig. 1 The study area with the two major routes, *Route A* and *Route B*, and the 10 *goals* (*Source* Bing map)

she continues the next route by following a "relative" instruction. In the current study, we added another sketch mapping task to externalize the routes travelled.

Participants and study areas A total of 18 university students (nine female) with an average age of 23.5 years ($SD = 2.28$) participated in the experiment. These students were from diverse disciplines. The study area (Fig. 1) is a residential district in the city of Muenster and we chose two routes in this area. Route A has a total length of 1,380 m while Route B is 920 m long. The blue route is Route A and the orange route represents Route B. Both routes have a total length of 2,300 m with each sub-route ranging from 62 to 400 m. There were 10 goals specified in the verbal instructions which the participant had to reach before going to the next route. All participants claimed that they were unfamiliar with the study area and the routes.

Material and procedure We provided each participant a set of 10 verbal instructions and asked them to perform a wayfinding task. The verbal instructions were in German and they describe two major routes, Route A and Route B, with each composed of connected sub-routes. Route A is composed of sub-routes from 1 to 5 and route B

Table 1 Route A verbal instructions in both reference frames

Absolute frame of reference	Relative frame of reference
Route 1. Walk straight to ESE for 400 m. You see a block of houses to the north and to the south of the street. You see a restaurant and a hair salon to the north. You see a bakery with an automobile shop beside it to the south. You see a traffic light. You see a park to the north. After the park, you see a kiosk. When you hit an intersection, you see the bakery located ESE. You see a Teleport telephone booth [goal] to the east	Route 1. Walk straight ahead. You see a block of houses on both sides of the street. You see a restaurant and a hair salon to your left. You see a bakery with an automobile shop beside it to your right. Then, you see a traffic light. You see a park to your left. After the park, you see a kiosk. When you hit an intersection, you see a bakery located in the corner of the other side of the street. You see a Teleport telephone booth [goal] to your left
Route 2. Turn to the south and cross the street. Walk straight for 62 m. Turn to the ENE when you hit a shared pathway for bicycle and pedestrian. Walk for 220 m. You pass through a residential area where you see a white building to the north. You pass by an open space to the south. You see a parking space of a restaurant to the south. When you hit a north-south running road, you see a pharmacy [goal] to the north and a restaurant to the south	Route 2.Turn to the right and cross the street. Walk straight. Turn left when you hit a shared pathway for bicycle and pedestrian. You pass through a residential area where you see a white building to your left. You pass by an open space to your right. Then, you see a parking space of a restaurant to your right. When you hit the end of the road, you see a pharmacy [goal] to your left and a restaurant to your right
Route 3. Walk 70 m north. You see the Ulf Import Farschule to the east. Cross the east-west running road, you see a Haus Nemann restaurant [goal] to the west. You see Johanniter Akademie Gästehaus to the north. You also see Schlecker store to the east and a church to the ESE	Route 3. Turn left from the pharmacy and walk straight. You see the Ulf Import Fahrschule to your right. Cross the street, you see the Haus Nehmann restaurant [goal] to your left. You see Johanniter Akademie Gästehaus in front of you. You also see Schlecker store and a church to your right
Route 4. Walk to the NNE for 40 m and then walk 290 m NNW. You pass by the whole block of Johanniter Akademie Gästehaus to the west. You see a shop for decorations to the east. You pass by the Johanniter-Stift Seniorenhäuser Münster to the west. You cross two streets to the west. You see a whole block of brick building [goal] to the west	Route 4. Turn to the right beyond the island. Turn to the left in the first street and walk straight. You pass by the whole block of Johanniter Akademie Gästehaus to your left. You see a shop for decorations to your right. You pass the Johanitter-Stift Seniorenhäuser Münster to your left. You cross two streets to your left. You see a whole block of brick building [goal] to your left
Route 5. Turn to the NE and walk for 150 m. When you hit a shared bicycle and pedestrian pathway to the north, turn to the NNW and walk for 120 m. You pass through a residential block with a white-colored house to the west. At the end of the pathway [goal], You see a brick building to the north with a kiosk and a shop for girls on both ends of the building	Route 5. Turn to the right and walk straight. When you hit a shared bicycle and pedestrian pathway to the left, turn to the left and walk straight. You pass through a residential block with white-colored house to your left. At the end of the pathway [goal], you see a brick building in front of you with a kiosk and a shop for girls on both ends of the building

Note Original verbal instructions were written in German language

Fig. 2 A schematic view of the classification of landmarks based on locations

is composed of sub-routes from 6 to 10. Table 1 shows an example of the verbal instruction of the first set of five routes in Route A in both absolute and relative frames of reference. The experimenter walked behind the participant without any conversation.

After completing the wayfinding task for Route A, we retrieved the verbal instructions from the participants and immediately asked them to draw sketch maps of the routes they had travelled. The participants were provided with an A3 sized chapter and a pen. We asked them to complete sketching the map within 15 min. The drawing activity was recorded using a video camera. After finishing the drawing task, the participants proceed with Route B. The same procedure was followed in the sketch mapping task.

Method of landmark classification Landmarks classification was based on their locations with respect to the route proposed by Lovelace et al. (1999). These are landmarks along the route, at decision points and landmarks off the route. Landmarks along the route (LAR) refer to the landmarks located on the route which the wayfinder encounters when he or she travels along the route. LAR type of landmarks is located at either one or both sides of the route being travelled. Landmarks at decision points (LDP) are landmarks that are located at junctions or intersections where a turning action is required. Landmarks off the route (LOR) are the landmarks that are not directly along the route and they can be visible or not. Oftentimes, LOR are distant landmarks but their locations relative to the participant are known. Figure 2 is a schematic view of the classification of landmarks: circle represents LAR, rectangle represents LDP and triangle represents LOR.

4 Results

We collected in total 18 sketched route maps. The average time spent to complete sketch maps was around 8 min for Route A and 6 min for Route B. We used mixed analysis of variance, with time as within subject factor and the route group as between subject factor. There was a significant effect of time in drawing the sketch maps for the two groups. This means that it took time for participants to draw longer routes, $F(1, 16) = 11.42$, $p < .01$. The changing reference frame was not a factor for the number of landmarks drawn.

Table 2 Percentage of the landmarks recalled and missed from verbal instructions

Route	LAR (%)	LAR (missed) (%)	LDP (%)	LDP (missed) (%)	LOR
Route A	37.58	62.42	64.82	35.18	35.19%
Route B	51.59	48.41	72.22	27.78	No data[a]

[a]No landmark off the route mentioned in the verbal instruction for Route B

Table 3 Percentage of all landmarks drawn in sketch maps

Routes	LAR (%)	LDP (%)	LOR (%)
Route A	42.81	37.33	19.86
Route B	41.53	42.62	15.85

4.1 Landmarks

Landmarks recalled and missed from verbal instructions In verbal instructions, Route A has 24 landmarks and Route B has 15 landmarks. The numbers of the different types of landmarks included in the verbal instructions differ for two routes. In Route A, there are 17 landmarks along the route (LAR), nine landmarks at decision points (LDP) and six landmarks off the route (LOR). In Route B, there are seven landmarks along the route (LAR), six landmarks at decision points (LDP) and no landmarks off the route (LOR). We compared the collected sketch maps with the verbal instructions in landmark information. Table 2 shows the percentage of landmarks that participants recalled and drew and the percentage of landmarks that were missed in sketch maps.

The above table shows that participants included all three types of landmarks in their sketch maps. It was not only the landmarks of LDP were frequently drawn but also the landmarks of LAR were recalled and sketched. For Route A, participants also recalled and represented around 35 % of LOR in their sketch maps. In Route B, there is a high percentage of landmarks recalled and drawn from the verbal instructions. The table also shows the information of missing landmarks. There is a high percentage of LAR missed in Route A and less percentage of LDP missed in Route B.

Types of landmarks drawn in sketch maps We examined now all types of landmarks including missing and extra ones drawn in sketch maps. Our aim was to explore what types of landmarks people commonly sketch in route maps. Table 3 includes all the landmark information generated from sketch maps regardless of being mentioned in route instructions. The table shows that landmarks along the route were drawn as frequently as landmarks at decision points. Landmarks off the route were also evident in both routes. It occurred that all participants have included landmarks off the route in Route A which comprised almost 20 % of the total landmarks drawn and almost 16 % in Route B.

Figure 3 shows all three types of landmarks in a sketch map from one of our participants. The figure shows the first five route segments of Route A with landmarks along the route (circle), at decision points (rectangle), and off the route (triangle) (refer to Fig. 1 for the map and Table 1 for the route instructions).

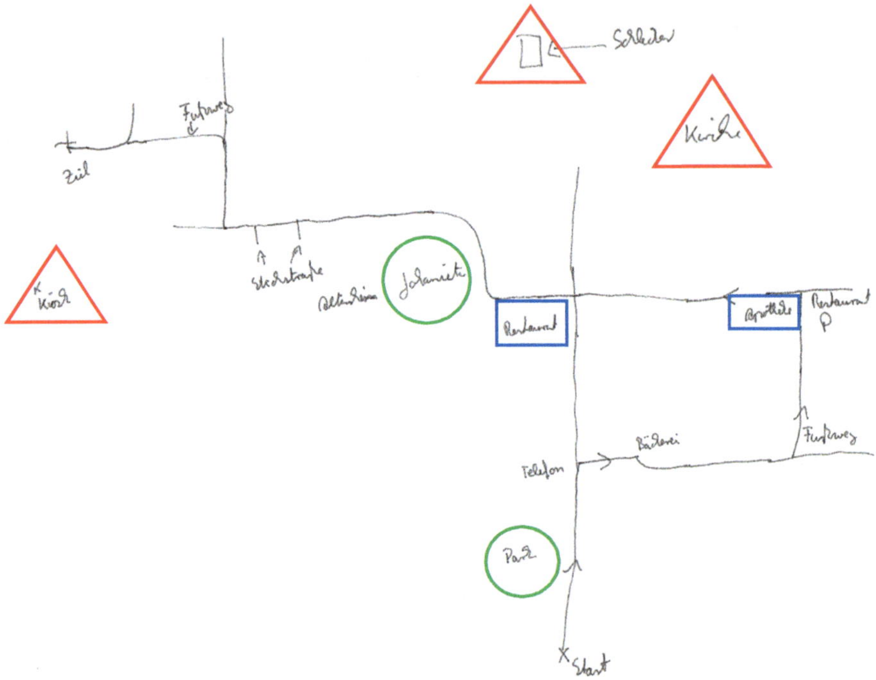

Fig. 3 Sketch map showing all types of landmark classification

Table 4 Percentage of all landmarks recalled in individual sketch maps

Landmarks	LAR (%)	LDP (%)	LOR (%)
Recalled from verbal instructions	45.89	48.02	6.08
Added in sketch maps	41.66	40.51	17.82

Landmarks drawn per participant In this analysis, we tried to examine landmarks drawn by each participant. We investigated all three types of landmarks in sketch maps comparing them with the verbal instructions. Similar to the previous results from Tables 2 and 3, Table 4 shows that landmarks along the route were drawn as often as landmarks at decision points in individual sketch maps for both routes. Regarding LOR type of landmarks, participants tended to draw extra ones that were not mentioned (17.82 %) than the ones that were mentioned in verbal instructions (6.08 %). Half of the participants added landmarks along the route for both routes.

Figure 4 shows two sketch maps with added landmarks. These added landmarks were commonly LOR type and were not mentioned in the verbal route instructions.

Taking all the landmarks drawn on sketch maps, there is a significant difference of the landmarks along the route, $F(1, 16) = 24.63$, $p < .001$; decision points $F(1, 16) = 19.06$, $p < .001$; and off the route $F(1, 16) = 19.06$, $p < .001$ using

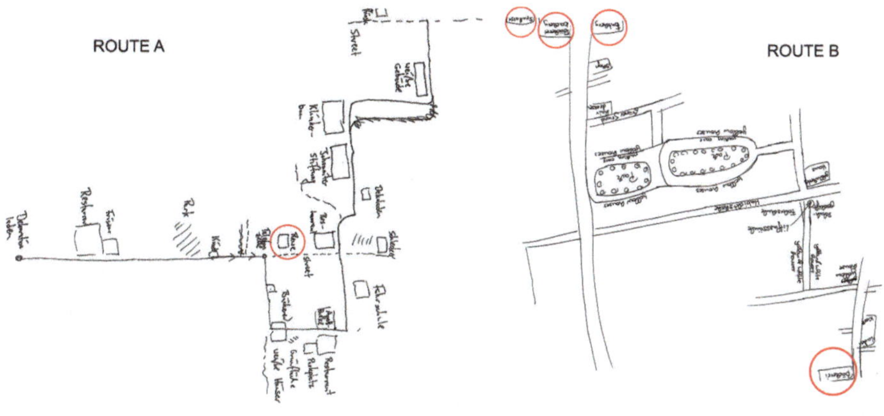

Fig. 4 Landmarks (*encircled*) are the ones added to sketch maps

Table 5 Percentage of streets from verbal instruction and extra streets

Route	From verbal instructions (%)	Not from verbal instructions (%)
Route A	79.91	13.95
Route B	85.35	12.13

mixed analysis of variance. The changing reference frame does not have an effect on the number of landmarks drawn on sketch maps.

Relationships of the types of landmarks A significant correlation occurred in the landmark types of Route B. LAR and LDP were positively correlated ($r = .66$, $p < .01$). This means that if participants draw LAR, they tend to also draw LDP. LOR, on the other hand, was negatively correlated with LAR, ($r = -.66$, $p < .01$) and LDP, ($r = -1$, $p < .01$). The more LAR and LDP were drawn, the less LOR was included in the sketch map. Regarding both Route A and Route B, there is a significant negative correlation between LOR and LDP ($r = -1$, $p < .01$). This means that if participants draw more LOR, they include less LDP in their sketch maps.

4.2 *Streets*

There were 13 streets mentioned in the verbal instruction of Route A and 11 streets mentioned in the verbal instruction of Route B. Table 5 shows that participants were able to recall and draw most of the streets mentioned in the route instructions. For Route A, almost 80 % of the streets were recalled and drawn. For Route B, 85 % of the streets were recalled and represented in sketch maps.

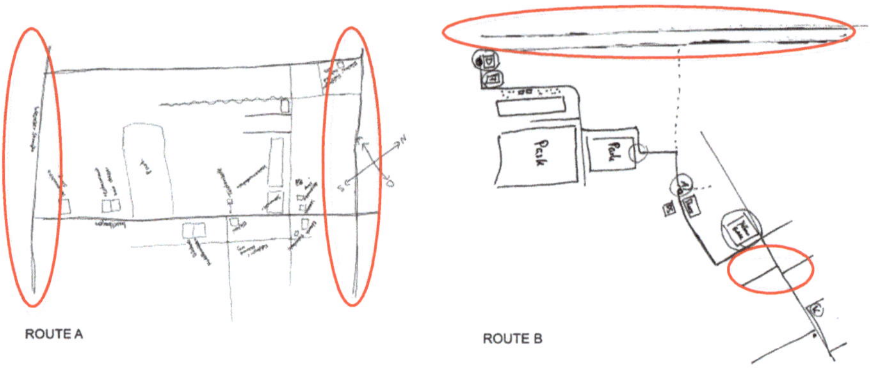

Fig. 5 Extra streets (*encircled*) added to sketch maps

Almost 78 % of all participants added streets that were not mentioned in verbal instructions in their sketch maps for both routes. Participants added and labeled mostly distant street they recognized (Fig. 5).

5 Discussion and Conclusion

This study shows what information people remember after following verbal instructions of a route. Sketch maps cannot tell everything what people learned from verbal instructions as well as travel experiences but they can provide insights into what people may recall from verbal instructions and other information of landmarks and streets that are excluded from verbal instructions. With regard to types of landmarks, people did not include all landmarks mentioned in the route instructions to their sketch maps. The importance of landmarks at decision points have been widely studied specifically in wayfinding. Denis (1997) stated that landmarks where action is required are considered the most essential type of landmark to be included in verbal route descriptions. However, we found that landmarks along the route were drawn as frequently as landmarks at decision points. This shows that participants found the landmarks without turning action made were also relevant in representing the study area. Schwering et al. (2013) also showed a similar result wherein landmarks along the route are frequently mentioned in the route descriptions and sketch maps. Our result of the types of landmarks agrees with Lovelace et al. (1999) stating that landmarks along the route are also important elements to be considered in route instructions. In the results, there was a positive correlation in one major route which suggests that when participants drew LDP, they also draw LAR. It shows that the landmarks that they mostly remember and represent in the sketch map were not only located at decision points but also along the route. When participants draw more LAR and LDP, they tend to draw less LOR. This shows that all these landmark types are commonly remembered and are considered important to be represented in the sketch maps. Participants also remembered landmarks that are not mentioned in the instructions as well as streets. This tells us that people are not confined in landmarks and streets that they read in verbal instructions but are also interested in providing added information they have seen along the route. The landmarks that were added vary among participants wherein most of these appeared to be visually recognizable. With regard to streets, participants recalled more than half of it and included recognizable streets that are distant from the study area. These were mostly major roads.

To conclude, this chapter addresses the following aspects. Firstly, we looked at the types of landmarks based on their locations on the route. Results showed that these landmarks were not only located at decision points but also along the route. Landmarks off the route were also common types of landmarks drawn. Secondly, this chapter addresses the added information found in sketch maps. Our results showed that participants included extra landmarks and streets to sketch maps, which were not mentioned in verbal instructions. This shows that participants paid attention to the study area during wayfinding and they were able to learn extra spatial knowledge directly from their travel experiences.

This study is helpful to cognitive psychologists who try to understand how human externalize spatial knowledge both from verbal instructions and real travel experiences. This provides an idea of which type of landmarks people remember considering the location on the route. When giving directions, landmarks at decision points may be more commonly used but in this study, landmarks along the route appeared to be important as well. Participants also recalled landmarks off the route which were included in the sketch maps. Providing better orientation in an unfamiliar environment could be one reason why participants remember adding information in the sketch maps especially those landmarks located off the route. In addition, this study showed that people are able to learn an unfamiliar environment which is evident in the sketch maps showing a natural way of how people represent the environment. For future work, we intend to investigate how participants sketched route maps to determine importance of landmarks drawn. Also, it would be interesting to know how participants provide verbal instructions after following a wayfinding task.

Acknowledgments We thank the support of Deutsche Akademischer Austauschdienst (DAAD) and the SketchMapia Project funded by Deutsche Forschungsgemeinschaft (DFG). Many thanks also go to all participants who took part in the experiment and Deborah Nyangulu for reviewing the final version of the chapter.

References

Albert WS, Reinitz MT, Beusmans JM, Gopal S (1999) The role of attention in spatial learning during simulated route navigation. Environ Plan A 31(8):1459–1472

Anacta VJA, Schwering A (2010) Men to the east and women to the right: wayfinding with verbal route instruction. In: 7th international conference on spatial cognition, Mt. Hood Oregon, USA, Aug 2010. Lecture notes in computer science, vol 6222. Springer, Heidelberg, pp 70–84

Appleyard D (1970) Styles and methods of structuring a city. Environ Behav 2:100–116

Couclelis H, Golledge RG, Gale N, Tobler W (1987) Exploring the anchor-point hypothesis of spatial cognition. J Environ Psychol 7(2):99–122. doi:10.1016/S0272-4944(87)80020-8

Denis M (1997) 'The description of routes: a cognitive approach to the production of spatial discourse. Cah de Psychol Cogn 16(4):409–458

Denis M, Zimmer HD (1992) Analog properties of cognitive maps constructed from verbal descriptions. Psychol Res 54(4):286–298

Garling T, Book A, Ergezen N (1982) Memory for the spatial layout of the everyday physical environment: differential rates of acquisition of different types of information. Scand J Psychol 23(1):23–35

Golledge RG, Stimson RJ (1997) Spatial behavior: a geographic perspective. The Guilford Press, New York

Heft H (1979) The role of environmental features in route-learning: two exploratory studies of way-finding. Environ Psychol Nonverbal Behav 3:172–185

Ishikawa T, Kiyomoto M (2008) Turn to the left or to the West: verbal navigational directions in relative and absolute frames of reference. In: 5th international conference on GIScience, Park City, UT, USA, Sept 2008. Lecture notes in computer science, vol 5266. Springer, Heidelberg, pp 119–132

Jansen-Osmann P, Wiedenbauer G (2004) Wayfinding performance in and the spatial knowledge of a color-coded building for adults and children. Spat Cogn Comput Interdisc J 4(4):337–358

Klippel A, Winter S (2005) Structural salience of landmarks for route directions. In: International conference on spatial information theory, NY, USA, Sept 2005. Lecture notes in computer science, vol 3693. Springer, Heidelberg, pp 347–362

Lovelace KL, Hegarty M, Montello DR (1999) Elements of good route directions in familiar and unfamiliar environments. In: International conference on spatial information theory, Stade, Germany, Aug 1999. Lecture notes in computer science, vol 1661. Springer, Heidelberg, pp 365–82

Lynch K (1960) Image of the city. MIT Press, Cambridge

MacEachren AM (1992) Application of environmental learning theory to spatial knowledge acquisition from maps. Ann Assoc Am Geogr 82(2):245–274

Magliano JP, Cohen R, Allen GL, Rodrigue JR (1995) The impact of a wayfinder's goal on learning a new environment: different types of spatial knowledge as goals. J Environ Psychol 15(1):65–75

Michon P-E, Denis M (2001) When and why visual landmarks used in giving directions? In: International conference on spatial information theory, CA, USA, Sept (2001). Lecture notes in computer science, vol 2205. Springer, Heidelberg, pp 292–305

Nothegger C, Winter S, Raubal M (2009) Selection of salient features for route directions. Spat Cogn Comput Interdisc J 4(2):113–136

Raubal M, Winter S (2002) Enriching wayfinding instructions with local landmarks. In: 2nd international conference on GIScience, Boulder, CO, USA, Sept 2002. Lecture notes in computer science, vol 2478. Springer-Verlag London, UK, pp 243-259

Rovine MJ, Weisman GD (1989) Sketch-map variables as predictors of way-finding performance. J Environ Psychol 9(3):217–232. doi:10.1016/S0272-4944(89)80036-2

Schwering A, Li R, Anacta VJA (2013) Orientation information in different forms of route instructions. In: Vandenbroucke D, Bucher B, Crompvoets J (eds) Proceedings on the 15th AGILE international conference on geographic information science, 2013

Siegel AW, White S (1975) The development of spatial representations of large-scale environments. Adv Child Dev Behav 10:9–55

Tom AC, Tversky B (2012) Remembering routes: streets and landmarks. Appl Cogn Psychol 26(2):182–193

Waller D, Lippa Y (2007) Landmarks as beacons and associative cues: their role in route learning. Mem Cogn 35(5):910–924

Walmsley DJ, Jenkins JM (1992) Tourism cognitive mapping of unfamiliar environments. Ann Tourism Res 19(2):268–286. doi:10.1016/0160-7383(92)90081-Y

Young M (1999) Cognitive maps of nature-based tourists. Ann Tourism Res 26(4):8179–8839

Part V
Geospatial Decision Support Services

Behaviour-Driven Development Applied to the Conformance Testing of INSPIRE Web Services

Francisco J. Lopez-Pellicer, Miguel Ángel Latre, Javier Nogueras-Iso, F. Javier Zarazaga-Soria and Jesús Barrera

Abstract The implementation of the INSPIRE directive requires to check the conformity of a large number of network services with the implementing rules of INSPIRE. The evaluation whether a service is fully conformant with INSPIRE is complex and requires the use of specialized testing tools that should report how verification has been made and should identify non-conformances. The use of these tools requires a high degree of technical knowledge. This fact makes very difficult for non-technical stakeholders (end users, managers, domain experts, etc.) to participate effectively in conformance testing, hinders stakeholders understanding of the causes and consequences of non-conformant results and may cause in some stakeholders disinterest in conformance testing. This work explores the suitability of a *behaviour-driven development* (BDD) approach to the conformance testing of OGC Web services in the context of the INSPIRE directive. BDD emphasizes the participation of non-technical parties in the design of acceptance tests by means of automatable abstract tests expressed in a human readable format. Using this idea as base, this work describes a BDD based workflow to derive abstract test suites and executable test suites from INSPIRE implementation requirements that can be

F. J. Lopez-Pellicer (✉) · M. Á. Latre · J. Nogueras-Iso · F. J. Zarazaga-Soria
Universidad Zaragoza, Zaragoza, Spain
e-mail: fjlopez@unizar.es

M. Á. Latre
e-mail: latre@unizar.es

J. Nogueras-Iso
e-mail: jnog@unizar.es

F. J. Zarazaga-Soria
e-mail: javy@unizar.es

J. Barrera
GeoSLab S.L, Zaragoza, Spain
e-mail: jesusb@geoslab.es

J. Huerta et al. (eds.), *Connecting a Digital Europe Through Location and Place*, Lecture Notes in Geoinformation and Cartography, DOI: 10.1007/978-3-319-03611-3_19,

written in the language used by non-technical stakeholders. This work also analyses if BDD and popular BDD tools, such as Gherkin and Cucumber, are compatible with ISO 19105:2000 testing methodology. As demonstration, we present an online conformance tool for INSPIRE View and Discovery services that executes BDD test suites.

1 Introduction

The implementation of the INSPIRE directive must undergo the implementation of a software testing infrastructure to verify the conformance of Web based Geographic Information (GI) services with the implementing rules of INSPIRE on in-teroperability. Some authors, such as Bertolino (2007), describe software testing as a task "*ad hoc, expensive and unpredictably effective*". In the opinion of Canfora and Di Penta (2009), software testing is even more costly and risky when services are involved. INSPIRE stakeholders are aware that conformance testing tools for INSPIRE Web services are necessary (Bernard et al. 2005). For example, the ACE-GIS testing suite is one of the earliest examples (Esbrí et al. 2004). However, it is really very difficult to ensure an effective participation of non-technical stakeholders (end users, managers, domain experts, etc.) in the conformance testing process due to its inherent complexity. A relevant symptom is that available INPIRE tools that automate total or partially such process (e.g. GDI-DE Test suite (Hogrebe 2012), INSPIRE Metadata Validator (JRC IES/SDI Unit 2011), NeoGeo WMS INSPIRE Tester (Chartier 2011)) are targeted to technically skilled endusers with deep knowledge of UML models and XML processing tools (testers, developers, Web services experts, etc.).

This work focuses on the suitability of a *behaviour-driven development* (BDD) approach to the conformance testing of Web based GI services against the requirements of INSPIRE stakeholders. These requirements are embodied in the documents that define the technical guidance for the implementation of INSPIRE Network Services (European Commission 2013). In Software Engineering, BDD is a lightweight and non-formal model-based software development process in which software developers and domain experts collaborate in developing a human readable model of a system for acceptance tests (North 2007).

The main contributions of this chapter are an analysis of the suitability of BDD techniques and tools for INSPIRE conformance testing, and the presentation of an application that implements such approach. To do so, we first discuss in Sect. 2 existing approaches to test the conformance of Web services applicable to Web based GI services. Next, in Sect. 3, we present how BDD can be applied to INSPIRE conformance testing, and, in Sect. 4, we confront the BDD approach against the ISO 19105:2000 testing methodology identifying similarities and differences. Following, we present in Sect. 5 an online test execution application for INSPIRE View and

Discovery Services based on BDD. In Sect. 6, we discuss the use of BDD for conformance testing of Web-based GI services. We conclude with some remarks on the use of a BDD approach for conformance testing of Web based GI services.

2 Related Works

Conformance testing is the process to determine the extent to which a product or system conforms to the requirements of a specification with the aid of testing (Gray et al. 2010). It is acknowledged in the GI domain that the availability of conformance tests for data, metadata and services promotes and eases the adoption of interoperability initiatives (Nebert et al. 2007). Conformance testing for data and metadata often focuses on syntactic and semantic validation against schemas and rules. There are available many works about conformance testing for data and metadata in very different scenarios (e.g. domain conformance (Martirano 2013), online validation tool (JRC IES/SDI Unit 2011), metadata edition (Nogueras-Iso et al. 2012)). Service conformance testing is different from data and metadata conformance testing. Service conformance tests are built for verifying if a service behaves as it is supposed to behave according to a specification. Survey papers (e.g. Canfora and Di Penta (2009); Bozkurt et al. (2013)) show that there are a multitude of tools, testing techniques and procedures that have been proposed for testing any kind of Web services. In Europe, thanks to the INSPIRE directive, the need for tools, testing techniques and procedures suitable for Web based GI services has soared across organizations and countries recently. The most outstanding examples are the discussion platform *Persistent Test Bed* (PTB) (Östman 2010) and the testing tools developed by the *Geo Data Infrastructure Germany* (GDI-DE) (Hogrebe 2012) and the European Commission's JRC Institute for Environment and Sustainability (JRC IES/SDI Unit 2011). INSPIRE conformance testing has become also a research area. For example, Horák et al. (2011) show how to analyse performance, capacity and availability of view services. Giuliani et al. (2013) perform a similar analysis for download services. Kliment et al. (2012) and Martirano (2013) are examples of recent efforts towards a methodology for conformance testing of INSPIRE Network Services. The industry, represented by the OGC, has a program named *OGC Compliance and Interoperability Testing and Evaluation* (CITE) (Bermudez and Bacharach 2013) that has developed tools to determine a product implementation of an OGC Web service standard fulfils all mandatory elements. The CITE tools are the *Compliance Test Language* (CTL) and the TEAM Engine tool. The CTL is an XML grammar for documenting and scripting test suites that embeds XML stylesheet transformations (XSLT) and calls to native code. The TEAM Engine is a test execution tool able to run CTL files. In addition, the CITE program has developed test suites for OGC standards following the ISO 19105:2000 testing methodology. Several INSPIRE conformance testing initiatives, such as the GDI-DE Testsuite, are based on these tools and test suites.

Fig. 1 Typical Gherkin template used in BDD for depicting the behaviour of a system

```
Feature [title]
   In order to [benefit]
   As [role]
   I want [feature]
   Scenario [title]
     Given [context]
     And [some more contexts]…
     When [event]
     And [some more events]…
     Then [outcome]
     And [some more outcomes]…
   Scenario [title]…
 Feature [title]…
```

3 BDD Applied to INSPIRE Conformance Testing

BDD is an agile software development process in which developers, domain experts, users and stakeholders collaborate to specify in a human readable model written in a ubiquitous language the expected behaviour of a system for acceptance testing purposes. The concept of ubiquitous language describes a language built to be shared and used by developers, domain experts, users and stakeholders to promote a common understanding of the business domain (Evans 2003). This concept is fundamental in BDD. The ubiquitous language used in BDD is often referred as the Gherkin language[1] and typically follows the template for describing the behaviour of a system presented in Fig. 1.

Corriveau and Shi (2013) classify BDD as a *model-based testing* (MBT) tool. MBT is a kind of black-box testing where tests cases are generated from a specification, and then executed (Utting and Legeard 2010). BDD is a special case of MBT because its ubiquitous language is not formal and the automatic derivation of test cases from BDD models only outputs test stubs. As many other MBT tools, BDD is supported by a set of tools able to execute the scenarios found in BDD models. RSpec, JBehave, StoryQ, SpecFlow, Behat and Cucumber are examples of those toolkits (Solis and Wang 2011). Compared with other MBT tools, BDD can be considered too simple. Corriveau and Shi (2013) express this concern when comparing BDD with other tools based on formal modelling languages such as Spec Explorer (Veanes et al. 2008). However, the simplicity of BDD is the most probable cause of its adoption by the industry (Lerner 2010).

The production of a test framework for INSPIRE conformance testing of network services based on BDD should follow the five main steps of MBT (see Fig. 2).

1. Selection of requirements.
2. Production of an *abstract test suite* (ATS).
3. Production of an *executable test suite* (ETS).

[1] Properly speaking, the Gherkin language is the ubiquitous language understood by the Cucumber and Behat test execution tools.

4. Execution of ETS against an *instance under test* (IUT).
5. Analysis of results.

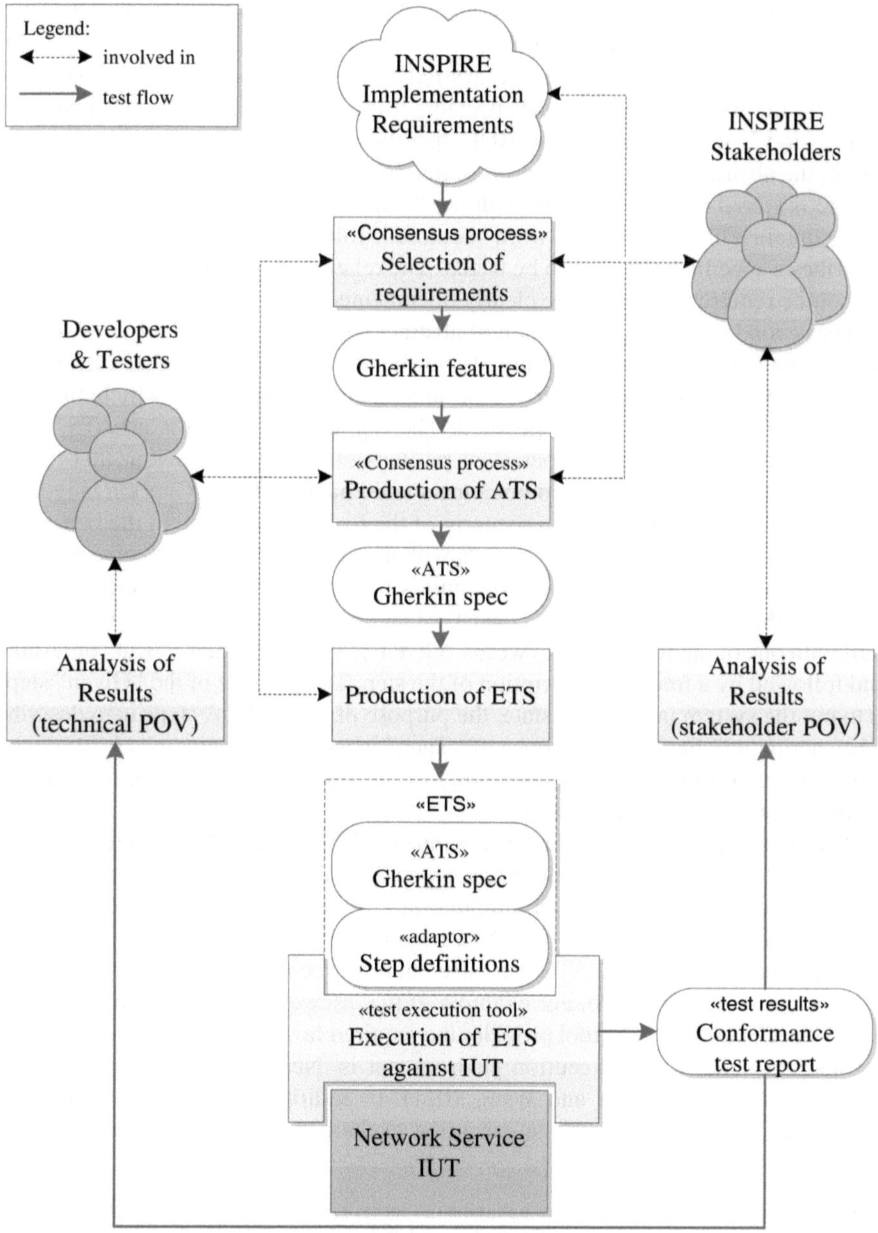

Fig. 2 BDD applied to INSPIRE Network Service conformance testing

The first step is the selection of requirements. It is a process based on consensus where stakeholders, developers and testers agree on a subset of INSPIRE implementation requirements for Web based GI services whose conformance must be tested. The outcome of this process should be a high abstraction model of the behaviour of the system expressed as a set of expected features. In this context, the term feature identifies a specific desired behaviour to be tested. In MBT, this model is known as abstract model because it must not refer to specific instances (Utting and Legeard 2010). If the BDD specification language is the Gherkin language, the abstract model will be represented in plain text files where each expected feature is stored in a separate file with the ".feature" filename extension. Each file must contain a line with the keyword "Feature"[2] followed by free indented text that describes a specific behaviour to be tested. The relationship between the feature and the source requirements must be clearly documented in text to support traceability.

The second step is the production of abstract tests from the model by consensus. Since resources for testing are limited and there is an infinite number of possible tests, involved parties should agree first on some criteria to decide which and how many abstract tests should be specified. The output of this step is an ATS. Each abstract test is a sequence of operations or steps related to a behaviour that put an IUT in a state where an expected outcome should happen. In the Gherkin language, each abstract test is encoded as a scenario of the feature associated to the behaviour that the test relates. Every scenario starts with the keyword "Scenario" on a new line after a feature or scenario declaration, and is followed by a free indented text that describes the test. Every scenario consists of an ordered list of steps. Each step must start with one of the following keywords "Given", "When", "Then", "But" or "And", and followed by a free text description of the step. The purpose of the "Given" steps is to put the system in a known state, the purpose of the "When" steps is to describe a key action and finally the purpose of "Then" is to observe outcomes.[3] "But" and "And" are used to increase the readability of the abstract test. Gherkin uses tags to group features and scenarios together.

The third step is to implement the ATS into an ETS. In BDD, this is done by coding some adaptor code that implements each step described in the ATS in terms of the Web service application-programming interface of the IUTs. The main advantage of this approach is the isolation of the ATS from the implementation details. The only requirement for reusing the ATS in a different test execution environment is to code an appropriate adaptor code. For example, if the test execution environment is Java based, the Cucumber-JVM tool provides the required Java artefacts for implementing the adaptor code; if the execution environment is .Net based, the SpecFlow tool can be used instead (Solis and Wang 2011). In addition, BDD assumes that test execution tools will automate the execution of the lists of steps found in the ATS.

---

[2] We assume in this section that the behavior model will be written in plain English although the Gherkin language supported by Cucumber and Behat provides keywords for more than 40 languages.

[3] Popular BDD tools, such as Cucumber, do not distinguish semantically among these steps. This behavior has practical, strong implications discussed in next sections.

These tools will look up the implementation of a step in the adaptor code by some matching procedure at runtime. For example, the Cucumber tool will look for a step definition annotated with a keyword, string or regular expression that matches the text of a Gherkin step and extract from the matching text parameters for invoking the code. That is, in BDD, an ETS for a specific test execution environment is a bundle composed by an ATS and an adaptor code for such environment.

The fourth step is to execute the ETS against an IUT with an appropriate test execution tool. In BDD, the reports of the test executions are generated from the ATS bundled in the ETS. That is, the reports are expressed in human readable terms that were agreed and written by one of the final recipients of these reports: the INSPIRE stakeholders. For example, this chapter presents a Web based testing tool able to execute ETS for OGC Web services. In this tool, users can select the ETS to be executed and the location of the capabilities XML of the OGC Web Service that they want to test. Moreover, each ETS is multilingual, that is, each bundle consists of an ATS written in English, an ATS written in Spanish and a shared adaptor code. Hence, the user can select which ATS drives the tests, and the INSPIRE conformance report produced by the tool will be written in the corresponding language.

Finally, the fifth step requires that involved parties analyse the human readable results of the ETS executions from their point of view. For example, when an IUT fails to pass, the main cause of the failure or error should be determined. Technical parties may use the reports to find faults in the IUT, in the testing execution tool, in the adaptor code, and even in the ATS. Non-technical parties may use reports to improve the communication with technical parties while the fault is fixed, to discover faults in the ATS that technical parties may not be aware of and, perhaps, to discover that a flawed implementation requirement is the main cause of the failure.

4 BDD and ISO 19105:2000 Testing Methodology

The ISO 19105:2000 testing methodology (ISO/TC 211 2000), which is based on testing methodology for software, is the conceptual framework for conformance testing in the domain of geographic information (Kresse and Fadaie 2004). Any testing framework intended to be used in the geographic information domain should be aligned to ISO 19105:2000 in order to detect its strengths and weaknesses. Table 1 maps key ISO 19105:2000 concepts to BDD concepts presented in the Sect. 3. In general, there is a recognizable correspondence between ISO 19105:2000 and BDD concepts. The mapping also reveals that BDD does not provide a robust mechanism for defining modules and suites yet.

The conformance assessment process in ISO 19105:2000 involves four phases: *preparation for testing*, *test campaign*, *analysis of results* and *conformance test report*. The first three phases of BDD applied to INSPIRE Network Service conformance testing (*selection of requirements*, *production of ATS*, *production of ETS*) fall within the scope of the *preparation for testing* phase. The *execution of ETS against an IUT* phase is equivalent to the *test campaign* phase as both are the process

Table 1 Mapping between ISO 19105:2000 concepts and BDD concepts

ISO 19105:2000	Definition	BDD	Definition
Abstract test case	Generalized test for a particular requirement	Scenario	A sequence of operations necessary to test for a particular feature
Abstract test method	Method for testing implementation independent of any particular test procedure	Step list	The sequence of steps that define a scenario
Abstract test module	Set of related abstract test cases	Set of tagged scenarios	Set of scenarios or features annotated with the same tag
Abstract test suite (ATS)	Abstract test module specifying all the requirements to be satisfied for conformance	Feature suite	All the scenarios specifying all the features to be satisfied for acceptance
Executable test case	Specific test of an implementation to meet particular requirements	Scenario and step definitions (adaptor code)	A sequence of operations necessary to test for a particular feature along with its adaptor code for a particular test execution tool
Executable test suite (ETS)	Set of executable test cases	Feature suite and step definitions (adaptor code)	All the scenarios specifying all the features to be satisfied for acceptance along with their adaptor code for a particular test execution tool

of executing the ETS against an IUT and recording in a log the observed test outcome and any other relevant information. The shared *analysis of results* phase presents a subtle difference. In ISO 19105:2000, it refers to the evaluation of the observed test outcome against the pass and fail criteria prescribed by the abstract test case. This analysis may overlap in time with the test campaign. In BDD, an automated execution tool computes during the execution of the ETS a pass or fail test verdict automatically. Hence, the evaluation also involves confirming or overturning the computed verdict. Finally, ISO 19105:2000 identifies a *conformance test report* phase where the results of the conformance assessment process are documented in a proforma conformance test report. This phase does not exist explicitly in the BDD approach because execution tools can generate automatically proforma test reports based on the ATS.

Although ISO 19105:2000 and BDD have similarities, BDD tools cannot be considered as mature tools yet. For example, the most popular BDD tools the Gherkin language and the Cucumber tool (Wynne and Hellesøy 2012) do not provide in its present state a complete support to the ISO 19105:2000 testing methodology. We can point out that the Gherkin language (Table 2) and the Cucumber tool (Table 3) do not support conditional requirements, inconclusive verdicts, hierarchical ATS, conformance levels and dependence between abstract tests methods. Such features can be emulated producing more complex ATS (pervasive use of tags and duplicate steps) and requires a careful analysis of results in some scenarios (risk of wrong computed verdicts).

Table 2 Issues found in the Gherkin language

ISO 19105:2000	Description	Issue	Consequences
Hierarchical abstract test modules	Abstract test modules may be nested in a hierarchical way	Lack of semantic relationship between tags	*Tag explosion* Nested modules can be implemented by tagging each feature or scenario belonging to these modules with tags that identify the respective container modules
Conformance clauses with levels	A conformance level is a special class of conformance class in which requirements of a higher level contain all the requirements of the lower levels	Lack of semantic relationship between tags	*Tag explosion* Lower conformance levels can be implemented by tagging each feature or scenario belonging to these levels with tags that identify the respective higher conformance levels
Dependence among abstract test methods	An abstract test method may depend on the outcome of other abstract test methods	No supported by the language The sequence of operations is specific to each scenario	*Step explosion* Increases the complexity of the production of ATS due to the risk of an explosion of duplicate sequences of operations

Table 3 Issues found in the Cucumber tool

ISO 19105:2000	Description	Issue	Consequences
Conditional requirements	Conformance requirements that shall be observed if the conditions set out in the specification apply	The tool does not distinguish semantically steps (e.g. Given steps do not have guard semantics)	*Wrong verdicts* Check for wrong verdicts in conditional requirements whose guard depends on an observable value known during the execution of the steps
Inconclusive verdict	Test verdict when neither a pass verdict nor a fail verdict apply	The tool only supports pass or fail verdicts	*Wrong verdicts* Check for false pass or fail verdicts

5 Test Execution Tool for INSPIRE Network Services

The approach presented in the Sect. 4 has been applied to develop a Web application able to perform an assessment on the conformity of both INSPIRE View and Discovery services.[4] The application is based on two of the most popular BDD software tools: the Gherkin language and the Cucumber-JVM test execution tool. The

[4] This system is planned to be publicly available at IDEE, the SDI of Spain. At the moment of the writing, the access to the development version is restricted. Readers can request the corresponding author access to the service.

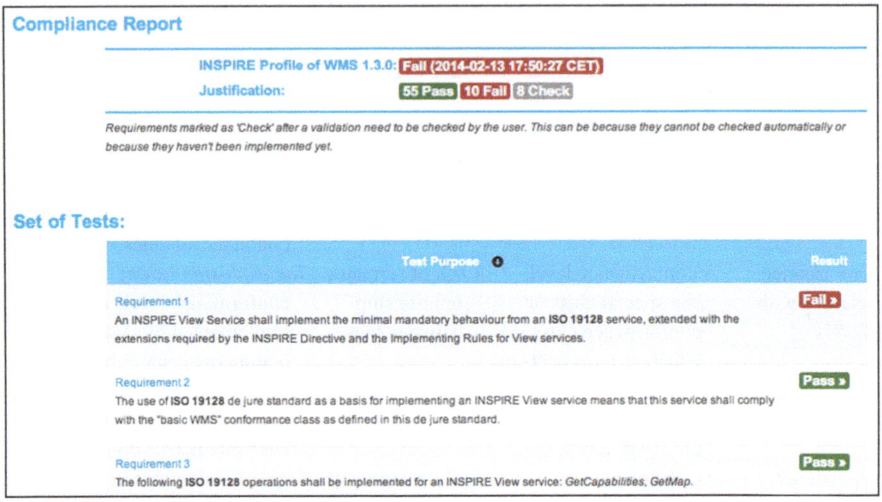

Fig. 3 Conformity test report

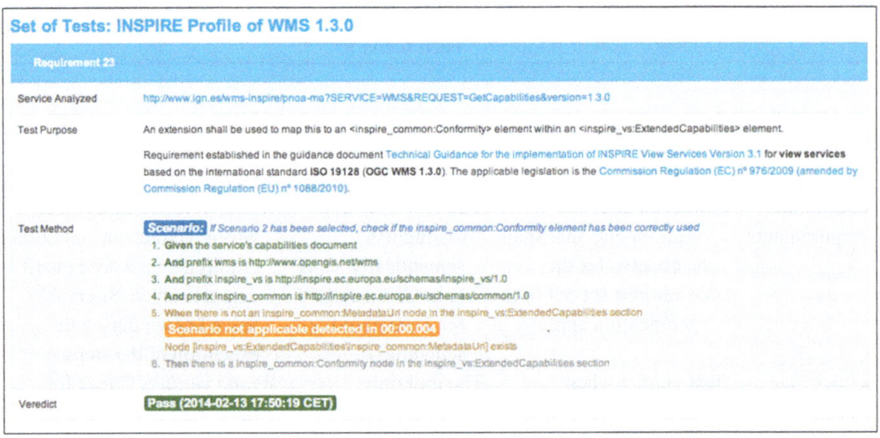

Fig. 4 Test case report

test execution tool was patched to solve the issues detected in the Sect. 4. Next, we describe how an end-user can interact with the application, its architecture, and the production of ATS and ETS.

This application has a landing page where the user fills in a form with the information related to an IUT, that is, a view or download OGC Web service under test: the online location of the capabilities XML document of an OGC WMS 1.3.0 or an OGC CSW 2.0.2 service, and the corresponding ETS. After the user sends the form, the application returns to the browser a master view of the conformance report labelled "in progress". In parallel, each executable test case is running or scheduled to run on the server. The application notifies the user in real time of each of the verdicts produced by the test cases and computes an overall verdict for the service (Fig. 3). The user can also request for a detailed view of each executable test cases. Each executable test view displays the abstract test case, the execution trace, the execution outcome and the test verdict (Fig. 4).

Figure 5 presents the architecture of this Web-based multilayer application. The presentation layer offers a landing page and master and detail views of live conformance test reports. The presentation layer depends on three services:

- **Conformance test builder**. Given the provided information related to the IUT, this service instantiates the appropriate ETS to be executed against the selected instance, schedules jobs to run its executable test cases (test jobs), and creates an empty conformance test report. The unique identifier of this report is returned to the user.
- **Test executor**. This service is invoked when a scheduler fires a test job. It instructs the Cucumber component to run an instantiated executable test case, records in a log its trace, its observed outcome and its verdict (pass, fail or inconclusive), and notifies the verdict to the user. However, if the test case depends on the finalization of other test cases, this service reschedules it.
- **Conformance test report**. This service provides access through unique identifiers to the conformance test reports that consist in an overall verdict and the log and the computed verdict of each test case.

As ETSs are decoupled from their adaptors, it is possible that the test executor discovers at runtime that a step is not implemented or that a feature has no scenarios. In such a case, the test case is ignored for the overall verdict and the user is notified that the test case requires human verification. The overall verdict is computed as follows:

- **Pass verdict**: a minimum number of tests are implemented and all return pass verdicts.
- **Fail verdict**: at least one implemented test returns a fail verdict.
- **Inconclusive verdict**: none of the above verdicts are met.

These services use the components exposed by the component layer. The core components of the application are a Gherkin pre-processor that detects explicit dependences between test cases marked with tags (i.e. test modules), a Cucumber-JVM testing that has been modified to support conditional "Given" and "When" clauses and adaptor code that throws inconclusive verdicts, and, as plugins, multilingual ETS bundles with shared adaptor code written in Java. Logs, computed verdicts and the test job queue are stored in the persistence layer in a relational database.

The approach described in Sect. 3 has been followed to produce the ETS from the most recent technical guidance documents for the implementation of INSPIRE view and discovery services. In a first stage, domain experts and test execution tool developers decided to select all implementation requirements and create a feature per implementation requirement (73 for view services and 32 for discovery services).

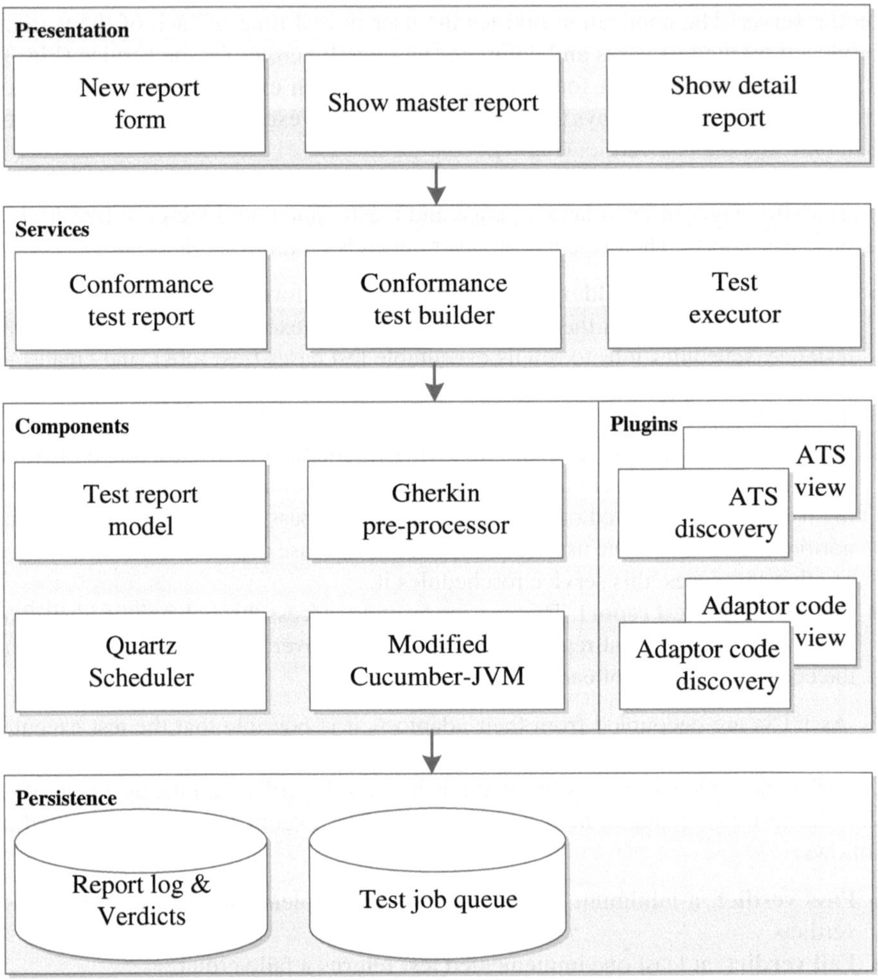

Fig. 5 Architecture diagram of a test execution tool for INSPIRE Network Services

Next, they started the process of the production of an ATS for view services and an ATS for discovery services. Both ATS were written in English. The outcome of this process was an ATS for view services with 60 well-defined features (7 test modules, 53 test cases with test methods), 61 scenarios and 270 steps, and an ATS for download services with 25 well-defined features (6 test modules, 19 test cases with test methods), 23 scenarios and 158 steps. During this procedure, it was detected that was non-feasible to devise fully automatable scenarios for some features (13 for view services and 7 for discovery services). They were kept as part of the ATS for documentation purposes although they were not automatable. The steps from both ATS were considered for the production of the adaptor code. As many steps were duplicated or matched by the same regular expression, the 428 steps were mapped

Table 4 Testing artefacts produced

Artefact	View services	Discovery services	Total
Implementation requirements	73	32	105
Features (test modules)	7	6	13
Features (test cases with test methods)	53	19	72
Features (human verification required)	13	7	20
Scenarios	61	23	84
Steps	270	158	428
Operations (annotated Java methods)	–	–	72

to 72 operations implemented as annotated Java methods. Once the adaptor code was ready, each ATS was translated to Spanish and the adaptor code was updated to match also the description of the steps in Spanish. Finally, all the ATS produced along with the shared adaptor code were deployed in the application. Table 4 presents a detailed summary of the testing artefacts produced.

6 Discussion

In this section, we discuss the use of BDD for conformance testing of Web-based GI services. Software testing intrinsically faces a lot of challenges but Web service testing faces additional issues that makes it a task of outstanding complexity. For example, Canfora and Di Penta (2009) highlight as key issues lack of observability of service code, lack of test data, complex or not fully specified input/output types, testing costs and side effects of testing. The use of a BDD-approach does not avoid dealing with such issues. For example, 20 requirements could not be implemented because it was no agreement on a suitable sequence of operations for the identified scenarios, or because such sequence was perceived as not automatable. Similar issues can be found in other conformance testing systems for INSPIRE and, although the technical guidelines are available, it is acknowledged that there are issues that have not been addressed yet (JRC IES/SDI Unit 2011).

Other aspect to analyse is if the use of a ubiquitous language helps a better understanding of the standards, the specifications and the test methods. BDD tests suites are written in a language that have no syntactic noise and is more readable. BDD practitioners claim this feature not only improves the understanding but also ease the participation of stakeholders. There is little empirical evidence available in the literature that supports this claim. Future research needs to evaluate to which extent BDD test suites are perceived as more understandable than test suites produced by alternative approaches.

Traceability helps to understand the test case and its execution, and thus to increase the confidence of stakeholders. Traceability is the ability to relate different items involved in testing, such as requirements and tests. BDD tools provide a quite simple and straightforward support for traceability between tested requirements (*features*), abstract test cases (*scenarios*), and test implementations (adaptor code) that interact with an IUT. Similar support can be found in CITE-based tools. However, an effective development environment for conformance testing needs to support not only traceability but also the debugging of test cases. Nowadays, integrated development editors (IDE) offer an extensive support to run and debug BDD specifications and the respective adaptor code by means of plugins (Chelimsky et al. 2010). The CTL, for example, lacks of such wide support.

7 Conclusions

We have presented the progress made in the investigation of novel procedures for INSPIRE conformance testing of Web based GI services. The use of BDD for conformance testing of Web based GI services is new in this domain. As other MBT approaches, it has as advantage that authoring ATS is truly independent of the implementation of the adaptor code. In addition, non-technical stakeholders can participate in authoring ATS and could gain insights on conformance process. This work also shows that BDD is partially compatible with the ISO 19105:2000 testing methodology and has desirable qualities such as traceability and readability. Therefore, in the INSPIRE context, the adoption of BDD could facilitate a wider participation of stakeholders in the development of ATS and ensure the effective understanding of INSPIRE implementation requirements and their consequences by both technical and non-technical INSPIRE stakeholders.

Acknowledgments This work has been partially supported by the Spanish Government (project TIN2012-37826-C02-01), the National Geographic Institute (IGN) of Spain and GeoSLab, S.L.

References

Bermudez L, Bacharach S (2013) Compliance testing program policies and procedures. Open Geospatial Consortium, Wayland

Bernard L, Kanellopoulos I, Annoni A, Smits P (2005) The European geoportal–one step towards the establishment of a European Spatial Data Infrastructure. Comput Environ Urban 29:15–31. doi:10.1016/j.compenvurbsys.2004.05.009

Bertolino A (2007) Software testing research: achievements, challenges, dreams. Future of software enginnering (FOSE'07), Minneapolis, 23–25 May 2007. doi:10.1109/FOSE.2007.25

Bozkurt M, Harman M, Hassoun Y (2013) Testing and verification in service-oriented architecture: a survey. Softw Test Verif Reliab 23:261–313. doi:10.1002/stvr.1470

Canfora G, Di Penta M (2009) Service-oriented architectures testing: a survey. In: De Lucia A, Ferrucci F (eds) Software engineering. Springer, Berlin, pp 78–105

Chartier B (2011) Vos services WMS sont-ils INSPIREd? In: Neogeo technologies. http://www.neogeo-online.net/blog/archives/1331/. Accessed 3 Dec 2013

Chelimsky D, Astels D, Dennis Z, Helmkamp B, Hellesøy A, North D (2010) The RSpec book. The Pragmatic Bookshelf, Dallas

Corriveau J-P, Shi W (2013) On acceptance testing. International conference on software engineering research and practice (SERP 2013), Las Vegas, 22–25 July 2013

Esbrí MÁ, Gould M, López ML (2004) Conformance Test Engines for quality assurance of INSPIRE Services. 10th EC-GI&GIS Workshop, Warsaw, 23–25 June 2004

European Commission (2013) Guidance documents. In: Network services: legislation. http://inspire.jrc.ec.europa.eu/index.cfm/pageid/5. Accessed 4 Dec 2013

Evans E (2003) Domain-driven design. Addison-Wesley Professional, Boston

Giuliani G, Dubois A, Lacroix P (2013) Testing OGC web feature and coverage service performance: towards an efficient access to geospatial data. J Spat Inf Sci (In press). doi:10.5311/JOSIS.2013.7.112

Gray M, Goldfine A, Rosenthal L, Carnahan L (2010) Conformance testing. In: Information technology laboratory, NIST. http://www.nist.gov/itl/ssd/is/conformancetesting.cfm. Accessed 4 Dec 2013

Hogrebe D (2012) GDI-DE Testsuite. Improving interoperability. INSPIRE Conference, Istanbul, 23–27 June 2012

Horák J, Ardielli J, Růžička J, (2011) Performance testing of web map services. In: Nguyen N, Trawiński B, Jung J (eds) New challenges for intelligent information and database systems. Springer, Berlin, pp 257–266

ISO, TC 211, (2000) ISO 19105:2000—Geographic information—conformance and testing. Switzerland, Geneva

JRC IES/SDI Unit (2011) INSPIRE geoportal metadata validator. In: INSPIRE geoportal. http://inspire-geoportal.ec.europa.eu/validator2/. Accessed 4 Apr 2013

Kliment T, Tuchyna M, Kliment M (2012) Methodology for conformance testing of spatial data infrastructure components including an example of its implementation in Slovakia. Slovak J Civil Eng XX:10–20, doi:10.2478/v10189-012-0002-y

Kresse W, Fadaie K (2004) ISO standards for geographic information. Springer, Berlin

Lerner RM (2010) At the forge: cucumber. Linux J 2010:7

Martirano G (2013) The eENVplus approach for data harmonization and validation. eENVplus workshop, INSPIRE conference, Florence, 24 Jun 2013

Nebert D, Reed C, Wagner RM (2007) Proposal for a spatial data infrastructure standards suite: SDI 1.0. In: Onsrud H (ed) Research and theory in advancing spatial data infrastructure concepts. ESRI Press, Redlands, pp 147–159

Nogueras-Iso J, Latre MA, Béjar R, Muro-Medrano PR, Zarazaga-Soria FJ (2012) A model driven approach for the development of metadata editors, applicability to the annotation of geographic information resources. Data Knowl Eng 81–82:118–139. doi:10.1016/j.datak.2012.09.001

North D (2007) Introducing behaviour driven development. In: Dan North & Associates. http://dannorth.net/introducing-bdd/. Accessed 25 Nov 2013

Östman A (2010) Network for testing GI services. GIS Ostrava, Ostrava

Solis C, Wang X (2011) A study of the characteristics of behaviour driven development. 37th EUROMICRO conference on software engineering and advanced applications (SEAA), Oulu, 20 Aug–2 Sept 2011. doi:10.1109/SEAA.2011.76

Utting M, Legeard B (2010) Practical model-based testing. Morgan Kaufmann, San Francisco

Veanes M, Campbell C, Grieskamp W, Schulte W, Tillmann N, Nachmanson L (2008) Model-based testing of object-oriented reactive systems with spec explorer. In: Hierons RM, Bowen JP, Harman M (eds) Formal methods and testing. Springer, Berlin, pp 39–76

Wynne M, Hellesøy A (2012) The cucumber book: behaviour-driven development for testers and developers. The Pragmatic Bookshelf, Dallas

Making the Web of Data Available Via Web Feature Services

Jim Jones, Werner Kuhn, Carsten Keßler and Simon Scheider

Abstract Interoperability is the main challenge on the way to efficiently find and access spatial data on the web. Significant contributions regarding interoperability have been made by the Open Geospatial Consortium (OGC), where web service standards to publish and download spatial data have been established. The OGCs GeoSPARQL specification targets spatial data on the Web as Linked Open Data (LOD) by providing a comprehensive vocabulary for annotation and querying. While OGC web service standards are widely implemented in Geographic Information Systems (GIS) and offer a seamless service infrastructure, the LOD approach offers structured techniques to interlink and semantically describe spatial information. It is currently not possible to use LOD as a data source for OGC web services. In this chapter we make a suggestion for technically linking OGC web services and LOD as a data source, and we explore and discuss its benefits. We describe and test an adapter that enables access to geographic LOD datasets from within OGC Web Feature Service (WFS), enabling most current GIS to access the Web of Data. We discuss performance tests by comparing the proposed adapter to a reference WFS implementation.

J. Jones · S. Scheider
Institute for Geoinformatics, University of Münster, Münster, Germany
e-mail: jim.jones@uni-muenster.de

S. Scheider
e-mail: simon.scheider@uni-muenster.de

W. Kuhn
Center for Spatial Studies University of California, Santa Barbara, USA
e-mail: kuhn@geog.ucsb.edu

C. Keßler (✉)
CARSI, Department of Geography, Hunter College, City University of New York,
New York, USA
e-mail: carsten.kessler@hunter.cuny.edu

J. Huerta et al. (eds.), *Connecting a Digital Europe Through Location and Place*,
Lecture Notes in Geoinformation and Cartography, DOI: 10.1007/978-3-319-03611-3_20,

1 Introduction

Linked Open Data (LOD) is an approach for creating typed links between data from different sources in the Web. These typed links are based on objects, which have their meaning explicitly defined by terms in shared LOD vocabularies (Heath and Bizer 2011). With the advent of LOD vocabularies, these objects and their links can be built in a machine-readable way, enabling computers to perform queries and reasoning on datasets. The LOD approach is based on the Linked Data Principles,[1] which define essential steps for publishing data in the Web and for making it part of a single global dataset (Bizer et al. 2009). These principles help to enable interoperability and discoverability of datasets, creating a rich network of information. Due to these characteristics, LOD has become a key solution when it comes to efficiently publishing data on the Web.[2]

The LOD cloud is growing very rapidly, and some of its most important central hubs contain vast amounts of geographic information. The DBPedia initiative,[3] for example, systematically extracts information from Wikipedia,[4] publishes it as LOD and links it to other datasets (Auer et al. 2007). Part of this information is a geo-coordinate for every localizable phenomenon described in Wikipedia. Successful efforts on implementing geographic LOD have also been carried out by government agencies, such as the Ordnance Survey of Great Britain,[5] which contributes significantly to the growth of the Web of geographic LOD based datasets (Goodwin et al. 2008).

Despite the benefits and efforts around LOD and also its inarguably increasing acceptance, the specific requirements of publishing geographic information on the Web have been addressed by standardized web services so far. An example is the Web Feature Service (WFS), a standard for providing geographic features on the Web, widely implemented in most Geographic Information Systems (GIS), but not supporting LOD. Despite their difference, both techniques, LOD and geographic web services, have their specific benefits and shortcomings for publishing and accessing geographic information on the Web. It has been argued before that combining both worlds has a great potential for boosting accessibility and interoperability of geographic information (Janowicz et al. 2010). For example, making Linked Open Data available in a geo service standard will turn all geo-service compatible GIS tools, whether they consist of simple desktop clients or distributed service implementations, into powerful geographic analysis tools of the LOD cloud. This combines the strengths of spatial data manipulation in a GIS with the potential of accessing datasets that are interlinked in the Web of Data.

This chapter addresses one of the open challenges for reaching this goal. We propose a way to efficiently access geographic LOD datasets via WFS. The main

[1] http://www.w3.org/DesignIssues/LinkedData.html

[2] http://lod-cloud.net

[3] http://dbpedia.org/About

[4] http://www.wikipedia.org

[5] http://www.ordnancesurvey.co.uk

idea is to use current Geographic Information Service standards and re-implement them in order to consume geographic LOD datasets published on the Web. The remainder of the chapter is structured as follows: Sect. 2 gives an overview of Linked Geographic Data, showing how it is described in different vocabularies. Section 3 describes the Web Feature Service standard, and explores its capabilities through its standard operations. Section 4 outlines the requirements and introduces our solution. Section 5 evaluates the performance of our implementation against the WFS reference implementation. Section 6 reviews related work, followed by conclusions and an outlook on future work in Sect. 7.

2 Linked Geographic Data

LOD datasets are described using the Resource Description Framework[6] (RDF), specified by the World Wide Web Consortium (Brickley and Guha 2004). RDF is a technology for describing resources and their interrelations in subject-predicate-object form. These so-called RDF Triples are commonly stored using an optimized storage and retrieval technology called Triple Store. Most Triple Stores organize RDF Triples in sub-sets called *Named Graphs*.[7] Named Graphs aggregate data, so that, for example, RDF Triples from distinct sources can be easily identified.

There have been several efforts to use LOD with geographic data. Suggestions include vocabularies for describing geographic data, together with storage and query techniques (Battle and Kolas 2011). Among the existing vocabularies for describing geographic LOD datasets is the Basic Geo Vocabulary[8] (WGS84 lat/long), which provides a namespace for describing geographic entities by coordinates pairs. This vocabulary is thus limited to points using WGS84 as a geodetic reference datum. Listing 1 shows an example using the WGS84 Vocabulary.

Listing 1 An example of a feature described with the WGS84 lat/long Vocabulary.

```
@prefix wgs84_pos: <www.w3.org/2003/01/geo/wgs84_pos#>.
@prefix my: <http://ifgi.lod4wfs.de/resource/>.
@prefix gn: <http://www.geonames.org/ontology#>.

my:GEOMETRY_1 a gn:Feature ;
        wgs84_pos:lat  "1.71389" ;
        wgs84_pos:long "69.3857" .
```

6 http://www.w3.org/RDF/

7 http://www.w3.org/TR/rdf11-concepts/#section-dataset

8 http://www.w3.org/2003/01/geo/

An alternative to describe geographic LOD is the GeoSPARQL Vocabulary,[9] defined by the Open Geospatial Consortium[10] (OGC). It offers not only classes and properties for describing geographic LOD, but also spatial relations for querying geographic datasets (e.g. intersects, touches, overlaps, etc.). Listing 2 shows an example of a geographic LOD dataset using the GeoSPARQL Vocabulary, with the same point as in Listing 1. Geometries are defined by the class `Geometry` and the coordinates can be encoded in an RDF literal of type Well Know Text (WKT) using a single RDF property, namely `asWKT`.

Listing 2 An example of a feature described with the GeoSPARQL Vocabulary.

```
@prefix geo: <http://www.opengis.net/ont/geosparql/1.0#>.
@prefix my:  <http://ifgi.lod4wfs.de/resource/>.
@prefix sf:  <http://www.opengis.net/ont/sf#>.

my:GEOMETRY_1 a geo:Geometry ;
      geo:asWKT "POINT (-69.3857 1.71389)"^^sf:wktLiteral .
```

Due to the use of WKT literals, which correspond to OGC simple features (Herring 2011), GeoSPARQL enables an efficient way to describe many different kinds of geometry (e.g. polygons, lines, points, multipoint, etc.). Another important aspect of the GeoSPARQL vocabulary is the flexibility regarding coordinate reference systems. The latter are encoded as a literal type. This enables the use of many different coordinate reference systems by adding their corresponding URI to the WKT literal (see Listing 3). If no specific reference system is provided in the WKT literal, the WGS84 Longitude-Latitude[11] reference system is assumed by default.

Listing 3 An example of a feature described with the GeoSPARQL Vocabulary stating a specific Coordinate Reference System.

```
@prefix geo: <http://www.opengis.net/ont/geosparql/1.0#>.
@prefix my: <http://ifgi.lod4wfs.de/resource/>.
@prefix sf: <http://www.opengis.net/ont/sf#>.

my:GEOMETRY_1 a geo:Geometry ;
       geo:asWKT "<http://www.opengis.net/def/crs/EPSG/0/4326>
       POINT (-69.3857 1.71389)"^^sf:wktLiteral .
```

GeoSPARQL also offers the possibility to use the Geography Markup Language (GML) to encode geometries. In this case, the data type (`GMLLiteral`), property (`asGML`) and the URL for the geometry type (e.g. `http://www.opengis.net/def/gml/ Polygon`) have to be changed accordingly.

---

[9] http://www.opengis.net/doc/IS/geosparql/1.0

[10] http://www.opengeospatial.org/

[11] http://www.opengis.net/def/crs/OGC/1.3/CRS84

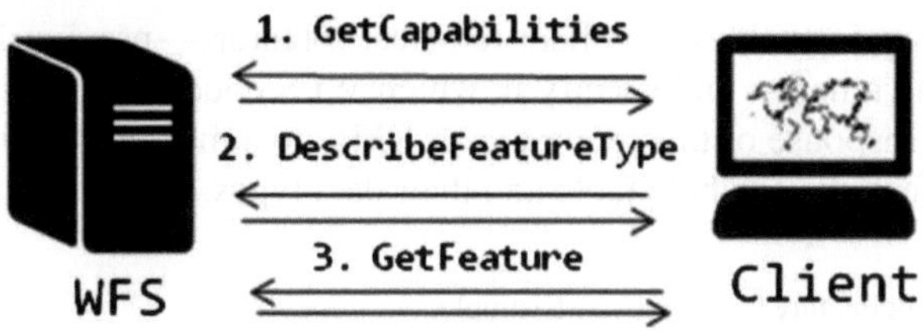

Fig. 1 Web feature service standard operations overview

3 Web Feature Service

The Web Feature Service[12] (WFS) is a platform-independent web service standard for vector-based geographic feature requests on the Web, defined by the OGC. A feature contains one or many geometries, optionally with attribute values. Its communication interface is established by HTTP requests encoded as key-value pairs, to which the server responds with XML documents. The standard operations of WFS are based on the `GetCapabilities`, `DescribeFeatureType` and `GetFeature` requests, as shown in Fig. 1.

3.1 GetCapabilities Request

The `GetCapabilities` request lists the WFS versions that the server can work with, the geometries available on the WFS server, together with their metadata (e.g. title, maintainers, abstract, provider's contact information, spatial reference system, etc.). It also informs the client which encodings are available for delivering the requested geometries (e.g. GML, GML2, JSON, CSV, etc.). Finally, the XML-based Capabilities Document also indicates which spatial functions are supported for each feature type. Listing 4 shows an example of how a GetCapabilities request can be sent to a WFS server.

Listing 4 GetCapabilities Request Example.

```
http://[SERVER_ADDRESS]/wfs?SERVICE=WFS&REQUEST=GetCapabilities
```

3.2 DescribeFeatureType Request

As shown in Fig. 1, the next step after receiving the Capabilities Document from the WFS server is to perform the `DescribeFeatureType` request. This request, as

[12] http://www.opengeospatial.org/standards/wfs

shown in Listing 5, enables the client to select a feature—previously listed in the Capabilities Document—and specify in which WFS encoding version it should be delivered. The response of this request is an XML document containing all fields of the requested features attribute table and their data types.

Listing 5 DescribeFeatureType Request Example.

```
http://[SERVER_ADDRESS]/wfs?SERVICE=WFS&VERSION=1.0.0&
REQUEST=DescribeFeatureType&TYPENAME=FEATURE_ID&SRSNAME=EPSG:4326
```

3.3 GetFeature Request

The last step to obtain features from a WFS is to perform the GetFeature operation. In this operation the client asks for a feature in a specific WFS encoding version, as shown in Listing 6. Finally, the client receives an XML document containing the feature and its attribute table.

Listing 6 GetFeature Request Example.

```
http://[SERVER_ADDRESS]/wfs?SERVICE=WFS&VERSION=1.0.0&
REQUEST=GetFeature&TYPENAME=FEATURE_ID&SRSNAME=EPSG:4326
```

Although the `DescribeFeatureType` and `GetFeature` requests syntactically only differ in the `REQUEST` parameter, they play different roles in the Web Feature Service standard, namely request information about a certain feature and retrieve the feature itself, respectively.

Another implementation of WFS—the Web Feature Service Transaction (WFS-T)—allows creating, deleting and updating features, but these functionalities are currently not addressed in this work. The WFS characteristics of: a) providing a platform-independent layer for querying geographic features requests on the Web, b) the capability of attaching attributes to the geographic features, and c) being a standard widely used as a vector data source, make WFS one of the most suitable standards for this work.

4 Linked Open Data for Web Feature Services (LOD4WFS Adapter)

Linked Open Data offers a structured approach to describe and interlink raw data on the Web, and the Web Feature Service standard offers a standardized and widely used way to deliver geographic features through web services. The union of these two technologies could increase the accessibility of geographic LOD datasets

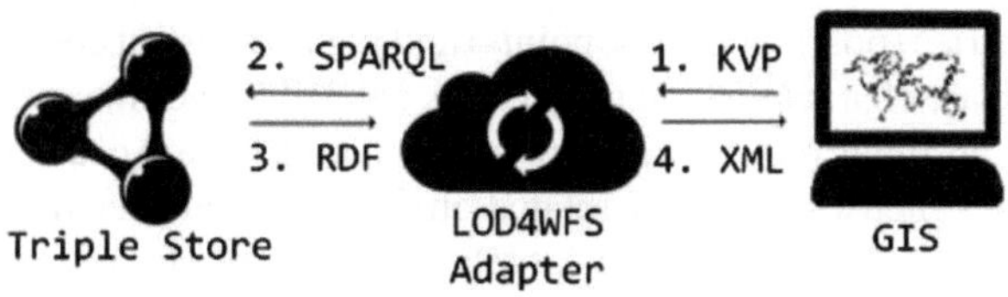

Fig. 2 LOD4WFS adapter overview

significantly. However, there is currently no common way for them to communicate. Filling this gap between LOD and WFS will allow current GIS to access geographic LOD datasets, thus enabling users to exploit the interactive tools of GIS to visualize and analyse them. Having LOD as a data source can also open new functionalities for WFS, namely the possibility of integrating different data sources, which is currently not supported by conventional WFS implementations that usually host their data sources in geographic databases or Shapefiles. This would enable, for instance, having access to the municipalities of a certain country from server A and having its river basins from server B in a single request. From this scenario emerged the idea of creating an adapter to enable access from WFS to LOD. Figure 2 gives an overview of how such an *LOD4WFS Adapter*[13] would enable access from GIS clients to geographic LOD datasets via WFS.

The adapter implements a service, compliant to the OGC WFS specification, which listens to WFS requests and converts these requests into the SPARQL Query Language for RDF.[14] After the SPARQL Query is processed, the LOD4WFS Adapter receives the RDF[15] result set from the Triple Store, encodes it as a WFS XML document, and returns it to the client (e.g. a GIS). This approach enables current GIS to transparently have access to geographic LOD datasets, using their implementation of WFS, without any adaptation whatsoever being necessary. In order to reach a higher number of GIS, the currently most common implementation of WFS has been adopted for the LOD4WFS Adapter, namely the OGC Web Feature Service Implementation Specification 1.0.0 (Vretanos 2002). The LOD4WFS Adapter enables access to geographic LOD datasets in two different ways, which we will call *Standard Data Access* and *Federated Data Accesses* in the following.

4.1 Standard Data Access

The Standard Data Access feature was designed in order to enable access to all geographic LOD datasets contained in a triple store. This feature basically works as an explorer, looking for geographic LOD datasets from a certain Triple Store and making them available via WFS. Due to the possibility of describing different

[13] https://github.com/jimjonesbr/lod4wfs

[14] http://www.w3.org/TR/rdf-sparql-query/

[15] http://www.w3.org/RDF/

types of geometries (polygons, lines, points) and many different coordinate reference systems, which are characteristic requirements of a WFS, we chose the GeoSPARQL Vocabulary as an input requirement for the Standard Data Access feature. Listing 7 shows how geometries and their related attributes are expected to be structured. The geometries are encoded as WKT literals and the attributes of features are linked to the instance of the `geo:Geometry` class via RDF Schema[16] and Dublin Core Metadata Element Set[17] vocabularies. However, there are no constraints on which vocabularies or properties may be used for describing attributes.

Listing 7 LOD dataset example: Turtle RDF encoding of a dataset, including ID and description.

```
@prefix geo:   <http://www.opengis.net/ont/geosparql/1.0#>.
@prefix my:    <http://ifgi.lod4wfs.de/resource/>.
@prefix sf:    <http://www.opengis.net/ont/sf#>.
@prefix dc:    <http://purl.org/dc/elements/1.1/>.
@prefix rdf:   <http://www.w3.org/1999/02/22-rdf-syntax-ns#>.
@prefix xsd:   <http://www.w3.org/2001/XMLSchema#>.

my:FEATURE_RECIFE a geo:Feature ;
   rdf:ID "2611606"^^xsd:integer ;
   dc:description "Recife"^^xsd:string ;
                  geo:hasGeometry my:GEOMETRY_REFICE .

my:GEOMETRY_RECIFE a geo:Geometry ;
   geo:asWKT "<http://www.opengis.net/def/crs/EPSG/ 0/4326> POLYGON ((
       -35.0148559599999984 -8.0564907399999992,
       -34.9939074400000010 -8.0493884799999993,
       ...
       -35.0148559599999984 -8.0564907399999992)) "^^sf:wktLiteral .
```

4.1.1 Required Metadata

In order to make the datasets discoverable via the Standard Data Access feature, additional metadata must be added to the datasets. We make use of Named Graphs for this purpose. Every Named Graph in the LOD data source must contain only objects of the same feature type. This approach facilitates the discoverability of Features, speeding up queries that list the Features available in the triple store. In case a Named Graph contains multiple feature types, the features can be split into different layers using the Federated Data Access (see Sect. 4.2). Finally, each Named Graph needs to be described by certain RDF properties, namely `abstract`, `title` and `subject` from the Dublin Core Terms Vocabulary.[18] This information helps the adapter to classify all Features available in a Triple Store, so that they can be further on discovered by the WFS client through the WFS Capabilities Document (see Listing 8). Alternatively, the LOD4WFS Adapter could also use a query based on other RDF types to construct the Capabilities Document.

[16] http://www.w3.org/TR/rdf-schema/

[17] http://dublincore.org/documents/dces/

[18] http://dublincore.org/documents/dcmi-terms/

Listing 8 Named Graph Example.

```
@prefix dct: <http://purl.org/dc/terms/>.
@prefix xsd: <http://www.w3.org/2001/XMLSchema#>.

<http://ifgi.lod4wfs.de/graph/municipalities>  dct:title "Brazilian
        Municipalities"^^xsd:string ;
        dct:abstract "Municipalities of the Brazilian Federal
        States."^^xsd:string ;
        dct:subject "municipalities boundaries"^^xsd:string .
```

It is important to emphasize that these RDF properties are used simply as a proof of concept for the proposed adapter, therefore other vocabularies and properties could be used instead.

4.2 Federated Data Access

The Federated Data Access feature offers the possibility of accessing geographic LOD datasets based on predefined SPARQL Queries. Differently than the Standard Data Access, the Federated Data Access feature is able to execute SPARQL Queries to multiple SPARQL Endpoints, thus enabling WFS features to be composed of data from different sources. As a proof of concept of what can be achieved, Listing 9 shows an example of a federated query, combining data from DBpedia and Ordnance Survey of Great Britain. The SPARQL Query is executed against the Ordnance Survey's SPARQL Endpoint,[19] retrieving the GSS Code[20] and geographic coordinates from districts of Great Britain—the coordinates are provided by the Ordnance Survey using the WGS84 lat/long Vocabulary, but this example converts them to WKT literals using the function CONCAT. Afterwards, the retrieved entries are filtered by matching the districts' names with DBpedia entries from the east of England, which are written in English language. The result of this SPARQL Query can be further on listed as a single WFS feature via the LOD4WFS Adapter, thereby providing a level of interoperability between datasets that is currently unachievable by any implementation of WFS, whether using Shapefiles or geographic databases.

Listing 9 Federated Data Access – SPARQL Query Example.

```
PREFIX xsd: <http://www.w3.org/2001/XMLSchema#>
PREFIX rdfs: <http://www.w3.org/2000/01/rdf-schema#>
PREFIX dbpo: <http://dbpedia.org/ontology/>
PREFIX dbp: <http://dbpedia.org/resource/>
PREFIX wgs84: <http://www.w3.org/2003/01/geo/wgs84_pos#>
PREFIX os: <http://data.ordnancesurvey.co.uk/ontology/admingeo/>

SELECT ?abstract ?name ?gss(CONCAT("POINT(", xsd:string(?long), " ",
```

[19] http://data.ordnancesurvey.co.uk/datasets/os-linked-data/explorer/sparql

[20] http://data.ordnancesurvey.co.uk/ontology/admingeo/gssCode

```
        xsd:string(?lat), ")") AS ?wkt)
WHERE
       {?subject rdfs:label ?name .
        ?subject wgs84:lat ?lat .
        ?subject wgs84:long ?long .
        ?subject os:gssCode ?gss .
        ?subject a os:District
        SERVICE <http://dbpedia.org/sparql/> {
                ?entry rdfs:label ?place .
                ?entry dbpo:abstract ?abstract .
                ?entry dbpo:isPartOf dbp:East_of_England
                FILTER langMatches(lang(?place), "EN")
                FILTER langMatches(lang(?abstract), "EN")
                FILTER ( STR(?place) = ?name )
                }
}
```

The LOD4WFS Adapter provides a web interface that allows users to write, validate and store SPARQL Queries (see Sect. 4.3.2).

4.3 LOD4WFS Software Architecture

The LOD4WFS Adapter, which was entirely developed in the Java programming language, is divided into 6 main system modules: *WFS Interface*, *Web Interface*, *Request Validator*, *Query Manager*, *Connection Manager* and *RDF2WFS Converter*. Figure 3 shows an overview of the application modules.

4.3.1 Web Interface

The Web Interface is responsible for receiving HTTP requests and translating them to the WFS Interface. It also provides access to a web-based system for maintaining SPARQL Queries created via Federated Data Access and changing the system's settings (e.g. default SPARQL Endpoint). This interface was developed using the Java-based HTTP server Jetty,[21] enabling the application to be deployed without the need of an external servlet container.

4.3.2 WFS Interface

The WFS Interface implements a listener for the standard operations defined in the OGC WFS Specification, namely `GetCapabilities`, `DescribeFeatureType` and `GetFeature`. Its main goal is to create an agnostic communication layer that enables any WFS client implementation to send requests and receive query results.

[21] http://www.eclipse.org/jetty/

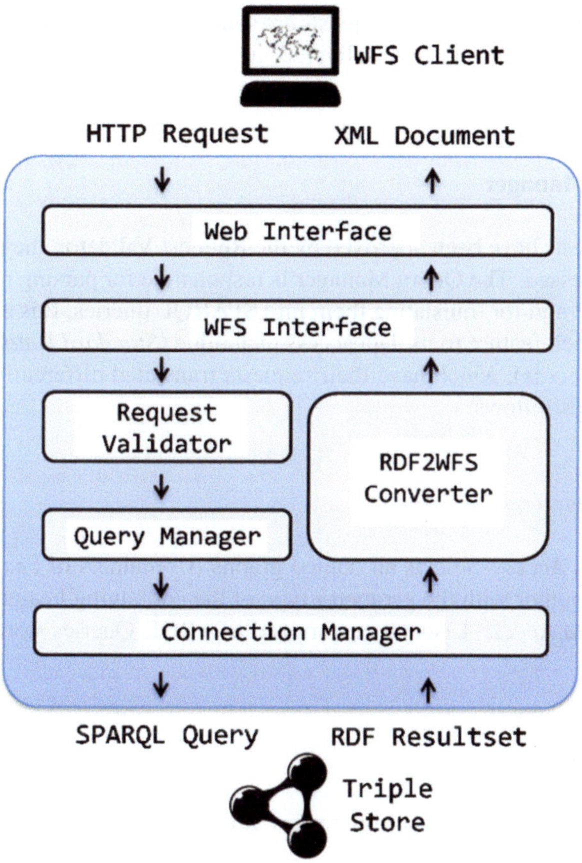

Fig. 3 LOD4WFS modules

Table 1 Validated WFS operations

Operation	Values
SERVICE	WFS by default
REQUEST	GetCapabilities, DescribeFeatureType or GetFeature
SRSNAME	Spatial reference system of a feature available in the system, e.g. EPSG:4326
TYPENAME	ID of a feature available in the system, provided by at the capabilities document
VERSION	1.0.0 by default

4.3.3 Request Validator

This module is responsible for validating the HTTP request received by the WFS Interface, making sure all operations sent by the WFS client are properly fulfilled. Table 1 shows the operations implemented by the Request Validator.

In case of invalid or unknown requests are sent (e.g. non-existing feature or wrong version), an exception report is delivered, according to the Web Feature Service Implementation Specification.

4.3.4 Query Manager

Once the requests have been approved by the Request Validator, they must be translated and processed. The Query Manager is responsible for parsing requests sent by the WFS client and for translating them into SPARQL queries. It is also responsible for mapping each feature to its data access technique (*Standard Data Access* or *Federated Data Access*), which have their requests translated differently. The requests are translated as follows:

GetCapabilities

Standard Data Access–Selects all named graphs (Containers of Features) from the triple store, together with the geometry type of the containing Feature.
Federated Data Access–Lists all customized SPARQL Queries stored via the Web Interface.

DescribeFeatureType

Standard Data Access–Lists all properties attached to a selected Feature together with their range.
Federated Data Access–Lists the variables expected from the customized SPARQL Queries.

GetFeature

Standard Data Access–Selects all geometries of a selected Feature together with the values of their related properties.
Federated Data Access–Executes the customized SPARQL Query of the requested feature to its predefined SPARQL Endpoint.

4.3.5 Connection Manager

The Connection Manager module is responsible for establishing communication from the LOD4WFS Adapter to Triple Stores. Its main goal is to execute SPARQL queries, previously composed by the Query Manager, and forwards its results to the

RDF2WFS Converter for further processing. It is based on the Apache Jena API[22] for building Semantic Web applications.

4.3.6 RDF2WFS Converter

Once the SPARQL Query has been processed and its results are returned to the system, the RDF2WFS module converts it to standard WFS documents. Depending on the request performed by the WFS client (`GetCapabilities`, `DescribeFeatureType` or `GetFeature`) it creates an XML document with the SPARQL Query result and delivers it back to the WFS client.

5 Solution Evaluation

In order to evaluate the performance of the proposed adapter, this section presents tests to compare it to the reference implementation of OGC WFS, namely the software server for geospatial data GeoServer.[23] The test compares the server response time for the `GetFeature` request in both LOD4WFS and GeoServer WFS implementations. Its main goal is to measure the time each of the services takes to process a GetFeature request, perform the query on the storage management system and send the XML document back to the client. For setting up GeoServer, the database PostgreSQL,[24] with its spatial extension PostGIS,[25] was chosen as feature storage for the WFS (Scenario A). For the LOD4WFS Adapter, three different Triple Stores were tested, namely Parliament,[26] Fuseki[27] and OWLIM-Lite[28] (Scenario B). The `GetFeature` requests were performed using the command line tool cURL.[29] The standard installations of all software involved in the tests were kept. Figure 4 shows an overview of how the test environment is structured.

5.1 Test Environment

All tests were performed using a virtual machine as specified in Tables 2 and 3.

[22] http://jena.apache.org/

[23] http://geoserver.org/display/GEOS/Welcome

[24] http://www.postgresql.org/

[25] http://postgis.net/

[26] http://parliament.semwebcentral.org/

[27] http://jena.apache.org/documentation/serving_data/

[28] http://www.ontotext.com/owlim

[29] http://curl.haxx.se/

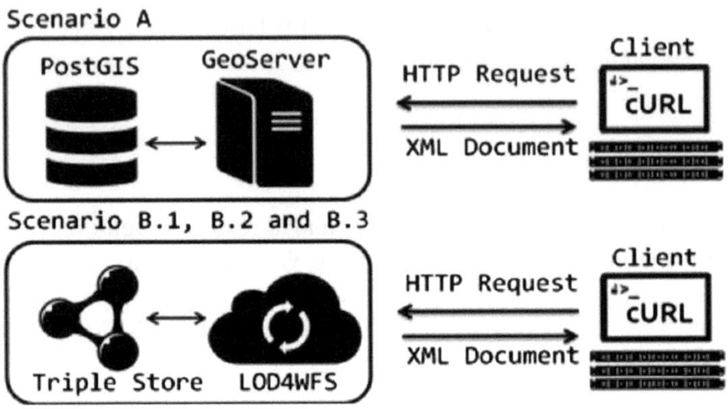

Fig. 4 Test environment overview

Table 2 Hardware environment

Processor	Intel(R) Xeon(R)
	CPU E5530 @ 2.40 GHz, dual core
Network card	82545EM Gigabit ethernet controller (copper)
	Capacity: 1 GB/s
Memory	Clock: 66 MHz
	8 GB

Table 3 Software environment

Software	Version
Operating system	Ubuntu server
	Linux 3.2.0-58-generic (amd64)
	Version 12.04 LTS
	File system: ext4
Apache tomcat	6.0.35
GeoServer[a]	2.4.4
PostgreSQL	9.1
PostGIS	1.5.3
OWLIM-Lite[a]	4.0
Java runtime	Sun microsystems Inc.: 1.6.0_27
	(OpenJDK 64-Bit server VM)
cURL	7.29.0

[a] Embedded at OpenRDF Sesame 2.7.0 and hosted with Apache Tomcat

Table 4 Test datasets

Brazilian municipalities dataset	
Number of geometries	5799
Dataset size	11.2 MB
Amazon rivers dataset	
Number of geometries	18690
Dataset size	45 MB
Amazon vegetation dataset	
Number of geometries	39082
Dataset size	173.2 MB

Table 5 Datasets overview for scenario A

Dataset	Table records	Table size[a]
Brazilian municipalities	5799	16 MB
Amazon rivers	18690	55 MB
Amazon vegetation	39083	183 MB

[a] Including indexes

5.2 Test Datasets

The datasets used for the tests (see Table 4) were created by the Brazilian Institute of Geography and Statistics[30] (IBGE). They all contain polygon geometries and are available in Shapefile format.[31]

To test *Scenario A*, the dataset was stored in the PostgreSQL database and further on added to the GeoServer as a data source for feature layers (see Table 5). This was necessary to enable access to the features through the GeoServer WFS interface.

In order to use the same dataset for *Scenario B*, the dataset had to be converted to LOD, fulfilling the characteristics previously discussed in Sect. 4.1. For this purpose, a script (*shp2rdf*) in the R programming language[32] was developed for reading Shapefiles and creating an LOD dataset in N-Triples syntax (Beckett 2014). The script uses the rgdal[33] and rgeos[34] packages.

After the conversion, the same RDF N-Triples files (see Table 6) were loaded into the Parliament (*Scenario B.1*), Fuseki (*Scenario B.2*) and OWLIM-Lite (*Scenario B.3*) Triple Stores. The datasets in all test scenarios could also be successfully

[30] http://ibge.gov.br/

[31] ftp://geoftp.ibge.gov.br/mapas_interativos/

[32] http://www.r-project.org/

[33] http://cran.r-project.org/web/packages/rgdal/rgdal.pdf

[34] http://cran.r-project.org/web/packages/rgeos/rgeos.pdf

Table 6 Datasets overview for scenario B

Dataset	Total triples	File size
Brazilian municipalities	86988	28.7 MB
Amazon rivers	359206	113.8 MB
Amazon vegetation	703497	416.1 MB

downloaded and displayed using the WFS clients of GIS QGIS[35] and ArcMap.[36] The converted datasets can be found at the following SPARQL Endpoint.[37]

5.3 Test Procedure

The loaded datasets were queried via HTTP `GetFeature` requests using cURL. The `GetFeature` request was performed 10 times in each test scenario for each dataset, afterwards the arithmetic mean value of the time elapsed was calculated. To avoid the network speed to affect the test results, the download speed was limited to 500 kilobytes per second, so that all test scenarios have the same download performance. Listing 10 shows an example of how the requests per cURL were sent to the test server. Table 7 summarizes the tests performed in each test scenario.

Listing 10 Sample HTTP Request Sent via cURL.

```
$ curl --limit-rate 500k 'http://[SERVER_ADDRESS:PORT]/wfs?SERVICE=
       WFS&VERSION=1.0.0&REQUEST=GetFeature&TYPENAME=FEATURE_ID'
       -o feature.xml;$
```

5.4 Results and Discussion

The results demonstrated a non-substantial efficiency difference between the test scenarios. Querying the Brazilian municipalities dataset, all tested scenarios showed a similar response time, having *Scenario A* as the most efficient one, being 1.33 % faster than the second fastest scenario, namely *Scenario B.1* (see Table 7-I). The efficiency difference querying this dataset was limited to the milli-second scale, though (see Fig. 5).

The tests querying the Amazon rivers dataset showed again a similar performance between the test scenarios using triple stores. Among them, Scenario B.3 had a slightly better performance than *Scenario B.1* and *B.2*. *Scenario A* had again the best

[35] http://www.qgis.org/

[36] http://esri.de/products/arcgis/about/arcmap.html

[37] http://data.uni-muenster.de/open-rdf/repositories/lod4wfs

Table 7 Performance of `GetFeature` requests

Test scenario	Avg. Time (mm:ss.ms)	Standard deviation
I. *Brazilian municipalities dataset*		
A – (GeoServer WFS with PostgreSQL)	00:38.217	0.1916
B.1 – (LOD4WFS with Parliament)	00:38.731	0.0961
B.2 – (LOD4WFS with Fuseki)	00:38.877	0.1283
B.3 – (LOD4WFS with OWLIM-Lite)	00:38.857	0.0762
II. *Amazon rivers dataset*		
A – (GeoServer WFS with PostgreSQL)	02:36.542	0.0802
B.1 – (LOD4WFS with Parliament)	02:38.110	0.1181
B.2 – (LOD4WFS with Fuseki)	02:38.150	0.2795
B.3 – (LOD4WFS with OWLIM-Lite)	02:38.076	0.0872
III. *Amazon vegetation dataset*		
A – (GeoServer WFS with PostgreSQL)	08:35.681	0.0642
B.1 – (LOD4WFS with Parliament)	08:44.013	0.2447
B.2 – (LOD4WFS with Fuseki)	08:44.771	0.1037
B.3 – (LOD4WFS with OWLIM-Lite)	08:39.079	0.0868

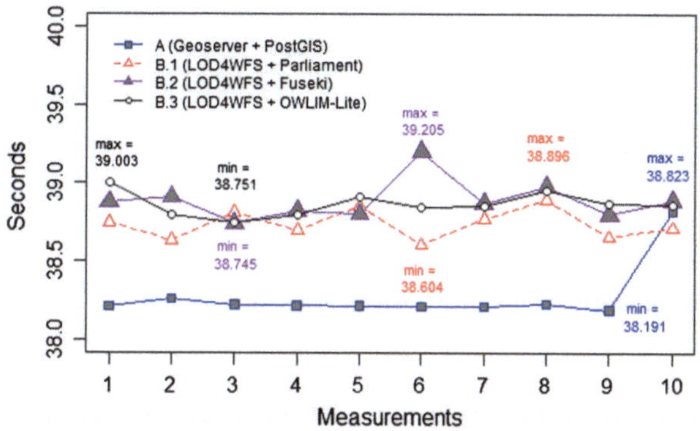

Fig. 5 Performance comparison for the Brazilian municipalities dataset

performance among all test scenarios (see Table 7-II), being 0.97 % faster than the second fastest test scenario, namely *Scenario B.3*.

Tests querying the Amazon vegetation dataset showed a bigger performance difference between the test scenarios involving triple stores (see Table 7-III). *Scenario B.3* demonstrated to have a more efficient response time than *Scenarios B1* and *B.2* when querying bigger datasets, being 0.94 % faster than the second fastest triple store based test scenario, namely *Scenario B.1*. Among all test scenarios, *Scenario A* demonstrated again a better performance than all others test scenarios (see Fig. 7), being 0.65 % faster than *Scenario B.3*.

Though the test results showed no expressive difference between the test scenarios, it demonstrated that the combination of GeoServer with the relational database PostgreSQL still provides a slightly faster platform for enabling access to geographic

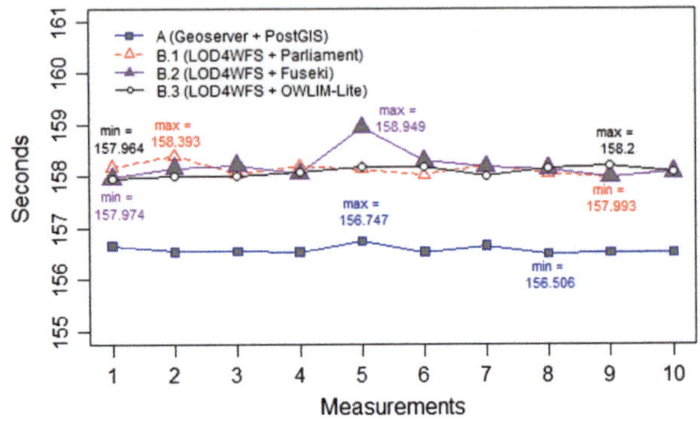

Fig. 6 Performance comparison for the Brazilian rivers dataset

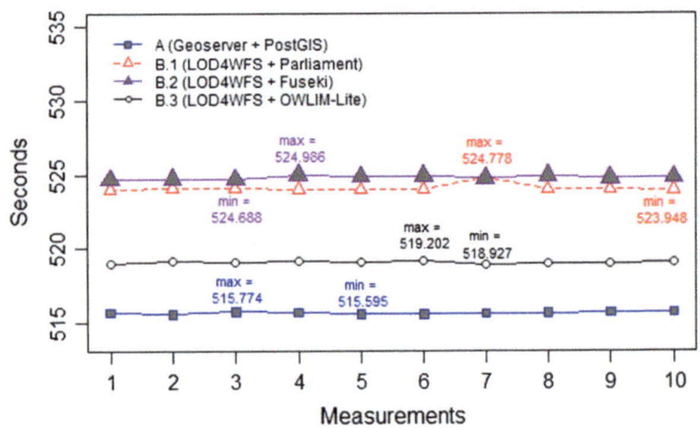

Fig. 7 Performance comparison for the Brazilian vegetation dataset

vector data. The results showed also, considering the given test environment, that the efficiency difference between the LOD4WFS approach and GeoServer with PostgreSQL gets smaller when bigger datasets are requested. The approach proposed by the LOD4WFS relies on the respective triple store's efficiency, which has been shown to be slower than a relational database in our test scenarios. However, the main point we want to stress in this work is the great benefit of having LOD datasets as data source for WFS. This approach provides not only an innovative and competitive way for serving data to current web service standards, but also offers the possibility of combining multiple data sources and creating new datasets on demand (see Sect. 4.2), which is currently not provided by any WFS implementation.

It is also important to mention that the results presented in these tests represent the performance of specific system versions in a single-user environment (see Sect. 5.1), therefore reproducing the tests with other releases will inevitably lead to different results.

6 Related Work

Significant efforts have been made to introduce and enhance the usage of semantics (Kuhn 2005) in geospatial information and web services. Among them are the works on geographical Linked Data (Goodwin et al. 2008), Semantic Geospatial Web services (Roman and Klien 2007), semantic enablement for spatial data infrastructures (Janowicz et al. 2010), structured alignment methods to geospatial ontologies (Cruz and Sunna 2008), semantic-based automatic composition of geospatial Web service chains (Yue et al. 2007) and a framework for semantic knowledge transformation of geospatial data (Zhao et al. 2009). The technological challenges and benefits of adding a spatial dimension to the Web of Data have been also discussed by Auer et al. (2009), where spatial data was systematically extracted from the collaborative project OpenStreetMap[38] and converted to RDF. Efforts on yielding geographic information in OGC web services and embedding them as LOD have been conducted by Roth (2011) with the Geographic Feature Pipes.

Other authors have suggested to use the OGC WFS standard as an interface for providing access to semantic data; Staub (2007) and Donaubauer et al. (2007) have proposed an extension of the existing WFS standard to create a model-driven interface. These works, however, require modification of the OGC WFS standard. In contrast, we use the WFS standard as it is specified by OGC, so that current GIS can access it without any modification.

7 Conclusions and Future Work

This chapter presents an alternative way of accessing geographic LOD datasets from current GIS. We have explored the possibility of using the OGC WFS standard as an intermediate layer between geographic LOD datasets and GIS. We developed an application (LOD4WFS Adapter) that acts as a service for: 1) listening to WFS requests and translating them to SPARQL Queries; and 2) transforming the RDF result set into WFS standard documents. Performance tests of the LOD4WFS Adapter against the reference implementation of OGC WFS (GeoServer) were conducted. The test environment involved three different triple stores and a relational database. The preliminary tests showed that our LOD4WFS Adapter can compete with the reference implementation for WFS services, while providing significantly larger flexibility in accessing and integrating data sources on the Web.

This chapter demonstrates that using LOD as data source for WFS is perfectly feasible and has a great potential. It combines the benefits of a widely used web service standard with the interoperability offered by LOD. This improves accessibility of geographical information on the Web of Data for GIS. Future work includes:

First, the implementation of WFS spatial operations. This would allow the LOD4WFS Adapter to translate supported WFS spatial operations (e.g. contains,

[38] http://www.openstreetmap.org/

intersects) to SPARQL using the Geographic Query Language for RDF (Geo SPARQL). Currently only a few Triple Stores implement GeoSPARQL (e.g. Parliament, Oracle Spatial RDF Semantic Graph,[39] Strabon[40]). This situation may improve once standard Triple Stores will adopt GeoSPARQL and corresponding OGC standards for spatial queries.

The second enhancement is the transaction operation (WFS-T). Currently, the LOD4WFS Adapter implements only requests of geographic information, and does not allow any data manipulation. Implementing the operations defined by WFS-T would enable WFS clients not only to query geographic LOD datasets, but also to insert, edit and delete existing features. The third enhancement we intend is the possibility of accessing geographic LOD datasets encoded as GML and other common formats, e.g. GeoRSS,[41] or GeoJSON.[42] Currently, only WKT is supported.

Finally, we intend to perform more detailed comparisons of the LOD4WFS Adapter and conventional WFS implementations. In order to achieve this, we plan to perform stress tests and to evaluate the application behavior in both single and multi-user environments using different operating systems.[43]

Acknowledgments This work is funded by the German Research Foundation (DFG) through the Linked Data for eScience Services (LIFE) Project, KU 1368/11-1.

References

Auer S, Bizer C, Kobilarov G, Lehmann J, Cyganiak R, Ives Z (2007) Dbpedia: a nucleus for a web of open data. In: The semantic web. Springer, Heidelberg, p 722–735

Auer S, Lehmann J, Hellmann S (2009) LinkedGeoData: adding a spatial dimension to the web of data. In: The semantic web—ISWC 2009. Springer, Heidelberg, p 731–746

Battle R, Kolas D (2011) Geosparql: enabling a geospatial semantic web. Semantic Web J 3(4):355–370

Beckett D (2014) N-Triples. A line-based syntax for an RDF graph. W3C proposed recommendation. http://www.w3.org/TR/n-triples/

Bizer C, Heath T, Berners-Lee T (2009) Linked data—the story so far. Int J Semantic Web Inf Syst 5(3):1–22

Brickley D, Guha RV (2004) RDF vocabulary description language 1.0: RDF schema. W3C recommendation. http://www.w3.org/TR/rdf-schema/

Cruz IF, Sunna W (2008) Structural alignment methods with applications to geospatial ontologies. Trans GIS 12(6):683–711

Donaubauer A, Straub F, Schilcher M (2007) mdWFS: a concept of web-enabling semantic transformation. In: Proceedings of the 10th AGILE conference on geographic information science

---

39 http://www.oracle.com/technetwork/database/options/spatialandgraph/overview/rdfsemantic-graph-1902016.html

40 http://www.strabon.di.uoa.gr/

41 http://www.georss.org/

42 http://geojson.org/

43 http://lodum.de/life

Goodwin J, Dolbear C, Hart G (2008) Geographical linked data: the administrative geography of great britain on the semantic web. Trans GIS 12(1):19–30

Heath T, Bizer C (2011) Linked data: evolving the web into a global data space. Synth lect semantic web theor technol 1(1):1–136

Herring JR (2011) Simple feature access—part 1: common architecture. OpenGIS implementation standard for geographic information, OGC 06–103r4. http://portal.opengeospatial.org/files/?artifact_id=18241

Janowicz K, Schade S, Bröring A, Keßler C, Maué P, Stasch C (2010) Semantic enablement for spatial data infrastructures. Trans GIS 14(2):111–129

Kuhn W (2005) Geospatial semantics: why, of what, and how. J Data Semant III. Lecture notes in computer science, vol 3534, pp 1–24

Roman D, Klien E (2007) Swing-a semantic framework for geospatial services. In: The geospatial web. Springer, Heidelberg, p 229–234

Roth M (2011) Geographic feature pipes. Diploma thesis, Institute for Geoinformatics, University of Münster, Germany

Staub P (2007) A model-driven web feature service for enhanced semantic interoperability. OSGeo J 3(1)

Vretanos PA (2002) Web feature service implementation specification. OpenGIS implementation standard for geographic information, OGC 02–058. http://portal.opengeospatial.org/files/?artifact_id=7176

Yue P, Di L, Yang W, Yu G, Zhao P (2007) Semantics-based automatic composition of geospatial web service chains. Comput Geosci 33(5):649–665

Zhao P, Di L, Yu G, Yue P, Wei Y, Yang W (2009) Semantic web-based geospatial knowledge transformation. Comput Geosci 35(4):798–808

CityBench: A Geospatial Exploration of Comparable Cities

Elizabeth Kalinaki, Robert Oortwijn, Ana Sanchis Huertas, Eduardo Dias, Laura Díaz, Steven Ottens, Anne Blankert, Michael Gould and Henk Scholten

Abstract In many city comparisons and benchmarking attempts, scores are purely one-dimensional with results split accordingly for each dimension. We propose a methodology for comparing cities on multiple dimensions implemented as a map-centric web tool based on a pairwise similarity measure. CityBench web tool provides a quick-scan geographical exploration of multidimensional similarity across European cities. With this dynamic method, the user may easily discover city peers that could face similar risks and opportunities and consequently develop knowledge networks and share best practices. This web tool is destined to provide economic/financing institutions', local governments' and policy makers' in Europe and beyond decision making support.

Keywords City comparison · Similarity measure · Geo-visualization

1 Introduction

Why do similarly structured cities behave differently in socio-economic terms? And which cities across Europe might be similar to my city of interest if I have a number of different life aspects? This chapter introduces the CityBench web tool an interactive, geospatial exploration of comparable (European) cities based on a pairwise similarity measure and a map-centric interface. The tool offers a quick and dynamic overview

E. Kalinaki (✉) · E. Dias · H. Scholten
FEWEB-RE-Vrije University Amsterdam, Rooms 4A-40/4A and Room 2G-30 (Filosofenhof), De Boelelaan 1105 Amsterdam, The Netherlands
e-mail: e.kalinaki@vu.nl

R. Oortwijn · E. Dias · S. Ottens · A. Blankert · H. Scholten
Geodan, Amsterdam, The Netherlands

A. S. Huertas · L. Díaz · M. Gould
Jaume I University, Castellon, Spain

J. Huerta et al. (eds.), *Connecting a Digital Europe Through Location and Place*, Lecture Notes in Geoinformation and Cartography, DOI: 10.1007/978-3-319-03611-3_21,

of the most similar cities for a user selected number of life aspects. The CityBench web tool can show cities apparently similar based on selected indicators and may help to identify geographic trends in the location of similar cities hence bringing GIS-like pattern analysis to an interactive web environment.

Often, economic/financial institutions such as the European Investment Bank Municipal and regional Unit (EIB-MRU) and EU policy makers wishing to obtain insight into regional trends in urban development or to explore effectiveness of regional urban cooperation programs, investors in search of a new business location as well as regional/urban practitioners searching for best practices or their own endogenous potentials and opportunities for co-operation, network or cluster forming carry out benchmarking studies. The studies in many cases like Helgason (1997), Lam et al. (2010) and Groenendijk (2004) involve identifying other 'peers' to compare a particular self with. The benchmarking and comparison process can benefit from technological developments offering new possibilities to create real-time, interactive map-centric visualization that give valuable insight.

For effective benchmarking, the regions or cities need to be clearly defined since comparing a capital city with a province or a municipality is not very useful due to the uniqueness of the different structures. Batty and Longley (1994a, b) explain that definitions of cities rely upon definitions of boundaries or delineations although such definitions are never comprehensive. If cities are highly dynamic landscapes according to Ramalho and Hobbs (2012), their definitions need regular revisions to include functional urban regions—the large areas around the proper boundary of the central city (Vasanen 2012; Hall 2009) and (Parr 2005). This and other similar descriptions highlight delineations such as 'inner city', 'metropolitan region' and 'central business district'. In this study, these delineations are normalized by aggregation or transformation into approximations of the recently proposed city delineation by Dijkstra and Poelman (2012) called a 'larger urban zone' (LUZ). In defining a LUZ, all grid cells covering the city with a density greater than 1,500 inhabitants per square km are selected. From these, clusters with a minimum of 50,000 inhabitants are defined as 'urban centers'. Then, all municipalities with at least half their population inside the urban center are selected as candidates to form the city. The LUZ is finally defined ensuring that there is a link to a political level with at least 50 % of the population living in an urban center and at least 75 % of the population of the urban center lives in a city (Dijkstra and Poelman 2012). Although different parts of the defined city boundaries can have very different patterns, European entities are proposing the LUZ as the most suitable city or region definition for comparability at European level and the backbone for city statistical data (EC 2004).

The CityBench web tool was proposed to efficiently integrate existing and new urban indicator data at the LUZ level available from the ESPON 2013 programme[1] and Eurostat[2] for a geospatial city exploration and identification of geographic patterns and to provide stakeholders the necessary functionality to visualize city status and perform benchmarking. The tool's purpose is to enable multi-scale,

[1] http://www.espon.eu/main/

[2] http://epp.eurostat.ec.europa.eu/portal/page/portal/eurostat/home/

multidimensional representation of cities for finding similarity and geographical trends for economic/financing institutions', local governments' and policy makers' decision making support.

2 Related Work

Ammons (1999) emphasizes that choosing an appropriate benchmarking technique and carefully applying it are both essential for benchmarking success. City benchmarking as defined by Luque-Martínez and Muñoz-Leiva (2005) should allow cities to implement the most effective practices and capacities learned from other cities in order to improve their actions in what they offer. This statement is in agreement with Ammons (1999) who suggests that the idea behind benchmarking should not focus on how an organization stacks up but should be captured by what the benchmarked cities learned and how they better themselves from the lessons learned from the benchmarking.

Many benchmarking and city comparisons studies have been performed. Luque-Martínez and Muñoz-Leiva (2005) offer ground work on both benchmarking in general and city benchmarking while Amelang (2007) combined two popular approaches by historians (the thematic approach, in which a single issue or series of issues are examined in relation to more than one urban area and another that focused on the cities themselves with an aim to identify similarities and divergences in individual urban trajectories by directly contrasting one city with another) to understand the metamorphosis of Barcelona.

Holloway and Wajzer (2008) explain that there is no single methodology for conducting city benchmarking exercises. The authors identified 'dimensions' (-the particular facet of a city to be compared e.g. quality of life or economic competitiveness), 'indicators' (-the measure of performance for each dimension of study e.g. for quality of life, a stable political environment could be considered as one dimension of quality of life and an indicator for this may be the level of crime) and 'scoring and ranking systems' (the methods used to analyze, compare or benchmark dimensions, indicators and cities) as the three elements evident in most studies.

Tuan Seik (2000) used many of these indicators such as health and family life, politics, religion and leisure to assess the quality of life in Singapore. In Luque-Martínez and Muñoz-Leiva (2005), approximately 180 indicators were gathered in different dimensions such as environmental management, housing, accessibility to evaluate the region of Andalusia, using Granada as the reference city. Baum (1997) performed a similar study with cities in Australia to test and explain the social polarization of Sydney. On a global scale, Lippman Abu-Lughod (1995) compared Chicago, New York and Los Angeles using a variety of indicators to specify a model for how forces generated at international level affect each city. Taylor and Walker (2001) used Friedmann's list of world cities (1986) and revised version Friedmann (1995) to produce a complex interweaving of hierarchical tendencies with distinct regional and interregional patterns from a multivariate analysis of their 'Service Complexes'.

The comparative study in Beaverstock et al. (2001) revealed a network model of inter-city relations which proved that there is more to benchmarking and city comparisons than which city comes first or last. It is noticeable in many studies that these comparisons are done to serve a specific purpose. European cases like Turok and Mykhnenko (2007) used city comparison and benchmarking to assess how 310 European cities had economically evolved over the period 1960–2005 revealing a general slowed growth over the last few decades at the time of their study. In Kasanko et al. (2006), a comparative analysis using 5 indicator sets of 15 European urban areas revealed that the structure of European cities had become less compact while Sager (2003) compared the performance of 'important' European cities in their function as tourist destinations to measure their gains and losses of the tourist market share in Europe.

Using indicators to benchmark cities or perform city comparisons has numerous advantages although as Holloway and Wajzer (2008) point out, city benchmarking has some limitations that undermine their validity for measuring and monitoring performance. These include the integrity and compatibility of data among cities, the overstatement of the cause and effect relationship between indicators and city performance and the subjectivity of the analysis and conclusions.

Most of the studies available use quantitative data from regional or governmental statistics offices and financial institutions and in most if not all the work involving indicators like in Luque-Martínez and Muñoz-Leiva (2005) and OECD (2012), the scores are purely one dimensional with results split accordingly for each dimension. If we want to determine the most similar cities to a reference city (LUZ), we need to supplement the above methodologies with a similarity measure.

There are various similarity measures as described in Deza and Deza (2013). Preoţiuc-Pietro et al. (2013) performed a physical city similarity analysis in which the authors represented each city as a point of vector space to compute 'pairwise' similarities between the cities using the cosine similarity and later quantified their results using a Kendall Tau rank correlation coefficient to find for each city the most similar cities. The cosine similarity is one of the techniques used in Seth et al. (2011) to determine similar cities that are not necessarily geographically close. In ecology and biology science, use of similarity measures like the Cosine similarity and Euclidean distance similarity are used successfully to determine similar entities. For example in (Luo et al. 2001) the similarity between gene expression patterns is measured by computing the Euclidean distances for each pair of samples based on log-transformed ratios across all of the genes. A similar example by Jain et al. (2000) uses the Euclidean distance between two corresponding "FingerCodes" for fingerprint matching and in Mane et al. (2010) for face recognition.

The choice between a Cosine and a Euclidean distance similarity is dependent on the purpose of the similarity measure. While the two are closely related, using 'Euclidean distance' is most useful to us when determining the most similar cities to our given reference city.

3 Methodology

3.1 CityBench Similarity Measure

One of the oldest and most influential theoretical assumptions is that perceived similarity is inversely related to psychological distance (Ashby and Ennis 2007). The authors explain that our mental representations of objects, concepts, positions on issues typically vary on a variety of psychological dimensions. The numerical values of a particular 'percept' on each of these dimensions can be interpreted as the coordinates of this percept in a psychological feature space.

The percepts close together are perceived as similar and percepts far apart dissimilar. Following the above principals, the CityBench similarity measure defines the current cities' multiple scores as geometric vectors and compares them across several indicator dimensions. It's not enough to rank the cities and their respective scores in each indicator dimension as in Luque-Martínez and Muñoz-Leiva (2005), what is important for CityBench are the cities that come closest to a specific city that scores k in any number of indicator dimensions.

We define k as the combined score of the reference city on the one, two or three indicators selected by the user. The specific city with score value k is most similar to itself and then similar to cities that score k + / − 1. The similarity value (Scv) for any city against the reference city is given by Eq. (1) where c is the score of any city other than the reference city in any indicator dimension i and v is the reference city value.

$$S_{cv} = \sqrt{\sum_{Ii=1}^{n} (C_{Ii} - V_{Ii})^2} \quad (1)$$

For example to find the most similar cities to Luxembourg as the reference city using 'GDP' as A and 'Ease of business' as B indicator dimensions, we normalize the values for all cities for both A and B by subtracting the minimum value in that indicator dimension from the original city indicator value and dividing that by the difference of the maximum and the minimum values.

With these normalized values, we then compute $(\text{CA-VA})^2$ and $(\text{CB-VB})^2$ for each city which later provides each city's Scv after computing the square root of their summation. If the list of cities and their respective Scv values provided by the equation above are displayed effectively, the values are capable of showing regional patterns based on the values of one indicator or a combination of indicators.

However, as with any statistical data, outliers cannot be avoided. Their sources can be traced to errors of measurement, faults in execution and or intrinsic variability (Woolley 2013). The presence of even just one outlier can offset the capacity to visualize variability of the remaining dataset therefore finding a best fit classification method can be used to observe the variability. Three data classification methods (Quintiles, Jenks natural breaks and Equal interval) and two alternative methods

Table 1 Overview of similarity maps derived from different classification methods (using GDP (€) per head as indicator)

Luxembourg (highest GDP)	Amsterdam (in-between GDP)	Plovdiv (lowest GDP)
Quintiles (5): 0 - .57,0.57 - 0.65, 0.65 - 0.72, 0.72 - 0.84, 0.84 - 1.0		
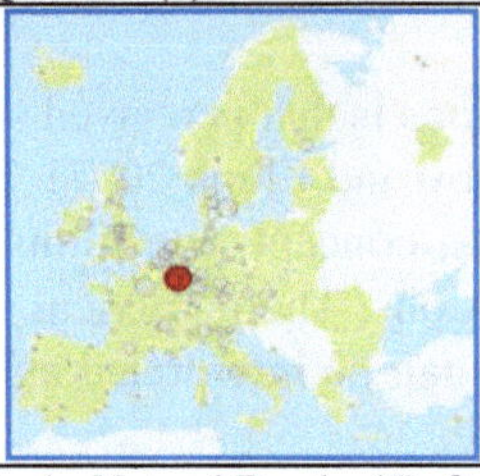	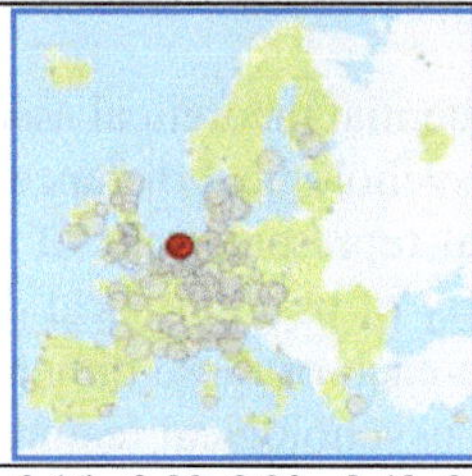	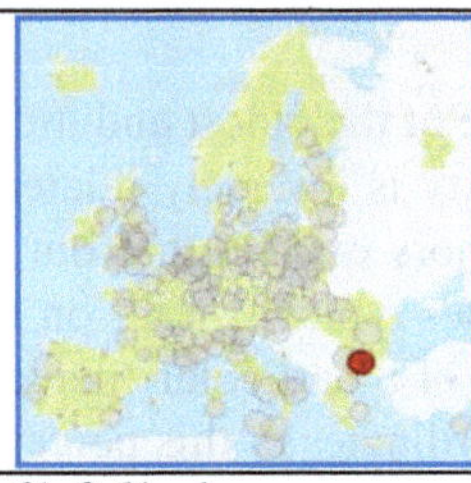
Jenks Natural Breaks (5): 0 - 0.16, 0.16 - 0.30, 0.30 - 0.43, 0.43 - 0.61, 0.61 - 1		
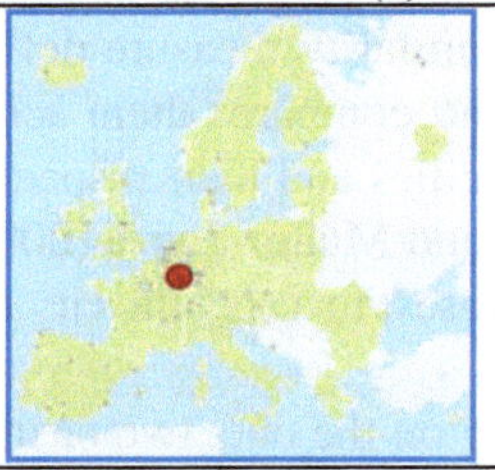	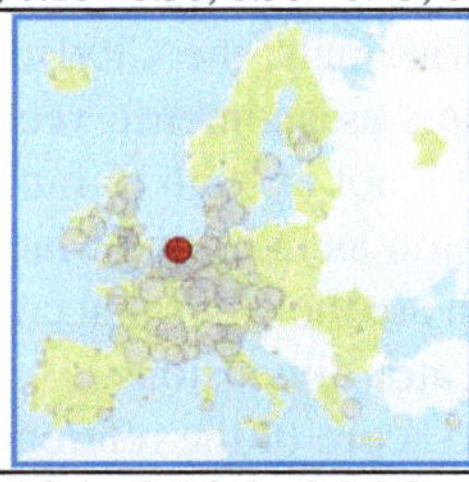	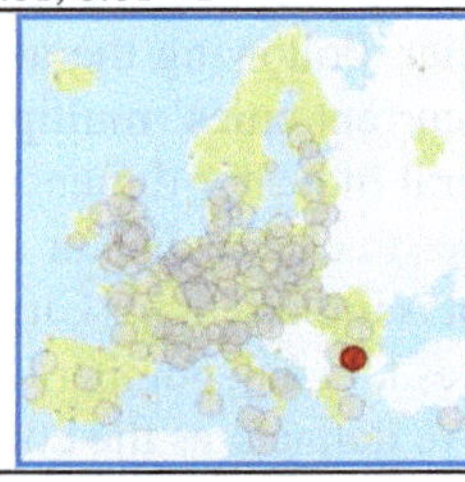
Equal Intervals (5):0 – 0.2,0.2 – 0.4,0.4 – 0.6,0.6 – 0.8,0.8 – 1.0		
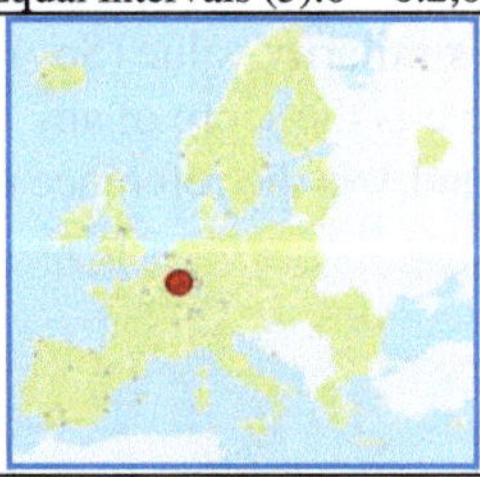	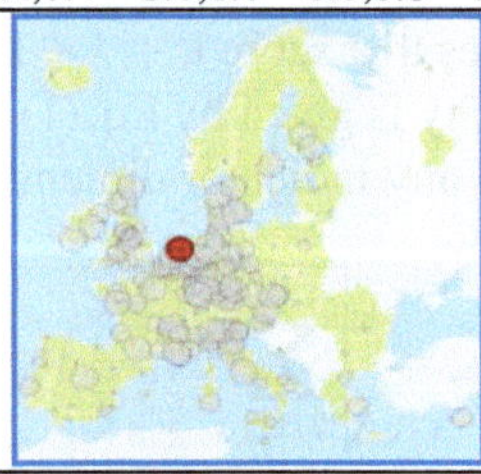	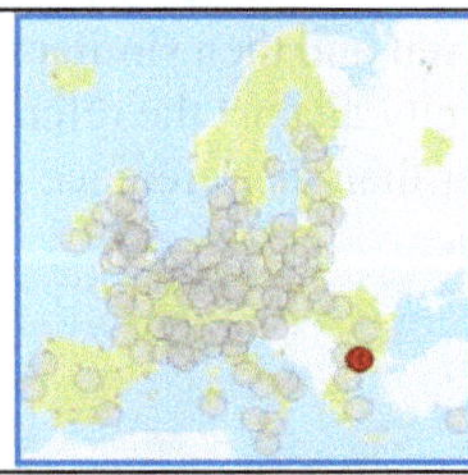
Based on ranking order (171)		
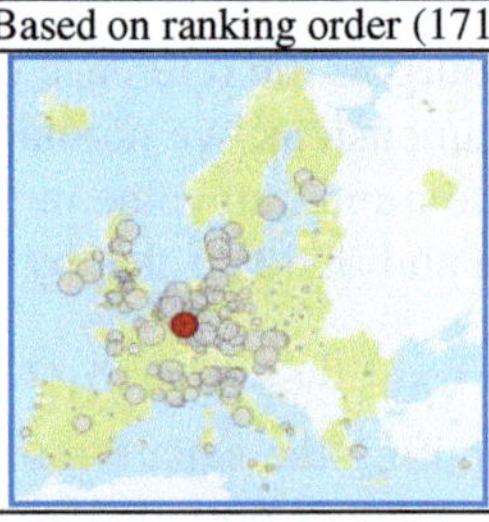	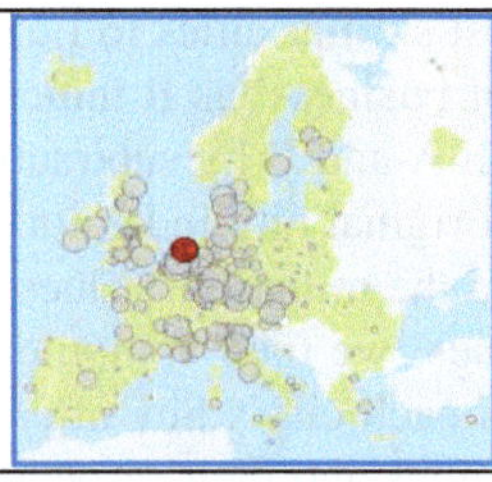	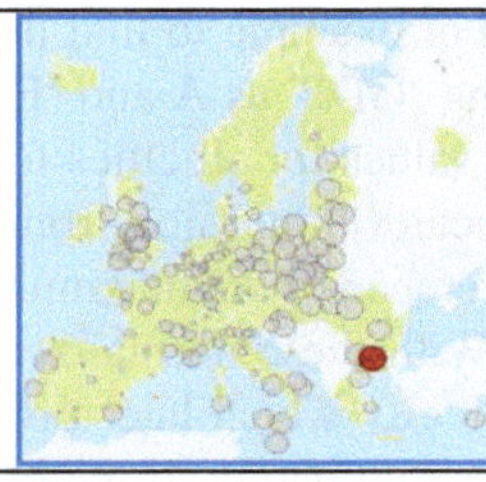

The red dot represents the reference city against which similarity is measured and shown

(ranking the values and using the original values) were explored to assess their effectiveness in displaying the similarity values.

In Table 1, samples of this exploration are shown for Luxemburg, Amsterdam and Plovdiv using original GDP (€) values (75,191, 40,568 and 3,728 respectively) as an indicator example. All the samples yield good visual regional patterns for cities with intermediate values (Amsterdam) while the high score outlier (Luxembourg) stands

Table 2 Similarity maps of the 3 cities using unclassified values from the pairwise measure

Luxembourg (highest GDP)	Amsterdam (in-between GDP)	Plovdiv (lowest GDP)
Unique values (Non-classified) 0.0-1.0		
	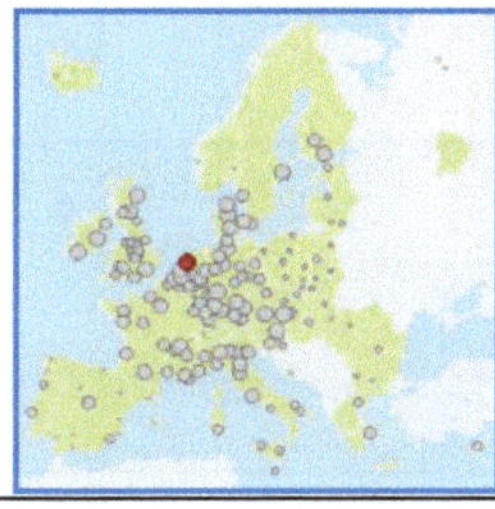	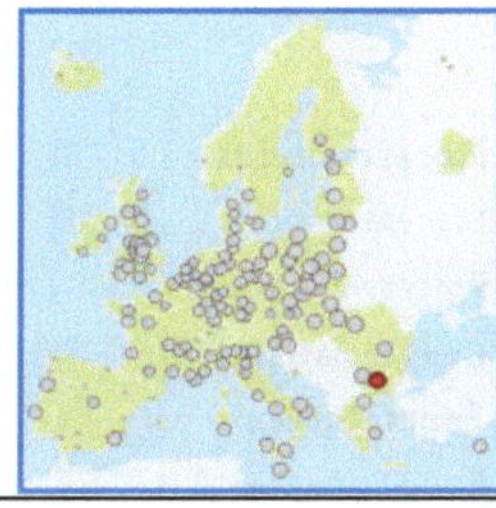

out in Quintiles and Jenks natural breaks but not as much in Equal intervals. For the low score outlier (Plovdiv), some regional patterns are only observable in the latter.

The high score outlier (Luxembourg) in the ranking method becomes part of the Central Europe, Ireland and Scandinavia cluster while the low score outlier (Plovdiv) joins the Eastern Europe cluster.

This classification does not distinguish between the high score outlier and the middle score cities implying that Luxembourg with original value 75,191 and Amsterdam with original value 40,568 are almost identical. While this is good for reflecting an order of magnitude common to Central Europe, Ireland and Scandinavia, it is difficult to visually 'explain' the difference in the original values. Based on the above findings, to allow for visualizing regional patterns and to visualize effects of outliers, we opted to use both position ranking and the unique values (Table 2).

4 Implementation

4.1 Data

An extensive data audit was performed especially focusing on ESPON[3] and Eurostat[4] data. The audit assessed whether existing datasets, collected at various geographic levels (LUZ, NUTS 3, and Metropolitan Region) could be used for city benchmarking. Other data termed licensed / commercial and open / volunteered was considered as well. The data audit resulted in a provisional list of indicators which served as a basis input for CityBench.

In order to enable integration of these indicators into the tool, an extraction, transforming and loading (ETL) module was created for the purpose. For new indicators

[3] http://www.espon.eu/main/

[4] http://epp.eurostat.ec.europa.eu/portal/page/portal/eurostat/home/

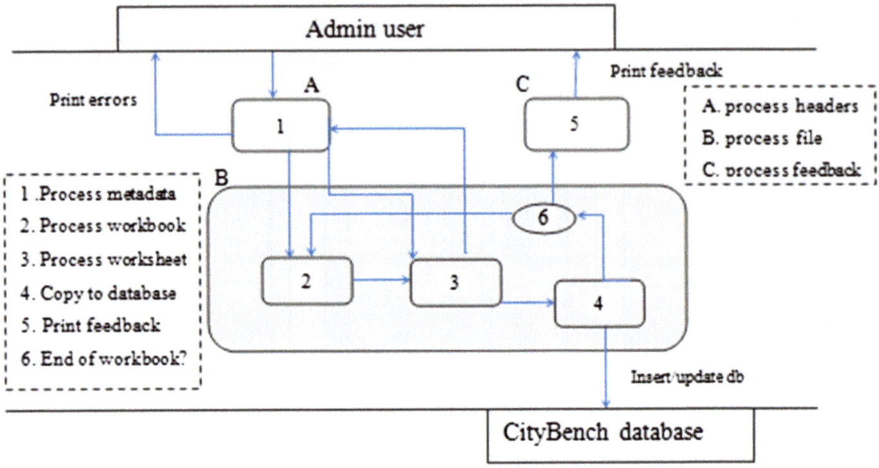

Fig. 1 CityBench ETL process flow

from other sources such as social media data, a module tailored to data of this nature was also developed. The 'CityBench indicators ETL' tool was developed to receive and consume data arranged in spreadsheets and load it into the CityBench database as new indicators, along with their 'meta' data. Figure 1 above is a schematic representation of the CityBench ETL tool.

The CityBench ETL tool contains several subroutines for processing an uploaded file. The file is first processed by the "process headers" which method ensures that the metadata for all indicators in the uploaded file is error free and will not add duplicates in the database. If the method finds no errors, the file is forwarded to the 'process file' method, which loops through the entire file reading data and creating new indicators for CityBench otherwise the user is asked to upload a new error free file.

Within 'process file', the 'process worksheet' method handles the individual processing of data for a single indicator. This process creates new unique indicator identification if the new indicator has no official id from the data provider and finally creates a new table in the CityBench database for the new indicator data while keeping track of the current process status. When all indicators in a file are loaded into the database, feedback is returned to the user with the results of the ETL process.

Using an ETL tool for data management contributes to the uniformity of the data to be used by the CityBench tool achieved by enforcing the usage of a data input template, which ensures that each indicator is uploaded in a standardized format and updates to indicators are uniformly performed and traceable.

4.2 Software Frameworks and Architecture

To achieve an interoperable, scalable and future extensible platform, CityBench employs the client-server architecture (Berson 1996). HTML5 and JavaScript at the client and Java on the server allow us to take advantage of the Rest model design.

Employing jQuery.js[5] and d3.js[6] provide us a set of curated user interface interactions like the scaling bar and the later, creation and control over our dynamic and interactive graphical charts and diagrams (Murray 2013). Java allows implementation of restful web services (Martin 2009; Burke 2009; Sandoval 2009).These serve Asynchronous JavaScript and XML (AJAX) requests from the CityBench client allowing seamless data exchange between the client and server.

Citybench follows and extends the European directive INSPIRE (EC 2007) which defines a classical service-oriented three-layer architecture to include the required functionality in the form of network services and the client applications as a web portal (Tatnall 2005). In Fig. 2, this architecture is described from a technical perspective to consist of the web client at the application layer, the services responsible for data retrieval, the database for storage of all CityBench related data and the ETL modules.

From bottom up these layers are; (A) the 'Data and Services' layer (external data fetching services and the ETL modules). This layer is configured to handle preprocessed data in Excel format from sources such as ESPON, EUROSTAT, European Environmental Agency (EEA), OpenFlights and social networks like Twitter (volunteered geographic information or VGI). (B) The 'Data' layer whose main component is a spatial database manages all web tool data needs. The database utilizes a custom data model well-adjusted to the needs of the CityBench tool and houses data from sources such as the ESPON database, metadata about the data and all indicators. (C) The 'Service' layer services implement well-known and standard-based interfaces which provide data discovery, view, and upload functionality for the CityBench tool. Together they constitute the accessible CityBench API available to other applications with needs similar to the CityBench tool. (D) The 'App' layer serves as the presentation layer comprising of a web portal which constitutes the visible CityBench tool and consists of a front-end user interface and a back-end login admin interface. Figures 5 and 6 are first CityBench prototype user interfaces with labeled functional areas.

5 Preliminary Results

The CityBench web tool utilizes both a map centric and a 'radial' display for the results of the pairwise similarity measure. The sectioned interactive interface (Fig. 3) is meant to be easy to use, above and below the main view section are city selectors for choosing cities.

---

[5] http://jquery.com/

[6] http://d3js.org/

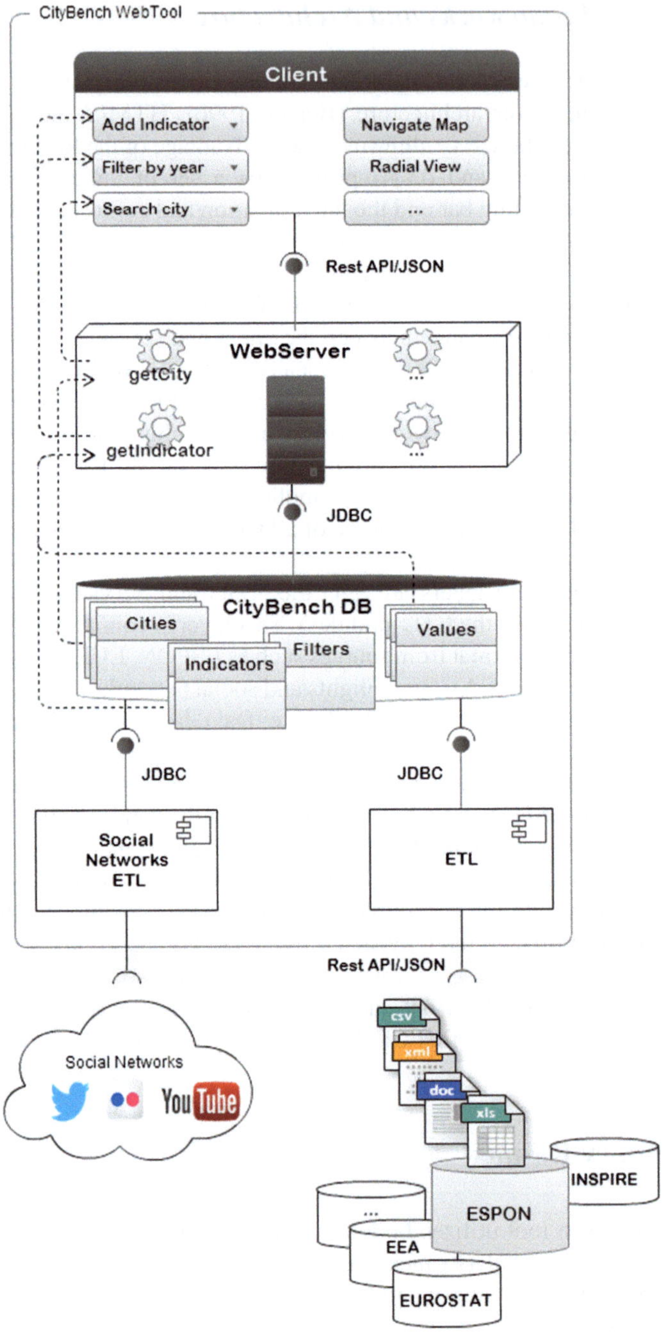

Fig. 2 Technical architecture of CityBench

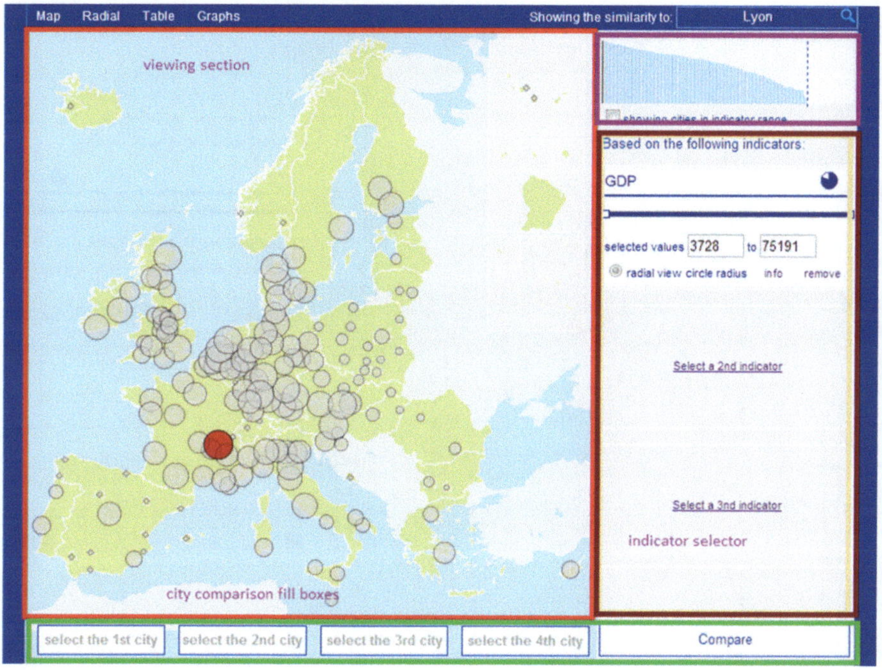

Fig. 3 CityBench interface

In both views (Figs. 3 and 4), there is a main view showing the results with each city represented as a circle whose radius is the pairwise measure score or a selected indicator value in Fig. 4. The reference city shown in red is naturally the most similar to itself and will always have the biggest radius in the main view and the highest bar in the list of cities to the right. Cities with no data in any of the selected indicators are shown with a 'cross' symbol in the map centric view (Fig. 3) and a white background in the bars widget located above the 'indicator selector' showing the distribution of values amongst the cities. In Fig. 3, the circle is placed in the middle of the geometry of the LUZ.

The radial view in Fig. 4 assumes an arrangement where the cities seem to converge or radiate from the reference city in the centre. The closest to the centre, the more similar the city is to the reference city. Each city in this view is positioned approximately according to its actual geographic location, but also relative to the reference city. This view was inspired by a similar visualization by Tulp (2012). The circle represents a city's normalized value in a chosen indicator. In Fig. 4, Luxemburg has the biggest circle meaning that it has the biggest GDP value although it is very dissimilar to Lyon (in red). Plovdiv has the smallest circle and therefore the smallest GDP but also dissimilar to Lyon.

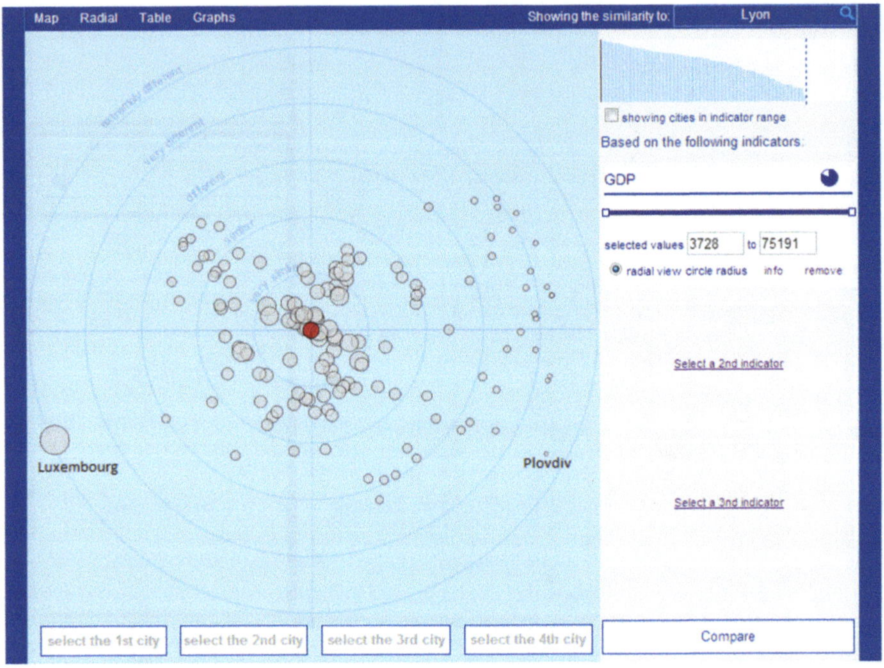

Fig. 4 CityBench radial view

5.1 Available Cities and Indicators

A number of criteria were used in selecting the CityBench LUZ (Cities). The first criterion included all European LUZ with corresponding metropolitan regions in the OECD Metropolitan Areas database (MAdb). The MAdb has 268 metropolitan regions whose performances on a set of key indicators are visualized in the OECD Metropolitan Explorer[7] of which 114 were selected as starting point for populating CityBench. The second criterion included all LUZ from EU27 + 4 countries (i.e. EU27 plus Iceland, Liechtenstein, Norway and Switzerland). The third and fourth criteria supplemented the Citybench city list with LUZ from underrepresented countries. These countries are either small or relatively sparsely populated, or have a population that is distributed relatively equally over the country and may have only one or even no LUZ of which the population exceeds 400,000. Using a second city provided its population exceeds 200,000 ensured a proper representation of these countries in CityBench (criterion 4). The current list of cities may be available upon request.

Consideration for alternative data sources for the web tool resulted in use of social media for populating the CityBench web tool. This arose due to the need for relevant

[7] (http://www.oecd.org/statistics/datalab/metro-explorer.htm)

and up-to-date indicators as at the time of data auditing, ESPON and Eurostat the main sources of LUZ data and therefore indicators were in the process of reviewing and updating their databases. This meant that for several relevant indicators, recent data was not available. In addition, these databases do not cover all indicators potentially relevant to the CityBench mission. Geo referenced user-generated content is acquiring a fundamental role in a wide range of applications even though it currently represents only a small percentage. Tweets from Twitter for example may play a major role in response actions to emergencies (Lanfranchi and Ireson 2009) and (Núñez-Redó et al. 2011). Social media inclusion in CityBench utilized a custom implementation of social media extraction module out of scope of this research.

The search for data resulted in a set of indicators such as population and unemployment in an economy dimension, aging and old age dependency in a demography dimension, out flight route in a connectivity dimension. A complete list of these indicators and methodology to create them can be available on request.

5.2 City Similarities

In both Sects. 1 and 3, we mentioned that the CityBench tool would be helpful in the identification of geographic patterns using its similarity measure, the results indeed support this statement. We see a geographic pattern in Fig. 3 when viewing similarity to Lyon and notice a high concentration of similar cities to Lyon in Western Europe and a strip of dissimilar cities in Eastern Europe.

In Fig. 5, a regional belt covering most of Eastern Europe down to Spain and Italy is visible when viewing similarity to Kielce using unemployment ratio and ease of doing business indicators. Figures. 6 and 7 continue to stress similarities among western European cities in respect to western European reference cities with exceptions here and there such as in Fig. 8.

The CityBench web tool results are very data dependent. As such the tool is not meant to provide a static evidence of similarities or benchmarking, this means that certain users (researchers, investors, city officials, citizens) with specific questions can select the suitable indicators for their questions, and CityBench allows them to explore, show clusters or spatial patterns not earlier anticipated.

5.3 Which City is Most Similar to Mine?

Using an example of 3 indicators (GDP, unemployment ratio, easy of doing business) with original values in Tables 3 and 4 lists the 10 cities most similar to each of the 6 cities Amsterdam, Cork, Helsinki, Luxemburg, Plovdiv and Madrid each used as a reference city as given by the CityBench similarity measure. We can already see relationships between cities in Table 4 whose top 4 are shown on the map centric in Fig. 8.

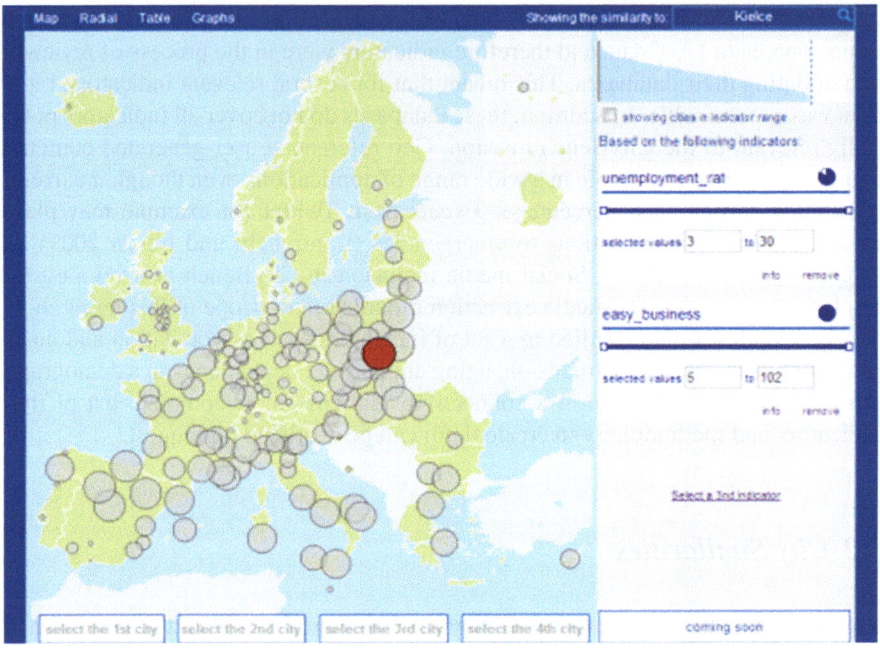

Fig. 5 Similarity to Kielce

The CityBench tool gives a quick scan into a city and cities that are most similar to it in respect to selected indicators. These results can be used by economists and policy makers to discover why similarly structured cities behave differently as it may be that cities that are similar may share similar risks and opportunities. These cities could learn from each other and also create a network of best practices.

6 Limitations, Future Works and Conclusion

6.1 Limitations

Some data from the vast collection of data from the main providers, ESPON and the EUROSTAT's Urban Audit (UA) collections is not yet available at LUZ geographical level revealing a shortage of directly comparable indicators across cities. Many indicators are collected at other regional scales other than the LUZ and conversion to LUZ level was not always possible.

In addition to the above, the latest edition of the UA uses different LUZ delineations resulting in the usage of different LUZ delineation version data when creating

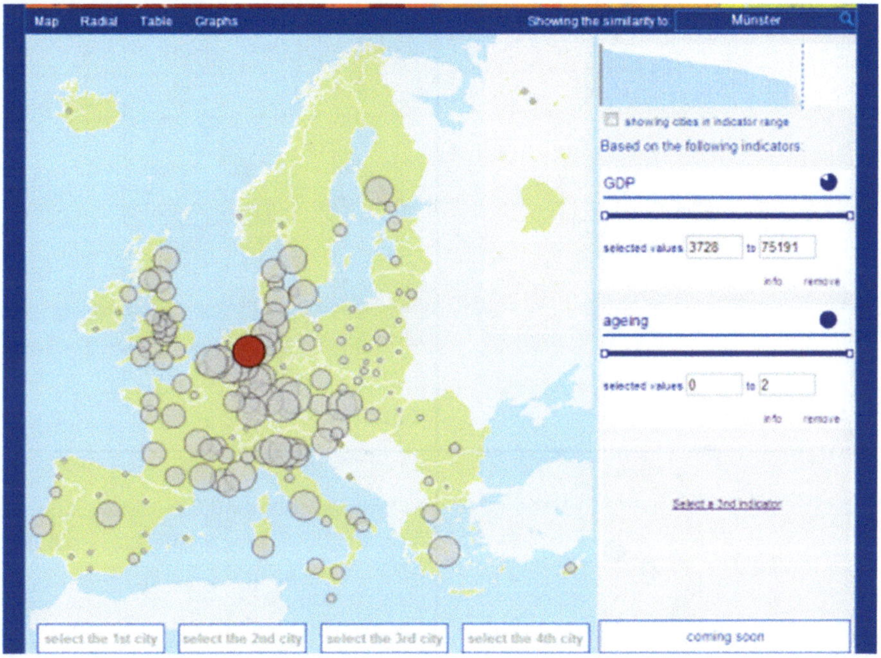

Fig. 6 Similarity to Munster

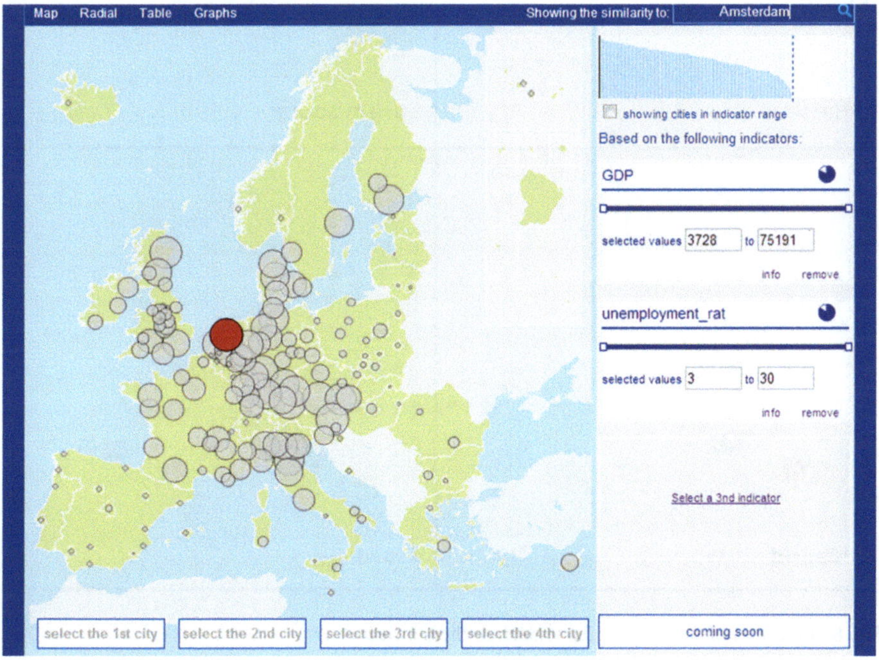

Fig. 7 Similarity to Amsterdam

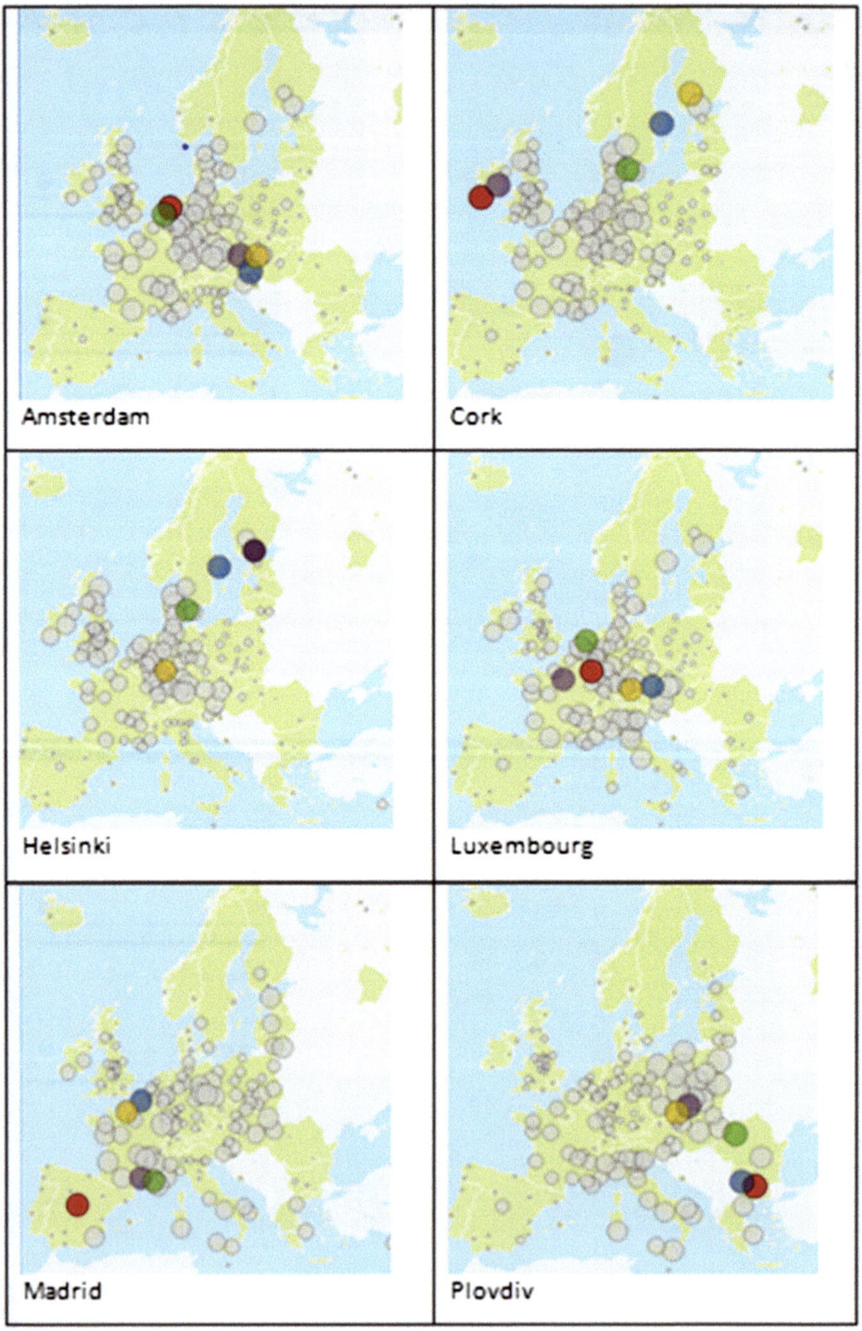

Fig. 8 Plotting *top* 4 similar cities to each of the 6 cities

Table 3 Original values of selected cities

	City	GDP	Unemployment ratio	Easy business
Max value		75191	30	102
	Amsterdam	40568	4.2	31
	Cork	44110	12.5	15
	Helsinki	44074	6.4	11
	Luxemburg	75191	4.4	56
	Plovdiv	3728	8.5	66
	Madrid	30040	16.1	44
Min value		3728	3	5

Table 4 Similarity results

Amsterdam		Cork		Helsinki	
Amsterdam	0.000	Cork	0.000	Helsinki	0.000
Linz	0.048	Dublin	0.029	Stockholm	0.037
Graz	0.055	Stockholm	0.203	København	0.088
Rotterdam	0.078	København	0.214	Frankfurt am Main	0.107
Wien	0.079	Tampere	0.224	Århus	0.125
Eindhoven	0.087	Helsinki	0.231	Aalborg	0.130
Karlsruhe	0.128	London	0.233	München	0.134
Frankfurt am Main	0.128	Düsseldorf	0.241	Hamburg	0.137
München	0.135	Aalborg	0.242	London	0.139
Regensburg	0.138	Paris	0.243	Karlsruhe	0.146
Luxemburg		Plovdiv		Madrid	
Luxembourg	0.000	Plovdiv	0.000	Madrid	0.000
Paris	0.476	Ostrava	0.121	Montpellier	0.121
Linz	0.523	Sofia	0.123	Lille	0.160
Amsterdam	0.549	Cluj-Napoca	0.126	Marseille	0.210
München	0.554	Brno	0.130	Rouen	0.240
Wien	0.575	Bialystok	0.132	Toulon	0.257
Lyon	0.584	Lódz	0.134	Nancy	0.280
Graz	0.592	Szczecin	0.134	Saint-Etienne	0.281
Bologna	0.603	Bydgoszcz	0.137	Nantes	0.284
Rotterdam	0.609	Kraków	0.137	Bordeaux	0.291

indicator time series from more than one UA edition. This may lead to errors in analysis due to the modifiable area unit problem (Openshaw 1984).

Even though we attempted to have the most complete indicator datasets, missing values for some cities in different indicators were unavoidable. Because of this, the presence of a city with no values in either one or more indicators means that this city is immediately dropped from the search or comparison operation resulting in an imperfect "image" of the current situation.

6.2 Future Work

We have used a similarity measure to provide a quick scan of cities (European LUZ) revealing the most similar cities of a selected city in respect to selected indicators. A next step would be to test the robustness of this similarity measure by performing sensitivity analysis and testing different similarity methods. It would also be interesting to quantify the geographical trends of the distribution of similarity. Future research will attempt to use spatial auto-correlation and cluster analysis to quantify and further help identify spatial patterns.

6.3 Conclusions

Comparing and benchmarking cities is an important activity in determining profitable or investment ready cities, for the above reason, many tools exist that display data and rank cities based on pre-defined indicators such as the OECD metropolitan explorer.

CityBench offers to supplement such tools with a methodology that offers a multifaceted insight into similarity of cities for variable multi-dimensional indicators.

By highlighting the cities that are performing similar and different to a compared city, CityBench can be used to create opportunities for cooperation in cases where cities share similar challenges/opportunities or syndicates where low performing cities make alliances with high performers. Preliminary results show that CityBench can show regional patterns when meaningful indicators are used together.

Prototype feedback from stakeholders suggested that while the concept is worthwhile and provides a new insight into city comparisons, results should be carefully communicate in a clear and transparent manner.

Acknowledgments We thankfully acknowledge the European Observatory for Territorial Development and Cohesion (ESPON) for funding this research and the team effort from both colleagues at Geodan Amsterdam, the Netherlands and Universitat Jaume I (UJI) Castellon, Spain.

References

Amelang J (2007) Comfigparing cities: a barcelona model? Urban History 34:173–189

Ammons DN (1999) A proper mentality for benchmarking. Public Admistration Rev 59(2):105–109

Ashby F, Ennis D (2007) Similarity measures. Scholarpedia 2(12):4116

Batty M, Longley P (1994a) Urban boundaries and edges. In: Batty M, Longley P (eds) Fractal cities: a geometry of form and function. Academic Press, San Diego, pp 164–198

Batty M, Longley PA (1994b) Fractal cities: a Geometry of form and function. Academic Press, San Diego

Baum S (1997) Sydney, Australia: a global city? Testing the social polarisation Thesis. Urban Studies, pp 1881–1902

Beaverstock JV, Hoyler M, Pain K et al (2001) Comparing London and Frankfurt as world cities: a relational study of contemporary urban change. Anglo-German Foundation for Study of Industrial Society, London

Berson A (1996) Client/Server architecture. McGraw-Hill, New York

Burke B (2009) RESTful Java with JAX-RS. O'Reilly Media. Inc.

Deza MM, Deza E (2013) Distances and similarities in data analysis. Encyclopedia of distances, pp 291–305

Dijkstra L, Poelman H (2012) Regional focus (RF 01/2012). Retrieved from http://ec.europa.eu/regional_policy/sources/docgener/focus/2012_01_city.pdf

EC (2004) Urban audit: methodological handbook. Office for Official Publications of the European Commission, Luxemburg

EC (2007) Directive 2007/2/EC of the European parliament and of the council of 14 March 2007 establishing an infrastructure for spatial information in the European community (INSPIRE). Official J Eur Union 50:1–14

Friedmann J (1986) The world city hypothesis. Development and change, 17:69–83

Friedmann J (1995) Where we stand: a decade of world city research. World cities in a World-System. Cambridge University Press, Cambridge, pp 21–47

Groenendijk NS (2004) The use of benchmarking in EU economic and social policies. In: Conference on the future of Europe. Odense, pp 24–25

Hall P (2009) Looking backward, looking forward:the city region of the mid-21st century. Reg Stud 43:803–817

Helgason S (1997) International benchmarking:experiences from OECD countries. In: Procedings of the Danish ministry of finance conference on international benchmarking. The Stationery Office, Copenhagen, pp 20–21

Holloway A, Wajzer C (2008) Improving city performance through benchmarking. In: International cities town centres & communities society conference, Sydney

Jain A, Prabhakar S, Hong L et al (2000) Filterbank-based fingerprint matching. IEEE Trans Image Process 9(5):846–859

Kasanko M, Barredo JI, Lavalle C et al (2006) Are European cities becoming dispersed? A comparative analysis of 15 European urban areas. Landscape Urban Plan 77(1–2):111–130

Lam EW, Chan AP, Chan DW (2010) Benchmarking success of building maintenance projects. Facilities 28(5/6):290–305

Lanfranchi V, Ireson N (2009) User requirements for a collective intelligence emergency response system. In: Proceedings of the 23rd British HCI group annual conference on people and computers: celebrating people and technology. British Computer Society, pp 198–203

Lippman Abu-Lughod J (1995) Comparing Chicago, New York and Los Angeles: testing some world city hypotheses. In: Knox P, Taylor P (eds) World cities in a World system. p 171–191

Luo J, Duggan DJ, Chen Yidong SJ et al (2001) Advances in brief: human prostate cancer and benign prostatic hyperplasia: molecular dissection by gene expression profiling. Cancer Research, pp 4683–4688

Luque-Martínez T, Muñoz-Leiva F (2005) City benchmarking: a methodological proposal referring specifically to Granada. Teodoro Luque-Martínez, Francisco Muñoz-Leiva, city benchmarking: a methodologCities 22(6):411–423

Mane AV, Manza RR, Kale KV (2010) The role of similarity measures in face recognition. Int J Comput Sci Appl (1):62–65

Martin K (2009) Java web services: up and running. O'Reilly Media Inc.

Murray S (2013) Interactivee data visualization for the web. O'Reilly Media Inc.

Núñez-Redó M, Díaz L, Gil J et al (2011) Discovery and integration of Web 2.0 content into geospatial information infrastructures: a use case in wild fire monitoring. Availability, reliability and security for business, enterprise and health, information systems, pp 50–68

OECD (2012) Redefining "Urban": a new way to measure metropolitan areas. OECD Publishing. doi:10.1787/9789264174108-en

Openshaw S (1984) The modifiable areal unit problem, Concepts and Techniques in Modern Geography Vol. 38 Geo Books, Norwich, England

Parr J (2005) Perspectives on the city-region. Reg Stud 39:555–566

Preoţiuc-Pietro D, Cranshaw J, Yano T (2013) Exploring venue-based city-to-city similarity measures. In: Proceedings of the 2nd ACM SIGKDD international workshop on urban computing. ACM, Chicago, pp 1–4

Ramalho CE, Hobbs RJ (2012) Time for a change: dynamic urban ecology. Trends Ecol Evol 27(3):179–188

Sager C (2003) European city benchmark study. Wonderful Copenhagen, Copenhagen

Sandoval J (2009) Master core REST concepts and create RESTful web services in java. Packt Publishing Ltd., Birmingham

Seth R, Covell M, Ravichandran D et al (2011) A tale of two (similar) cities-inferring city similarity through geo-spatial query log analysis. KDIR, pp 179–189

Sikos L (2011) Web standards: mastering HTML5, CSS3, and XML. Apress, Berkeley

Tatnall A (2005) Web portals: the new gateways to internet information and services. Idea Group Inc (IGI)

Taylor PJ, Walker D (2001) World cities: a first multivariate analysis of their service complexes. Urban Stud 38:23–47

Tuan Seik F (2000) Subjective assessment of urban quality of life in Singapore (1997–1998). Habitat Int 24(1):31–49

Tulp JW (2012) Close votes. Retrieved Nov 15, 2013, from Tulp interactive: http://tulpinteractive.com

Turok I, Mykhnenko V (2007) The trajectories of European cities, 1960–2005. Cities 24(3):165–182

Vasanen A (2012) Functional polycentricity: examining metropolitan spatial structure through the connectivity of urban sub-centres. Urban Stud 49:3627–3644

Woolley T (2013) An investigation of the effect of the swamping phenomenon on several block procedures for multiple outliers in univariate smaples. Opne J Stat 3(5):229–304

A GIS-Based Process for Calculating Visibility Impact from Buildings During Transmission Line Routing

Stefano Grassi, Roman Friedli, Michel Grangier and Martin Raubal

Abstract Planning linear infrastructures can be a tedious task for regions characterized by complex topography, natural constraints, high density population areas, and strong local opposition. These aspects make the planning of new transmission lines complex and time consuming. The method proposed in this work uses Multi-Criteria Analysis and Least-Cost Path approaches combined with a viewshed analysis in order to identify suitable routes. The visual impact is integrated, as a cost surface, into the process and combined with natural and anthropological constraints. The cumulated visibility of each raster cell is estimated as the sum of the weighted distance between buildings and the cell itself. In order to reduce the typical zig-zags resulting from Least-Cost Path methods, a weighted straightening approach is applied. A sensitivity analysis of the weights of the visibility and the straightening is carried out in order to assess different scenarios and to compare the existing TL path to the proposed ones. The method is applied to a case study where an old transmission line needs to be replaced by a new one and the local grid operator needs to identify feasible routes. A set of 30 routes is identified and most of them have a lower visibility that the existing path but, only some of them present a comparable complexity to be realized.

Keywords GIS · Multi-criteria analysis · Least-cost path · Transmission line routing · Visual impact from buildings

S. Grassi (✉) · R. Friedli · M. Raubal
ETH Zurich, Institute of Cartography and Geoinformation, 8093 Zurich, Switzerland
e-mail: sgrassi@ethz.ch

M. Grangier
Groupe-e, 1763 Granges-Paccot, Switzerland

J. Huerta et al. (eds.), *Connecting a Digital Europe Through Location and Place*, Lecture Notes in Geoinformation and Cartography, DOI: 10.1007/978-3-319-03611-3_22,

1 Introduction

Planning and enhancing new Transmission Lines (TLs) consists of finding suitable land to install masts, estimating the investment costs, avoiding protected and dangerous areas, settlements and forests, and assessing the visual impact of masts on the surroundings. In complex environments, characterized by a high density population, complex topography and the presence of several protected areas, the development of linear infrastructures leads to several challenges for planners.

Multiple issues have been identified in previous work when planning new TLs. The legal constraints and urban planning guidelines must be followed and the technical limitations of TLs have to be taken into account (Hirst and Kirby 2001). The integration of renewable energy resources into the power grid raised several issues (Mills et al. 2012) such as grid stability. Many aspects related to nature and are also considered critical (Bevanger et al. 2010). In addition, the visual impact on the landscape has always been a major issue for public acceptance (Cotton and Devine-Wright 2012; Furby et al. 1988). Building or replacing new TLs is capital intensive (McCalley and Krishnan 2014) and time consuming in particular due to social opposition (Towers 2000) which may delay new projects.

One of the challenges for grid operators is to find a trade-off between a cost-effective project, mainly as a function of overall length, and the minimization of the visual impact on the surrounding buildings in order to reduce local opposition (Marshall and Baxter 2002).

In this chapter a GIS-based methodology to support grid operators in selecting a TL path among different possible routes is proposed. The visibility of TLs from the surrounding buildings is introduced in order to determine suitable routes. A sensitivity analysis of the weighted visibility is proposed to show its impact on the TLs route and the results are compared to the outcome of an approach when only land features are considered.

A review of previous work in the field of transmission line planning is presented in Sect. 2. In Sect. 3 the geodata set used here is described. Section 4 continues with a discussion of the applied methodology. In Sect. 5 the structure of the sensitivity analysis is presented and applied to a case study. Section 6 presents conclusions and directions for future work.

2 Related Work

Work in the last decade has shown how GIS can support planners with respect to many aspects related to the routing of linear infrastructures such as roads (Sessions et al. 2006) and pipelines (Zhenpei et al. 2010).

Multi-Criteria Analysis (MCA) and Least-Cost Path (LCP) have been widely applied to support decision makers in assessing the optimal routing of linear infrastructures such as TLs using an Analytic Hierarchy Process (AHP) (Husain et al. 2012) and such as roads based on multiple environmental and economic criteria (Atkinson et al. 2005) or using anisotropic accumulated-cost surface (Yu et al. 2003). Other work described the theoretical basis to implement efficient methods for decision-aid (Malczewski 1999).

LCP is typically used to identify the shortest path between two points and different approaches have been developed in the last decades (Dijkstra 1959; Eppstein 1998) and applied for suitable corridor evaluation (Church et al. 1992).

To find a least-cost path between a starting point and a destination in a digital terrain model (DTM), there are two main steps. The first step is aimed at creating an accumulated cost surface with respect to a set of cost factors. The second step is to construct the least-cost path on the accumulated cost surface using the back-link mechanism (Xu and Lathrop 1995) to connect the departure and destination locations (this is the approach used in common GIS-software).

The accumulated cost surface represents the spatial distribution of the resistance of a region to be crossed (Douglas 1994) in terms of cost as result from the structure of the weighted factors defined by the user. The weights of each factor of the cost surface can vary as a function of the boundary (Atkinson et al. 2005) of the selected area (topography, land cover, etc...). These criteria take into account the impact of a given infrastructure on the studied area. A first study that integrated the viewshed to find LCP was carried out with a coarse digital elevation model (Stucky 1998).

In TL planning, GIS has been used to study the impact of local opposition when planning new electric infrastructure and to show that routing is affected by political decision (Towers 1997). Other work showed how the involvement of local people in the TL planning has decreased the risk of project failure (Jewell et al. 2010). Theoretical studies addressed the issue of path distortion using raster data in roads and TL planning (Huber and Church 1985). In recent work alternative LCP scenarios have been identified using MCA and LCP (French et al. 2008) when considering different groups of feature classes. Other studies addressed the issue of acceptable power line paths and the diverse socio-economic interests of the different groups involved in the planning process (Monteiro et al. 2005).

The issue of visual impact on the environment and the landscape of TLs corridors was also considered where the visual effect decreases linearly with the distance and using Digital Terrain Model (DTM) (Hadrian et al. 1988).

Nevertheless, in most of the previous studies the visibility of TL routes from settlements and individual buildings has not been taken into account when defining potential routes and corridors. In addition, the DTM, used in some previous studies does not represent the surface of a given region but only the terrain, and thus the elements that can obstruct the line of sight are not included. This is nowadays a critical issue to be addressed compared to decades ago when the first lines were built.

The aim of our work is to design an integrated workflow based on MCA and LCP that processes multiple linear elements and surfaces and determines a set of optimal paths to support planners in TL routing in a complex environment, where the visual impact plays a critical role. In order to assess the distribution of the visibility within a region, the existing buildings are used as observer points and a Digital Surface

Table 1 Geodata used to identify accumulated cost surface and avoidance areas

Primary area (vector)	Protected areas (vector)
Swisstopo VECTOR25-layer primary areas	BAFU-Landscapes and natural monuments of national importance (BLN)
Swisstopo swissBUILDINGS3D	
Terrain (raster)	BAFU-Alluvial zones
DTM25 (Digital terrain model)	BAFU-Low and high mires, mire landscapes
DSM25 (Digital surface model)	BAFU-Dry grasslands
Linear infrastructure (vector)	BAFU-Amphibian spawning areas
	BAFU-Aquatic and migratory bird reserves
Road and rail network	Groupe E-groundwater protection zones
Power grid network	
Danger areas (vector): cant. Fribourg	*Danger areas (vector): cant. Fribourg*
Danger zones landslide	Alluvial protection zones
Danger zones flood	Natural reserves
Danger zones avalanche	Mire protection zones
	Dry grassland protection zones
Danger areas (vector): cant. Neuchâtel	*Danger areas (vector): cant. Neuchâtel*
Danger zones landslide	Alluvial protection zones
Danger zones flood	Natural reserves
	Mire protection zones

Model (DSM) is used instead of a DTM. The distance of the buildings from the TL is taken into account and weighted when creating the new accumulated cost surface. A sensitivity analysis is carried out to show the variation of the TL route when the visual impact is assigned different weights compared to the shortest path defined using the global cost surface. The outcome is a set of routes whose parameters are correlated to each other using a standardization process that enables the grid operator to evaluate different possible paths.

The practical advantage of this methodology is the opportunity for the power grid planner to quickly reduce the time required to identify possible TL paths during the preliminary stage of a new project. As experienced with the local grid operator, the preliminary stage of a new TL route needs several months in order to identify some corridors as possible solutions.

3 Geodata

The geodata used for this research are divided into five categories according to the specific conditions present in the studied region: primary areas, linear infrastructure, terrain, protected areas and danger areas (Table 1).

Primary areas include land cover and land use; e.g., settlements, pasture, forest, water bodies, building footprints, etc. The group of linear infrastructure, as a separate dataset, includes roads, rails and the existing power grid. The terrain group includes

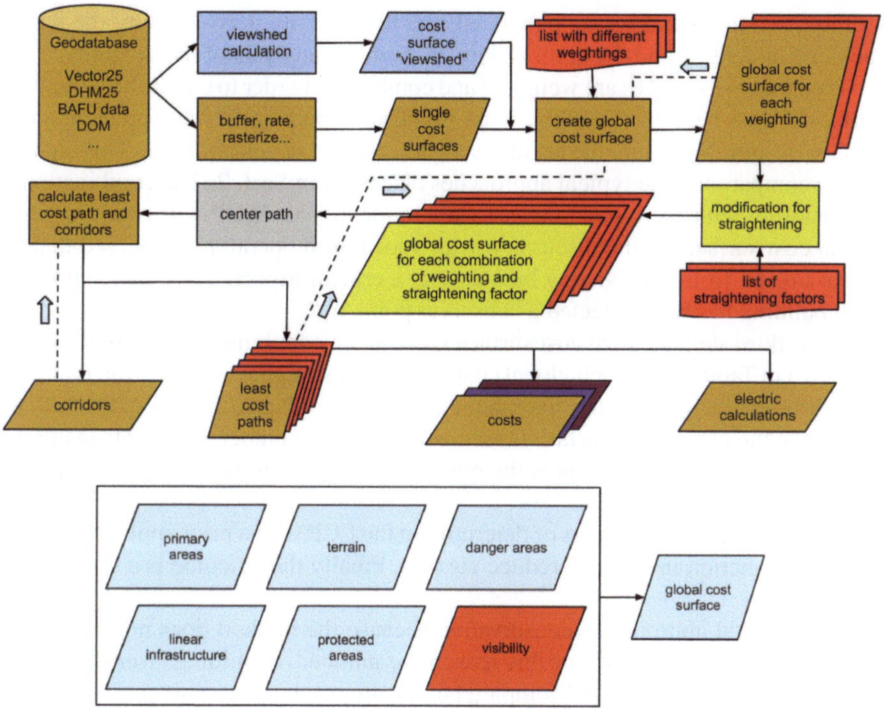

Fig. 1 *Above* schematic representation of the implemented workflow and *below* the factor groups

the topography, such as the 2 m resolution DTM and the DSM. Protected areas are crucial components in the multi-criteria analysis and must be avoided. Danger areas mainly consist of regions with a high risk of landslides, avalanches, or floods and thus not very suitable to install HV poles. The case study focuses on a region in the western part of Switzerland.

4 Methodology

A GIS-based model for transmission line siting can be implemented using different well-established GIS algorithms. Determining the most suitable path and corridors through a landscape is one of the oldest problems in spatial suitability analysis (Jankowski and Richard 1994). Corridors are areas where the traversing cost from the source to the destination lies below a certain threshold. They are useful in providing surfaces within which the optimal route can be identified by narrowing down the suitable areas. The path finding procedure used in this work consists of the following steps as shown in Fig. 1.

First, six cost surfaces are generated: the first five are derived using the factors included in Table 1 and the sixth is derived from the analysis of the visibility.

Then the cost surfaces are weighted and combined in order to create the global cost surface (Fig. 1 below). Visibility, as an influencing parameter when creating a cost surface, is introduced into the set of layers describing the land features. This element is new compared to the typical approaches discussed in Sect. 2. The combination of the visibility with the typical dataset used to identify the LCP will result in a new global cost surface. In the discussion with a local grid operator, it emerged that the visual impact from the existing buildings is a critical aspect. Thus, the centroid of each building has been selected as observer point.

In the third step, a global cost surface is created by applying the different weights as shown in Table 4. For each global cost surface the straightening function is applied with different factors in order to obtain a new set of global cost surface. Overall, the amount of the global cost surface is $m*n$ where n is the number of weights in the 'list of different weightings' and m is the number of weights in the 'list of straightening factors' in Fig. 1.

The following step consists of determining the LCP with a prior application of the centering function in order to reduce zigzags. Finally the outcome is a set of $m * n$ LCPs.

The adopted approach is conservative because the method does not distinguish between cables and masts. For this reason the number of buildings from which the TL is visible is overestimated compared to the approach where only masts are taken into account.

In previous work, the calculation of the viewshed and the cost surface has been carried out using the DTM. In this study the digital surface model (DSM) is used: it includes all visible objects such as buildings and forests built over the terrain surface that can block the line of sight from a given cell to the TL path and can thus reduce the number of buildings from which the TL is visible. The resolution of the selected DSM is $2\,\mathrm{m}^2$ which has been resampled to $5\,\mathrm{m}^2$ when calculating the visibility map in order to reduce the computational time without making the raster resolution too coarse.

4.1 Straightening

A linear approach for straightening is derived from previous work (Berry 2004): it modifies the cost surface by increasing the lower values of the cost surface and therefore reducing the length of the LCP. The straightening is applied by using the following relation:

$$AdjustedCost\ Surface = i + \frac{\max(costSurface) - i}{\max(costSurface)} * costSurface \tag{1}$$

where max(costSurface) is the highest value of the global cost surface.

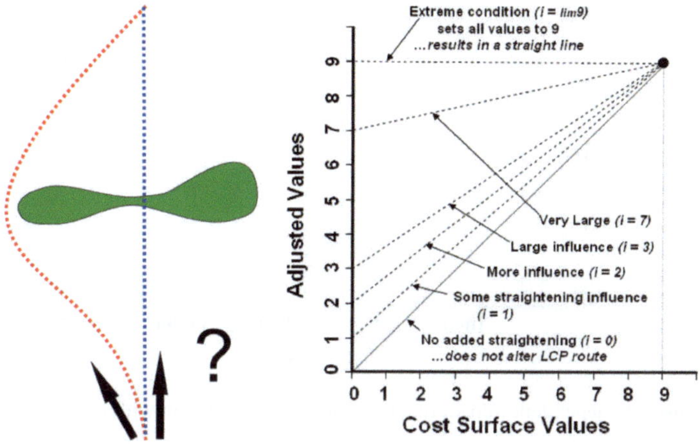

Fig. 2 *Left* visualization of a straightening effect depending on the factor *i:* (the *red line* is not straightened, the *blue line* is straightened), *right* impact of the *i* value on the straightening of the LCP (Berry 2004)

The example displayed in Fig. 2 (left) shows that a straightened path (blue line) that crosses areas with higher costs (green areas corresponding to vegetation) which is not the case of the unstraightened path (red line). The cost by crossing this area in reality does not necessarily increase if the distance is short, because most of the small areas with higher costs (unsuitable for masts) can be overflown with an OH line. With an increasing value of the straightening factor *i*, larger unsuitable areas are crossed. Practically, the challenge for the planner is to figure out the optimal *i* value. When the straightening function is applied, first, the maximum value of the global cost surface is determined and then the cost surface is accordingly modified.

Figure 2 (right) shows the impact of the *i* value on the cost surface and thus the LCP. A value *i* equal to zero does not affect the original cost surface and therefore the LCP is characterized by multiple turns, while, on the other hand, a value of *i* equal to 9 significantly modifies the cost surface and turns the LCP between two points into a straight line. In between, the *i* values comprised between 1 and 8 modifies the cost surface as a function of the selected value: the higher the *i* value, the higher is the impact on the cost surface and the straightening of the LCP. The maximum *i* value can be arbitrarily defined by the user.

The definition of a suitable straightening factor is a critical part when defining the LCP and for this reason it will be investigated later in the sensitivity analysis section.

4.2 Visibility Cost Surface

The assessment of the visual impact of the most common GIS software is mainly based on the cumulative viewshed analysis that uses multiple observers (Lake et al.

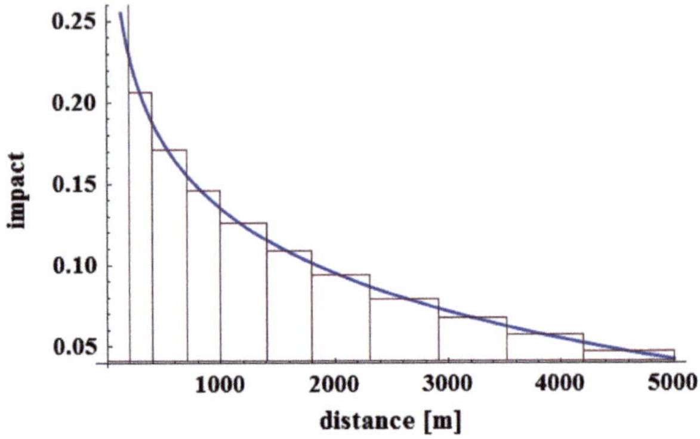

Fig. 3 The distance dependent impact function and its discretization with histograms (Paul et al. 2004)

Table 2 Inner and outer rings and corresponding weights resulted from the discretization of the distance dependent impact function

Inner radius [m]	0	200	400	700	1000	1400	1800	2300	2900	3500	4200
Outer radius [m]	200	400	700	1000	1400	1800	2300	2900	3500	4200	5000
Weight	0.286	0.21	0.17	0.145	0.125	0.108	0.09	0.08	0.068	0.057	0.05

1998). Nevertheless ArcGIS (the software suite used in this work) does not take into account that objects located in the space have a different impact on the observers as function of the reciprocal distance (Ogburn 2006).

In this work the visibility of a TL from the observer points is investigated and thus the distance from the observer point is taken into account. In previous work (Nohl 1993; Paul et al. 2004) the perceived impact w_i of TLs as a function of the distance x from the TLs (limited to 5,000 m) has been defined with the following relation, whose curve is shown in Fig. 3:

$$w_i = \frac{-0.0638 \times \ln x + 0.59}{1.105} \tag{2}$$

In order to overcome the current GIS limitation, i.e., not relating the viewshed impact to the distance, an inner and outer radius are defined in order to discretize the function shown in Fig. 3. With this approach the area around an observer is divided into multiple concentric rings and to each of them a weight is assigned depending on the distance from the center.

The weight of a circular ring (Table 2) equals the mean value of the impact between the inner and the outer radius and is calculated with the following relation:

$$w_i = \frac{\int_{x_i}^{x_{i+1}} \frac{-0.0638 \times \ln x + 0.59}{1.105} dx}{x_{i+1} - x_i} \tag{3}$$

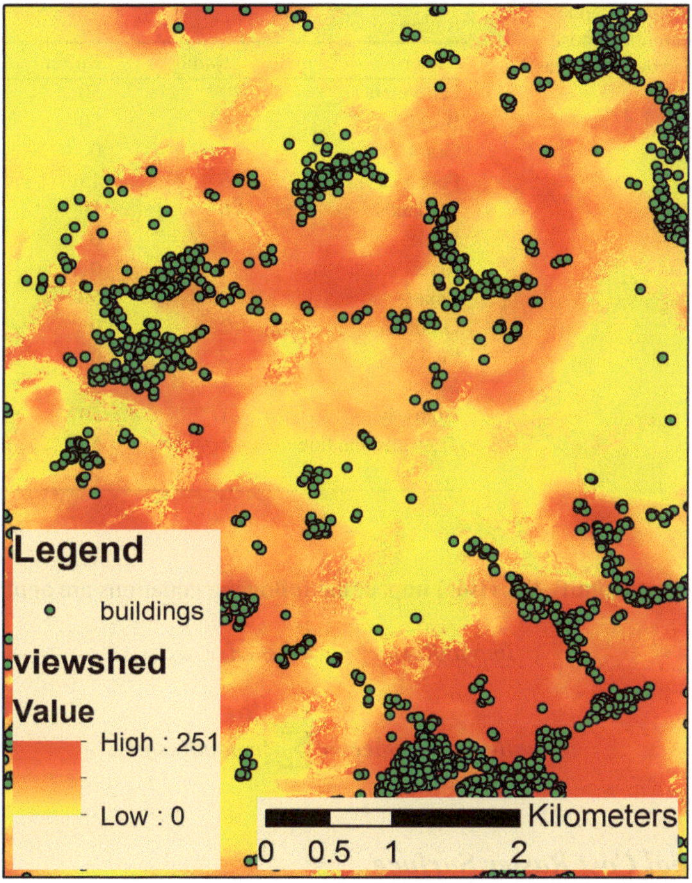

Fig. 4 Viewshed calculation with an *inner radius* of 700 m and an *outer radius* of 1,000 m

In Fig. 3 the width of the histograms corresponds to the width of the rings and is determined by discretizing the function expressed by the Eq. 3.

The difference between the inner and the outer radii increases with the distance whilst the corresponding weight decreases. The weight of each ring is roughly 0.85 smaller than the weight of the previous ring.

An example of the assessment of the visibility from the buildings, considering only one ring, can be seen in Fig. 4: red cells correspond to a high number of visible buildings from a given point 25 m (the selected height of a TL mast suggested by the grid operator) over the ground; on the other hand the yellow color cells correspond to a lower number of buildings from a cell. The figure corresponds to the ring whose inner radius is 700 m and the outer radius 1,000 m and it is called "$ImpactRaster_i$". The visibility analysis is repeated for all rings in order to create the weighted global visibility cost surface, here below called "*VisualImpactSurface*". The visible dark-green points are the centroids of the buildings.

Table 3 Geodata used to identify accumulated cost surface and avoidance areas

Rating table			
Factor	Rating	Buffer	Rating 2
Settlement	99	50	0
Building	99	50	0
Grass	1	0	0
Forest	5	15	0
Wetlands	6	10	0
Rock	1	0	0
River	1	0	0
Lake	99	50	0
Orchard	2	0	0
Pit	2	0	0
Street	1	50	3
Highway	1	50	3
Transmission line	1	50	1
...	...	...	...

In order to quantify the visual impact the following equations are applied:

$$ImpactRaster_i = viewshed_i * w_i \tag{4}$$

$$VisualImpactSurface = \sum ImpactRaster_i. \tag{5}$$

4.3 Global Cost Raster Surface

The five cost raster surfaces of the five categories listed in Table 1 are combined with the visibility cost surface to form a unique global cost surface. Table 3 shows the ratings and the buffers applied according to the local regulations of the studied area and based on the experience of the local grid operator.

Linear infrastructures have a double rating because the suitable lands lie along them. The process creates two buffers along linear infrastructures that define two different areas: the inner one, which is not very suitable and the outer one, which is more suitable according to the route classification. Unsuitable land for TLs routing are assigned a value of 99.

Figure 5 (left) shows the spatial distribution of the visibility from the buildings on each cell after having combined all rings using Eq. 5. The red colors correspond to weighted number of buildings that are visible from a given cell.

Figure 5 (right) shows the distribution of visibility of the map on the left: on the x-axis are all values from 0 to 322.369 shown and on the y-axis the corresponding number of cells containing each value.

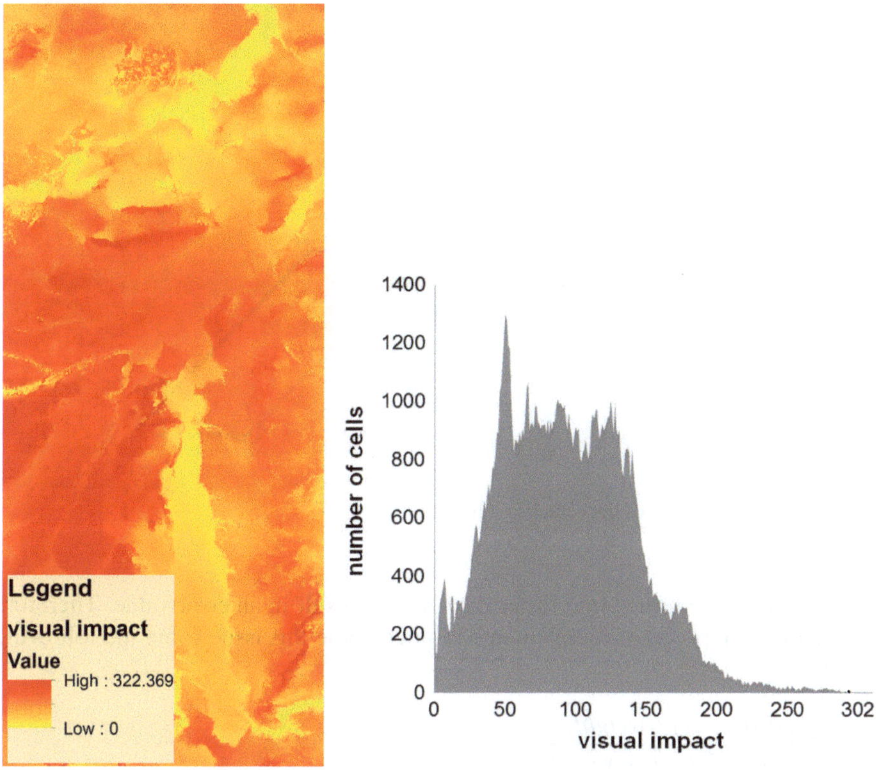

Fig. 5 *Left* the spatial distribution of the visibility from the buildings on the cells, *right* statistical distribution of visibility versus frequency

In Figure 5 (right) it appears that only few values have a high impact (high values on x-axis), which would significantly impact the reclassification. Thus if a straight linear reclassification was applied, the highest x-value would have a stronger impact compared to the other lower values that occur more frequently. Therefore a quantile subdivision is applied in order to take into account the frequency of the values. With this method the higher values (with lower frequency) have a lower impact on the reclassification.

A set of 100 classes is selected enabling to smooth the impact of cells with higher values. Finally the values from 0 to 99 are transformed into 6 classes with the following relation:

$$y = \frac{x}{99} * 5 + 1 \tag{6}$$

By applying this formula, all cost surfaces have values in the same interval.

The presence of linear elements does not enable a straight combination with the areal elements because the cells overlaying the 'NoData' value when the linear

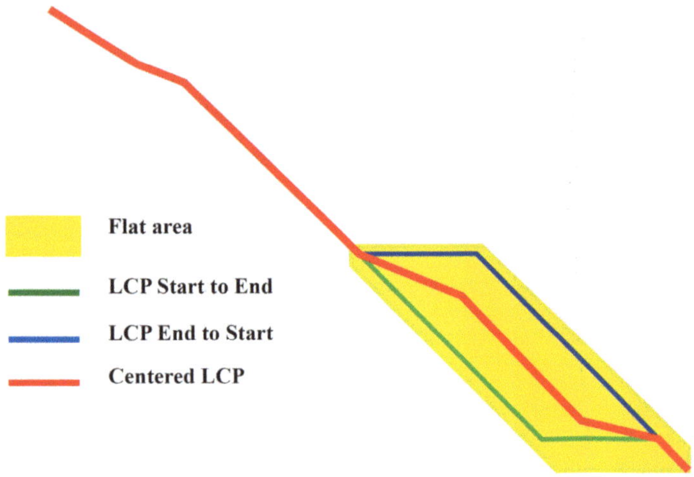

Fig. 6 Visual description of the centering function

infrastructures are turned into raster data, would result in an error value. Therefore a normalization process is implemented to overcome this issue using the following relation:

$$\frac{\sum costRaster_i * weight_i}{\sum normCostRaster_i * weight_i} = \frac{costRaster_a * weight_a + costRaster_b * weight_b + \cdots}{normCostRaster_a * weight_a + normCostRaster_b * weight_b + \cdots} \quad (7)$$

where:

costRaster$_i$: cost surface of a factor i.
weight$_i$: weight corresponding to a factor i.
normCostRaster$_i$: normalizing cost surface of a factor i.

The straightening function is then applied to the global cost surface.

Before determining the LCP, a centering function (Berry 2006) is applied aimed at reducing the zigzags and centering the LPC in "flat area" (Fig. 6) (regions with same cost surface) using the following relation:

$$Weight = \frac{maxProxValue - ProxValues}{maxProxValue} + 0.01 \quad (8)$$

Then a new global cost surface is obtained by multiplying each cell by its corresponding "*weight*" (Eq. 8). The new determined global cost surface is then used to determine the LCP of TLs.

Table 4 Five scenarios considered in the sensitivity analysis

Topic/percentage of visibility	0	0.25	0.5	0.75	1
Visibility	0	3	6	9	12
Linear infrastructure	3	2.25	1.5	0.75	0
Primary area	4	3	2	1	0
Terrain	1	0.75	0.5	0.25	0
Protected area	3	2.25	1.5	0.75	0
Danger area	1	0.75	0.5	0.25	0

5 Sensitivity Analysis

A sensitivity analysis has been carried out in order to assess the impact of the variation of the straightening factors and of the visibility weights. The weights are divided into five classes (Table 4). The sum of the weights of all five groups is equal to 12 and it equals the weight of the visibility when it is considered having the impact equal to 1. Practically, depending on the scenario, the parameters affect the LPC as function to their weights.

The five scenarios listed in Table 4 are combined with six scenarios of the straightening factor whose integer values are comprised between 0 and 5. The approach has been applied to a real case study in Western Switzerland where a new route between two substations needs to be identified in order to replace an outdated TL.

A set of 30 scenarios has been identified as combination of the 5 scenarios of Table 4 and the 6 straightening factors.

In Fig. 7, all LCPs resulted from the sensitivity analysis and the existing TLs (red line) are shown: three main corridors can be clearly identified.

In Table 5 all scenarios are numerically shown. For each scenario, three parameters (LCP columns) are estimated and normalized in order to be compared

In the column of the "length", is the normalized geometrical length of the LCP between the source and destination.

In the column "LCP Cost on Original Cost Surface (CoOCS)", is the normalised cost of the LCP calculated on the global cost surface obtained without considering the visibility in the starting dataset in Fig. 8 (below).

In the column "visibility", is the normalised cost of the LCPs on the "visual impact" cost raster.

The existing TL to be replaced presents length equal to 0.92, CoOCS equal to 1.38 and the visibility equal to 1.

The CoOCS is higher than the route with straightening factor and visibility equal to 1 because it crosses regions nowadays forbidden or with a relatively high cost.

In Fig. 8, the extreme scenarios are shown: straightening factor equal to zero on the left and visibility factor equal to zero on the right. On the left side of Fig. 8, the visibility weights equal to 0.25 and 1 are not shown because they match the LCP with weight visibility equal to 0 and 0.75, while in the right figure, the LCP with straightening factor equal to 1, 3 and 4 match respectively the LCP equal to 0, 2 and 5.

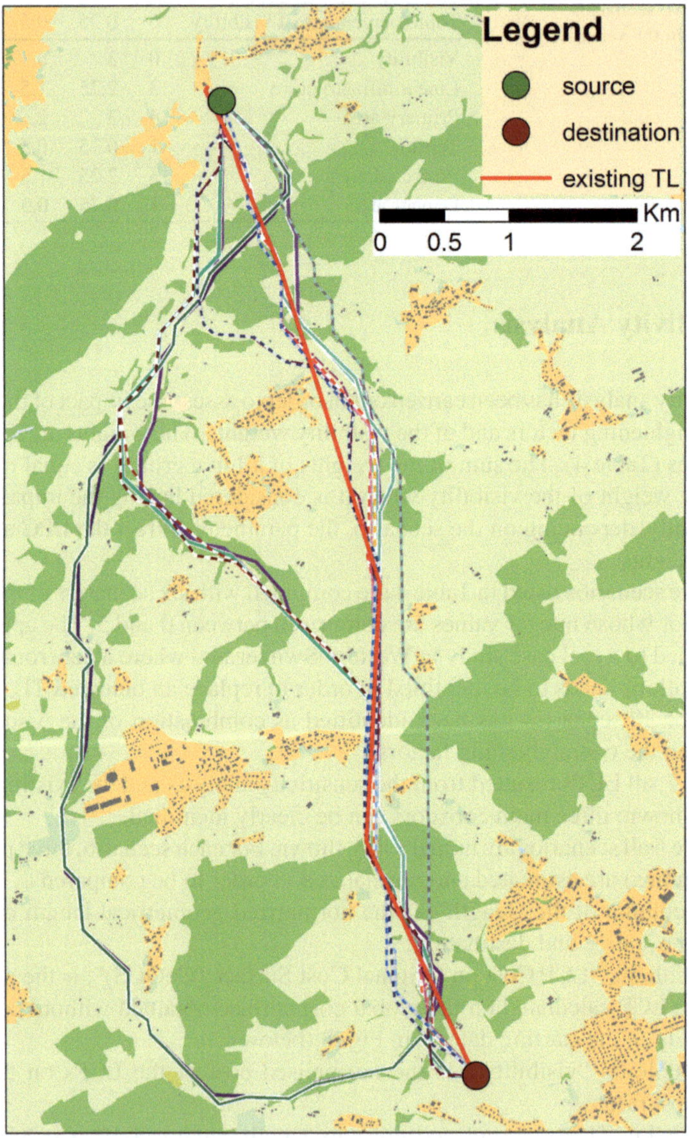

Fig. 7 Visualization of all LCPs resulting from sensitivity analysis

The LCP with visibility weight equal to 0.75 is longer and has a completely different path, because the process identifies those cells which have a low number of visible buildings and thus, using the DSM, the height of masts is obstructed by the vegetation or other objects. On the other hand, the other LCPs partially follow the path of the existing line, but they avoid the settlements built after the construction of

Table 5 Normalized LCP costs

Visibility Weight	Straightening factor	LCP Length	LCP cost on original Cost surface (CoOCS)	LCP Visibility
0	0	1	1	1
0	1	1	1	1.01
0	2	0.95	1.1	1
0	3	0.94	1.14	1.01
0	4	0.94	1.23	1.01
0	5	0.94	2.4	1.14
0.25	0	1.01	1.08	0.92
0.25	1	0.95	1.17	0.93
0.25	2	0.94	1.17	0.95
0.25	3	0.94	1.19	0.95
0.25	4	0.94	1.37	0.96
0.25	5	0.94	1.37	0.96
0.5	0	1.06	1.42	0.7
0.5	1	1.02	1.43	0.73
0.5	2	0.94	1.22	0.9
0.5	3	0.94	1.31	0.93
0.5	4	0.94	1.41	0.93
0.5	5	0.94	1.41	0.93
0.75	0	1.31	2.11	0.43
0.75	1	1.04	1.63	0.68
0.75	2	0.94	1.64	0.84
0.75	3	0.94	1.61	0.89
0.75	4	0.94	1.61	0.89
0.75	5	0.94	1.61	0.89
1	0	1.29	2.79	0.42
1	1	1.04	2.26	0.65
1	2	1.01	2.09	0.68
1	3	0.94	1.83	0.86
1	4	0.94	1.81	0.88
1	5	0.94	1.81	0.88

the old TL (black line in the figure). In addition all LCPs avoid crossing the vegetation or simply exploit the Right-of-Way (RoW) of the existing TL line which is a strip of land purchased by a grid operator from landowners to install the lines including all equipment (e.g., poles, towers, wires, appliances).

The right side of Fig. 8 shows overall straighter LCPs crossing relatively large forested areas (green regions) when a high straightening factor is applied (straightening factor of 5). The blue routes in the two figures basically correspond. Nevertheless it is difficult for a planner to compare the existing route to the proposed TLs and to make a pre-selection of a reduced set of routes.

The data shown in Table 5 are normalized but they have different units, therefore, in order to be compared to each other, they need to be standardized by calculating the z scores, with the z-transformation (Bahrenberg et al. 1999):

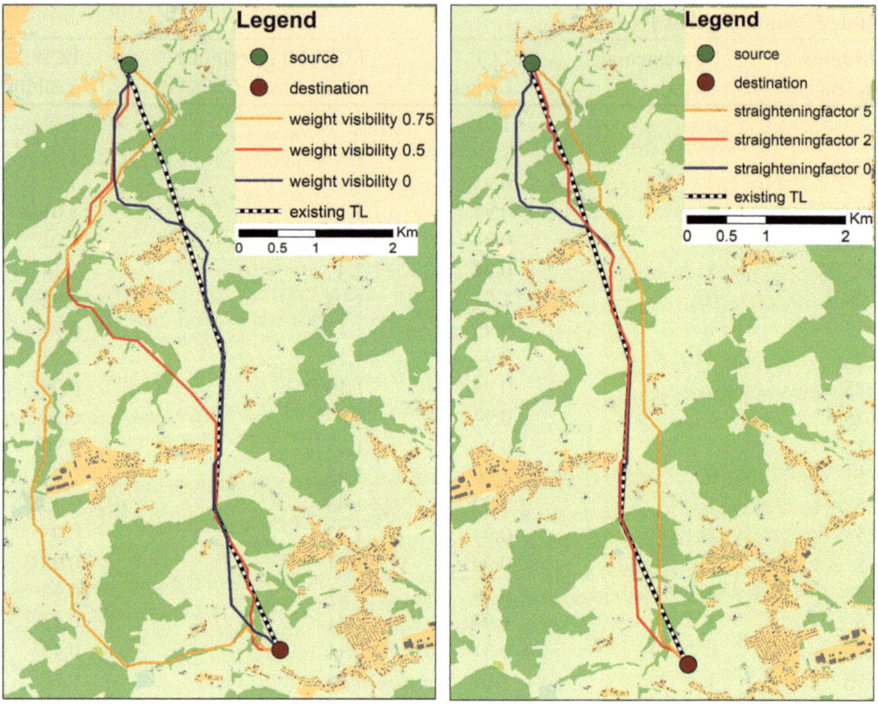

Fig. 8 *Left* the LCPs with straightening factor equal to 0 and varying visibility, *right* the LCPs with visibility factor equal to 0 and varying straightening

$$z_i = \frac{x_i - \mu}{\sigma} \tag{9}$$

where:

- x_i: raw value
- μ: mean value of the population
- σ: standard deviation

The obtained values are then plotted into a chart (Fig. 9) where the x-axis represents the standardized relation between the length and the CoOCS, while the y-axis represents the standardized visibility.

The relation between the x values and the visibility is a logarithmic formula shown in Fig. 9 which shows a correlation of 0.809. The input value in the formula is a standardized coefficient representing the complexity of a given project (the red triangles) made up of the length and the cost of the LCP_i on the original cost surface and representing the resistance of the land to be crossed between the source and the destination, linked by the following relation:

$$X_i = e^{Z^i_{vis.}} * \log\left(maxZ_{cost} + Z^i_{cost}\right) \tag{10}$$

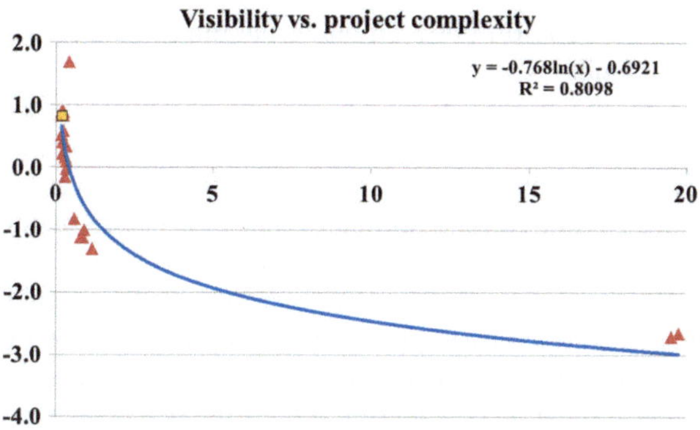

Fig. 9 Correlation between the project complexity and the visibility

The plot shows that an increase of complexity corresponds to a lower visibility and thus a potential lower local opposition; on the other hand a lower complexity corresponds to a much higher visibility.

The existing route is shown in the plot with an orange square. Most of the identified new routes have a lower visibility and the complexity remains rather constant until the curve crosses the x-axis when the proposed routes show a higher complexity.

6 Conclusions and Future Work

This paper addressed the issue of routing new TL routes through a complex environment where the visibility from buildings plays a critical role and needs to be minimized. In reality a trade-off between the costs and the visibility is more likely and realistic but different options need to be identified first. A set of 30 possible routes were identified in a sensitivity analysis where the visibility and the straightening factor have been assigned different weights. Finally the results have been standardized in order to be compared as they have different units and a relation between the visibility and the other two parameters (resistance of a raster cell to be crossed and the length) that describe the complexity of a given project have been identified with a correlation of 0.809. Most of the proposed new routes have a lower visibility of the existing TL but some of them appear unlikely to be realised.

The integration of the DSM allowed a more realistic assessment of the visibility of masts from the buildings which has been used as a parameter to quantify the spatial distribution of the visibility over a region and thus a new cost surface to be integrated in a typical MCA process.

The results are encouraging in the perspective of an increased reality when planning TLs using viewshed analysis compared to the classical methodologies. Our approach is able to support the grid planner in proposing different routes and providing a tool to find a trade-off between the visual impact and the complexity when crossing a piece of land.

The approach has the disadvantage of a long computational time that, in this case study, is about 10 h[1] which is also due to the resolution of the raster data used, to the extent of the selected region and the number of scenarios. Nevertheless, it is much shorter than the time required for a similar prefeasibility assessment which can last up to a few months.

The process can be further improved in order to conform to reality. The location of the masts can be identified by implementing an optimization workflow instead of considering the cable having the same impact of the masts. The visual impact of a mast can be further refined by considering the visible percentage of each mast. The investment costs of each single TL can be estimated as a function of the crossed land, the terrain features, the pole type and the amount and the angle of the turns along the route. The straightening of the route needs to be further improved in order to make the route more realistic. The methodology needs to be tested with more case studies to assess the applicability in transmission line routing.

References

Atkinson DM, Deadman P, Dudycha D et al (2005) Multi-criteria evaluation and least cost path analysis for an arctic all-weather road. Appl Geogr 25(4):287–307. doi:10.1016/j.apgeog.2005.08.001

Bahrenberg G, Giese E, Nipper J (1999) Statistische methoden in der geographie 1. Stuttgart

Berry J (2004) Beyond mapping III 'straightening' conversions improve optimal paths. Geo World 17:18–19

Berry JK (2006) Beyond mapping III-Use LCP procedures to center optimal paths. http://www.innovativegis.com/basis/BeyondMappingSeries/BeyondMapping_III/Topic8/FurtherReading_Topic8.htm#Section2

Bevanger K, Bartzke G, Clausen S et al (2010) Optimal design and routing of power lines; ecological, technical and eco-nomic perspectives (OPTIPOL) Kjetil Bevanger edn. NINA Publications, Norwegian Institute for Nature Research, Trondheim, NINA Report 619, pp 51

Church RL, Loban SR, Lombard K (1992) An interface for exploring spatial alternatives for a corridor location problem. Comput Geosci 18(8):1095–1105. doi:10.1016/0098-3004(92)90023-K

Cotton M, Devine-Wright P (2012) Putting pylons into place: a uk case study of public perspectives on the impacts of high voltage overhead transmission lines. J Environ Plan Manage 56(8): 1225–1245. doi:10.1080/09640568.2012.716756

Dijkstra EW (1959) A note on two problems in connexion with graphs. Numerische Mathematik 1:269–271

Douglas DH (1994) Least-cost path in GIS using an accumulated cost surface and slopelines. Cartographica: Int J Geogr Inf Geovisualization 31(3):37–51. doi:10.3138/D327-0323-2JUT-016M

Eppstein D (1998) Finding the k shortest paths. SIAM J Comput 28(2):652–673. doi:10.1137/S0097539795290477

[1] ArcGIS installed on a desktop computer, RAM 8GB, CPU 8 core 3.3GHz

French S, Houston G, Johnson C et al (2008) EPRI-GTC tailored collaboration project: a standardized methodology for siting overhead electric transmission lines. In: Goodrich-Mahoney JW, Abrahamson LP, Ballard JL, Tikalsky SM (eds) Environment concerns in rights-of-way management 8th international symposium. Elsevier, Amsterdam, pp 221–235. doi:10.1016/B978-044453223-7.50029-0

Furby L, Slovic P, Fischhoff B et al (1988) Public perceptions of electric power transmissionlines. J Environ Psychol 8(1):19–43. doi:10.1016/S0272-4944

Hadrian DR, Bishop ID, Mitcheltree R (1988) Automated mapping of visual impacts in utility corridors. Landscape Urban Plan 16(3):261–282. doi:10.1016/0169-2046(88)90073-4

Hirst E, Kirby B (2001) Key transmission planning issues. Electr J 14(8):59–70. doi:10.1016/S1040-6190(01)00239-1

Huber D, Church R (1985) Transmission corridor location modeling. J Transp Eng 111(2):114–130. doi:10.1061/(ASCE)0733-947X(1985)111:2(114)

Husain F, Sulaiman NA, Hashim KA et al (2012) Multi-criteria selection for TNB transmission line route using AHP and GIS. In: Paper presented at the international conference on system engineering and technology (ICSET), Bandung, 11–12 Sept 2012

Jankowski P, Richard L (1994) Integration of gis-based suitability analysis and multicriteria evaluation in a spatial decision support system for route selection. Environ Plan B: Plan Des 21(3): 323–340

Jewell W, Grossardt T, Bailey K et al (2010) A new method for public involvement in electric transmission line routing. In: Transmission and distribution conference and exposition, 2010 IEEE PES, 19–22 April 2010, pp 1–1. doi:10.1109/TDC.2010.5484237

Lake MW, Woodman PE, Mithen SJ (1998) Tailoring gis software for archaeological applications: an example concerning viewshed analysis. J Archaeol Sci 25(1):27–38. doi:10.1006/jasc.1997.0197

Malczewski J (1999) GIS and multicriteria decision analysis. Wiley, New York

Marshall R, Baxter R (2002) Strategic routeing and environmental impact assessment for overhead electrical transmission lines. J Environ Plan Manage 45(5):747–764. doi:10.1080/09640560220000013101

McCalley JD, Krishnan V (2014) A survey of transmission technologies for planning long distance bulk transmission overlay in us. Int J Electr Power Energ Syst 54:559–568. doi:10.1016/j.ijepes.2013.08.008

Mills A, Wiser R, Porter K (2012) The cost of transmission for wind energy in the united states: a review of transmission planning studies. Renew Sustain Energy Rev 16(1):1–19. doi:10.1016/j.rser.2011.07.131

Monteiro C, Miranda V, Ramirez-Rosado IJ et al (2005) Compromise seeking for power line path selection based on economic and environmental corridors. IEEE Trans Power Syst 20(3):1422–1430. doi:10.1109/TPWRS.2005.852149

Nohl W (1993) Beeinträchtigungen des Landschaftsbildes durch mastenartige Eingriffe. München

Ogburn DE (2006) Assessing the level of visibility of cultural objects in past landscapes. J Archaeol Sci 33(3):405–413. doi:10.1016/j.jas.2005.08.005

Paul H-U, Uther D, Neuhoff M et al (2004) GIS-gestütztes Verfahren zur Bewertung visueller Eingriffe durch Hochspannungsfreileitungen. Naturschutz und Landschaftsplanung: Zeitschrift für angewandte Ökologie, vol 36

Sessions J, Akay A, Murphy G et al (2006) Road and harvesting planning and operations. In: Shao G, Reynolds K (eds) Computer applications in sustainable forest management, vol 11. Springer, Netherlands, pp 83–99. doi:10.1007/978-1-4020-4387-1_5

Stucky JLD (1998) On applying viewshed analysis for determining least-cost paths on digital elevation models. Int J Geogr Inf Sci 12(8):891–905. doi:10.1080/136588198241554

Towers G (1997) Gis versus the community: siting power in southern west virginia. Appl Geogr 17(2):111–125. doi:10.1016/S0143-6228(97)00001-5

Towers G (2000) Applying the political geography of scale: grassroots strategies and environmental justice. Prof Geogr 52(1):23–36. doi:10.1111/0033-0124.00202

Xu J, Lathrop RG (1995) Improving simulation accuracy of spread phenomena in a raster-based geographic information system. Int J Geogr Inf Syst 9(2):153–168. doi:10.1080/02693799508902031
Yu C, Lee JAY, Munro-Stasiuk MJ (2003) Research article: extensions to least-cost path algorithms for roadway planning. Int J Geogr Inf Sci 17(4):361–376. doi:10.1080/1365881031000072645
Zhenpei L, Ping L, Ming W et al (2010) Application of ArcGIS pipeline data model and GIS in digital oil and gas pipeline. In: 18th international conference on geoinformatics, 18–20 June 2010. pp 1–5. doi:10.1109/GEOINFORMATICS.2010.5567619

MIX
Papier aus verantwortungsvollen Quellen
Paper from responsible sources
FSC® C105338

If you have any concerns about our products,
you can contact us on
ProductSafety@springernature.com

In case Publisher is established outside the EU,
the EU authorized representative is:
Springer Nature Customer Service Center GmbH
Europaplatz 3, 69115 Heidelberg, Germany

Printed by Libri Plureos GmbH
in Hamburg, Germany